Maths for Map Makers

2nd edition

Maths for Map Makers

2nd edition

by

Arthur Allan

Whittles Publishing

Published by
Whittles Publishing,
Dunbeath Mill
Dunbeath
Caithness, KW6 6EG
Scotland, UK
www.whittlespublishing.com

First edition 1997
Reprinted 2011

ISBN 1-870325-99-0
978-1-870325-99-8

Printed by Bell & Bain Ltd., Glasgow

Preface

This book is intended to help students to understand the elementary mathematics involved in map making. Such students may know very little mathematics and its notation, or they may be quite competent but ignorant of some requisite topics no longer treated in secondary school syllabuses, or they may simply require to brush up previous knowledge.

The first five chapters adopt an elementary approach in which no previous knowledge of the subject is assumed: the text then progresses to give the student a sufficient level of understanding to cope with most topics confronting the map maker, and finally it seeks to lead readers to a level of understanding which will enable them to make full use of the wide range of more traditional textbooks available on the market and to understand the various articles to be found in technical journals.

The assumption is made that the reader is already motivated and is prepared to work hard to carry out the exercises suggested. There are no quirky examples to test the reader's abstract reasoning powers, such as one finds in traditional text books and examination questions. If the concept is relevant, it is explained. A large number of worked exercises is given for the reader to perform as they arise.The experience so gained is vital for an understanding of the next stage in the argument. To provide some means of self testing, lists of keywords and formulae are given at the end of each chapter.

It is assumed that a hand calculator is available. Access to computer spreadsheets, though encouraged, is not essential, except for the new Chapter 13 where it is impracticable to carry out all but the simplest of calculations by hand. Even spreadsheets have their limitations here. Algorithmic forms of formulae are discussed, and computer applications are never far from sight.

The selection of topics and the level of treatment have been severely restricted to contain the size of the book within reasonable limits, and to avoid confusing the beginner with too much detail. It is also hoped that some readers may become so keen on mathematics that they will be able to cope with the rigour required by more advanced texts, some of which are listed in the reference section.

The material is arranged with careful cross-references. Initially basic mathematical topics and concepts are presented in a problem-oriented manner: that is, in the logical sequence which arises in dealing with the making of a map or other graphic product, such as a three-dimensional visualisation. In this way it is hoped to make clear the purposes and relevance of each topic, and to maintain the interest of the reader. To assist the visualisation of three-dimensional problems, the use of paper or other models is strongly advocated, because experience shows them to be effective with beginners. Strangely enough, once the ability to visualise three-dimensional objects has been acquired, physical models are no longer required and conventional drawings serve adequately.

Where possible the mathematical notation used in the book follows that of the pamphlet, *Formulae for Advanced Mathematics with Statistical Tables* published by the Cambridge University Press for the Schools Mathematics Project.

References listed at the end of the book are those I have found useful. Because texts for further reading are continually appearing, I feel it would be invidious to select some to the exclusion of others. Current teachers will be more competent to do this.

Arthur Allan

Acknowledgements

I am indebted to many people for assistance and encouragement with this book. My former students were particularly influential in suggesting the need for the book in the first place and for fashioning my approach to the text. Although some specific names come to mind, it would be invidious to mention some and not others. Many graduates in geography, archaeology and geology often struggled with the mathematics needed to become surveyors and cartographers. For their collective role in stimulating me with questions, comments and criticisms, spread over the years, I express my grateful appreciation.

Professor Ian Harley, of University College London, has given me much assistance with the intricacies of word-processing and its attendant software. He has also given some technical advice and encouragement about detailed aspects of the text, especially with Chapter 12.

I am most grateful to Dr R. Graham, Mr J.R. Smith and Dr L. Brooks for reading much of the text in its draft form and for their many constructive suggestions. The latter two also checked many of the arithmetic exercises. Professor M.A.R. Cooper made some helpful early suggestions about the layout and content, and Mrs Margaret Tomlinson made some useful comments on the initial chapters of the book.

I also owe a great debt to the authors of the texts which I have cited as references, and to all other authors not mentioned by name, who have assisted my understanding in some way. I am also indebted to the publisher Keith Whittles for his efficient and cordial support in converting the draft to its final form and to J.R.Smith for proof reading Chapters 1–12. In compiling the new Chapter 13 I received helpful advice from Professor P.A.Cross and Dr J.C.Iliffe, both of University College London, and from Mr N.Atkinson to whom I convey my gratitude.

Whilst I acknowledge assistance from many named and unnamed persons, I accept full responsibility for the text, and trust that any criticisms will be directed in my direction only. Inevitably some errors will have crept into the text for which I apologise in advance. I hope that I can be informed of them for future use in any reprint or errata sheet. The new Chapter 13 has been inserted in response to pleas from several reviewers of the first edition to whom I am grateful for their considered advice

A.L. Allan

Contents

How to use this book

The beginner is invited to read Chapters 1 to 5 in strict order, and to carry out the exercises also in strict order, before moving on to a new topic. A small calculator is required for most exercises. At the end of each chapter a list of key words is given so that the reader can reflect on their meanings to see if they are properly understood before proceeding further. If in doubt at any stage return to the text and repeat the work. A list of formulae is also given, in development order, at the end of each chapter. These formulae are all recast as an index at the back of the book, with the order changed to suit ready access to the text. Many users will use this mode of entry if the book is to serve as a refresher course.

Key paragraphs are numbered in sequence for cross-reference. Each exercise is numbered within these main paragraphs and indicated by the graphic shown below. A complete reference to an exercise requires both numbers, for example. 4.6 Exercise 6.

All key equations and formulae are numbered within braces uniquely throughout the book. Thus, formula 6 of Chapter 1 is labelled (1.6).

References are given towards the end of the book. An appendix of useful information is located also at the end of the book together with the index of formulae and the word index.

indicates an exercise to be undertaken

Common mathematical operators

$+$	addition or plus $2 + 3 = 5$
$-$	subtraction or minus $5 - 3 = 2$
$\pm$	plus or minus $5 \pm 3 = 8$ or 2
$\times$ or $*$ or $.$	multiplication $1 \times 2 \times 3 = 1 * 2 * 3 = 1.\,2.\,3 = 6$
$\div$ or $/$	division $6 \div 2 = 6/2 = 3$
$1/x$	reciprocal $1/2 = 0.5$
$\lvert x \rvert$	modulus of x = positive value of x: $\lvert -3 \rvert = 3$

$\frac{dy}{dx}$ or $\dot{y}$	first differential of y with respect to x
$\frac{d^2y}{dx^2}$ or $\ddot{y}$	second differential of y with respect to x
$\frac{\partial y}{\partial x}$	partial differential of y with respect to x
$\int$	integral (see integration)
Σ	sum of
Σn	sum of first n integers (whole numbers) $= 1 + 2 + 3 \ldots n$
x^n or $x \wedge n$	x to the power n $\quad 2^3 = 2.2.2 = 8$
$x^0 = 1$	by definition
$\sqrt{x}$	square root of x $\quad \sqrt{4} = 2$
$\sqrt{-1}$	not defined in ordinary arithmetic, but usefully denoted by i, then $i^2 = -1$
()	braces
{} []	brackets (curly) (square)
3!	factorial three $\quad 3! = 3 \times 2 \times 1 = 6$
$n!$	factorial n $\quad n! = n(n-1)(n-2) \ldots (3)(2)(1)$

Common mathematical symbols

$<$	less than $\quad 2 < 3$
$>$	greater than $\quad 3 > 2$
$=$	equal to $\quad 2 = 2$
$\neq$	not equal to $\quad 2 \neq 3$
$\leq$	less than or equal to $\quad x \leq 5$
$\geq$	greater than or equal to $\quad x \geq 5$
$\approx$ or $\doteqdot$	approximately equal to $\quad 3.1 \approx 3$
$\tilde{A}$ or $\angle A$	angle A
∞	infinity
δx	small change or increment (increase) in x $\quad \delta x = 0.02$
Δx	large change or increment (increase) in x $\quad \Delta x = 2$
ΔABC	triangle ABC
∂	del (see partial differentiation)
$\boldsymbol{A}$	matrix
$\boldsymbol{A}^T$	transposed matrix
$\boldsymbol{A}^{-1}$	inverted matrix
$\boldsymbol{x}$	vector

Note on common usage

The Roman and Greek alphabets are both used in mathematics. Coefficients and constants are usually represented by small letters (lower case letters) taken from the beginning of the alphabet. As a rule, those letters coming towards the end of the alphabet are used for variables or parameters and are printed in *italic* or sloping form. Capital letters (upper case letters) are normally used to describe points or matrices. Matrices and vectors are always printed in heavy type, thus ***A*** or ***u***, or if written by hand, they should be underlined.

Depending on circumstances, these general rules will be broken or appear to be broken. For example, the coefficients of linear equations can be either variables or constants according to circumstances. We encourage the reader always to question which entities are constants and which are variables irrespective of the notation used. For example in the triangle ABC, the sides a, b, and c and angles A, B C are often used as variables in surveying because they are measured quantities, and the letters x, y and z are required for the coordinates of points derived from these measurements.

The Greek alphabet – lower case

α	alpha	ν	nu
β	beta	ξ	xi
γ	gamma	o	omicron
δ	delta	π	pi
ε	epsilon	ρ	rho
ζ	zeta	σ	sigma
η	eta	τ	tau
θ	theta	υ	upsilon
ι	iota	ϕ	phi
κ	kappa	χ	chi
λ	lambda	φ	psi
μ	mu	ω	omega

The Greek alphabet – UPPER CASE

A	ALPHA	N	NU
B	BETA	Ξ	XI
Γ	GAMMA	O	OMICRON
Δ	DELTA	Π	PI
E	EPSILON	P	RHO
Z	ZETA	Σ	SIGMA
H	ETA	T	TAU
Θ	THETA	Y	UPSILON
I	IOTA	Φ	PHI
K	KAPPA	X	CHI
Λ	LAMBDA	ϑ	PSI
M	MU	Ω	OMEGA

Chapter 1
Numbers and Calculation

1.1 The language of mathematics

In mathematics, technical or specialised words are used which have to be learned so that ideas can be expressed easily. Problems can arise with the study of maths if the student does not absorb all the new meanings quickly enough. As with a foreign language, the vocabulary must be learned somehow, preferably by repeated use in practice. In this book, the first time such words are introduced they are *in italics*.

There is a further problem with jargon. In mathematics quite ordinary words are given special meanings. For example the word *normal* is given the special meaning 'at right angles to', or 'perpendicular to'. After a while the particular meaning of a word should appear from the context in which it is used. For example adjacent edges of this sheet of paper are normal to each other. Opposite edges are *parallel* to each other because they do not meet.

Mathematical modelling

We start by considering what happens when we look at a solid object, such as the room we sit in, or when we look at a drawing or map we wish to make of it. Assuming that the room we sit in is a *rectangular box*, how does it appear to us? What we see is a *perspective view* of the room and its contents. How does the rectangular sheet of paper appear to our eyes? If we hold the paper straight in front of us, it looks like a *rectangle,* i.e. its opposite sides are *parallel* to and equal to each other, and adjacent sides are *perpendicular* to each other.

If we place the paper on the table below our eyes, it no longer looks like a rectangle. Opposite sides no longer look parallel, and adjacent ones do not appear perpendicular to each other. What we see is a *projected view* of the rectangle. An image of the sheet of paper is projected on to the retina of our eyes. In a similar way, a photograph or television scene is a projected view of the scene we wish to view.

Look again at the room; the floor, the ceiling, and walls, all appear as projected views. If we use each eye separately, these views jump about slightly. Using both eyes enables us to see stereoscopically or in genuine three dimensions. Look with only one eye open and verify that the three-dimensional effect vanishes; all we see is a single perspective view of the room without depth.

As will be seen later, many of these effects can be explained by considering straight lines, or *rays*, or *vectors*, which are capable of mathematical treatment and calculation in a computer: a *mathematical model* is created for this to be done. For example, in this book you will be shown how to create a stereoscopic (three-dimensional) picture, such as we see in popular 'Magic Books', as well as the more traditional processes used in map construction and surveying.

1.2 Numbers and your calculator

The ability to count is essential to any civilisation and to scientific development. Historically many systems of counting were evolved to suit the needs of an emergent civilisation. The system we use today is based on early Arabic ideas with a zero and nine other symbols 1 to 9. Because the Romans had no concept of zero, they found multiplication and division almost impossible. The concept of zero to indicate multiples of the number base is vital. We are accustomed to use the number base of ten in everyday work. Other bases, such as two (binary) or sixteen (hexadecimal) are used by computers. We deal only with the base of ten. There are two kinds of number in use today, the integer and the real number.

In the following sections we will explain some of the arithmetic functions which arise in map making. They may be performed on a hand calculator or computer system, such as a spread sheet, or high level language. These are

addition +	subtraction –	multiplication × *	division ÷ /
exponent forms	integers 5	real numbers 1.23	truncation
rounding	significance	powers 4^3	log and 10^x
ln and e^x	square roots	reciprocals $1/x$	braces { }
brackets ( ) [ ]	stores K in/ K out	change sign ±	factorials $x!$

It is assumed that you have a hand calculator available. This calculator should have a memory M+, should preferably have six stores, the forward trigonometrical functions, sin, cos, tan, and their inverses $\sin^{-1}$, $\cos^{-1}$, $\tan^{-1}$, and polar to rectangular keys R–P and P–R. Other functions are usually supplied. Some of these functions will be explained in this chapter; others will be left until later.

Before reading any further you should be sure that you can *add* (+), *subtract* (–), *multiply* (×), and *divide* (÷) with your calculator. These symbols such as (+) are called *mathematical operators* which tell us what to do with the numbers, in this case to add them together.

The *brace* { }, or pair of square *brackets* [ ] or *parentheses* ( ), are also operators used to group numbers together. They are always used in pairs. Loosely they are all just referred to as 'brackets'. For example in *addition*

$$(2 + 3) + 7$$

The brackets indicate that we add the numbers within them, then add 7 to the result. Or again we might have

$$2 + (3 + 7) \quad \text{or} \quad 2 + 3 + 7$$

In these cases the results are all identical using the '+' operator. When using the '–' operator, more care is needed. Consider the operation

$$(2 - 3) - 7 = 2 - 3 - 7 = -8$$

But $$2 - (3 - 7) = 2 - (-4) = 2 + 4 = 6$$

This is because the negative sign outside the brackets applies to all numbers inside them. Also in the case of *multiplication* the position of the brackets alters the result: for

$$(2 + 3) \times 7 = 5 \times 7 = 35 \text{ while } 2 \times (3 + 7) = 2 \times 10 = 20$$

The general rule is that

brackets and multiplication are acted upon first

Thus, using the '×' operator, the results differ. This also applies to the '÷' operator.

Exercise 1 Show that

$$(1 + 3) \div 2 = 2 \quad \text{and} \quad 1 + (3 \div 2) = 2.5$$

Exercise 2 Evaluate the following

$$2 + [(3 + 5) \times 2 + (5 + 1)] \times 6 \tag{1.1}$$

The calculation stages are: carrying out operations from inside in order, as follows:

$$2 + [(8) \times 2 + (6)] \times 6$$
$$2 + [16 + 6] \times 6$$
$$2 + [22] \times 6$$
$$2 + 132$$
$$134$$

If your calculator is provided with brackets and braces, you can do this sum in the order expressed in line (1.1) pressing each number and operator in turn from left to right.

Normally the '×' signs are omitted from such an expression, but don't forget them when calculating.

Exercise 3 Evaluate the following

$$(((1 + 2)5 + 3)5 + 4)5 + 1$$

You will key in

$$1 + (((1 + 2) \times 5 + 3) \times 5 + 4) \times 5$$
$$= 471$$

Notice that there is always the same number of brackets facing inwards as there is facing outwards.

Exercise 4 Evaluate the above expression in separate stages.

$$(((1 + 2) \times 5 + 3) \times 5 + 4) \times 5 + 1$$
$$((3 \times 5 + 3) \times 5 + 4) \times 5 + 1$$
$$(18 \times 5 + 4) \times 5 + 1$$
$$94 \times 5 + 1$$
$$471$$

Exercise 5 Show that, when $a = \frac{1}{5}$

$$(((1 + 2)\,a + 3)\,a + 4)\,a + 1 = 1.944$$

The sequence of key strokes is

$$(((1 + 2) \div 5 + 3) \div 5 + 4) \div 5 + 1 = 1.944$$

So the brackets matter for some operations but not for others. We shall see later, when dealing with matrices, that the *order* in which the operations is done can also be important.

Exercise 6 Verify that

$$123\,456\,789 + 987\,654\,321 = 1\,111\,111\,110$$

Notice that we display these large numbers in groups of three to make them easier to read. In calculations, however, no spaces should be left between digits.

Exercise 7 Verify that

$$987\,654\,321 - 123\,456\,789 = 864\,197\,532$$

that

$$987\,654\,321 \div 123\,456\,789 = 8.000\,000\,073$$

and that

$$987\,654\,321 \times 123\,456\,789 = 1.219\,326\,311 \times 10^{17}.$$

The last number is so large that it cannot all be displayed on the calculator directly. The method is to convert it to the *exponent form* as explained next.

1.3 Numbers in exponent form

It is often convenient to express a very large number, such as the nine-digit number 123 000 000, in exponent form such as

$$123 \times 10^6$$

This means 123 × 1 000 000.

The number $1\,000\,000 = 10 \times 10 \times 10 \times 10 \times 10 \times 10$ is expressed as 10^6. Similarly a small number such as 0.000 000 123 can be written as

$$0.123 \times 10^{-6}$$

in which the symbol 10^{-6}, means 1/1 000 000. This exponent method allows you to concentrate on *significant figures* separately from the *order of magnitude* or size of the final answer.

Exercise 1 Multiply 2 000 000 by 0.000 003. Expressing each number in exponent form we have

$$2 \times 10^6 \times 3 \times 10^{-6} = 2 \times 3 \times 10^6/10^6 = 6$$

Most calculators have a key to convert a number to exponent form automatically. (In some it is labelled ENG). Most calculators give the power of ten, called the *index,* in groups of three, i.e. as

$$10^3,\ 10^6,\ 10^9, \text{ and } 10^{-3},\ 10^{-6},\ 10^{-9} \text{ and so on.}$$

Exercise 2 Verify that the exponent forms of the following numbers are respectively

$$123 = 123 \times 10^0, \quad 1\,234 = 1.234 \times 10^3, \quad 12\,345 = 12.345 \times 10^3.$$

To save space, only the powers are given in a window. An example of the display is

$$2^{\,03} \text{ or } 2^{\,-03}$$

To obtain the *magnitude* of an answer, in powers of ten, we add or subtract the indices separately from the exponents.

Exercise 3 Calculate the following

$$2 \times 0.005 \times 0.000\,007$$

Expressing the numbers in exponent form we have

$$2 \times 5 \times 10^{-3} \times 7 \times 10^{-6} \text{ and re-ordering we have}$$

$$2 \times 5 \times 7 \times 10^{-3} \times 10^{-6} = 70 \times 10^{-9} = 7 \times 10^{-8}$$

Normally the result is left in this form, but it can be written as a decimal as

$$0.000\,000\,07$$

1.4 Integers: sequence and series

An integer is a whole number. For example, the following is a *sequence* of the first five *integers*

$$1, 2, 3, 4, 5$$

It is called a *sequence* because it gives the integers in some order. A different sequence of the same digits is

$$5, 4, 3, 2, 1$$

A *series* shows a *relationship* between numbers of a sequence, such as their sum

$$1 + 2 + 3 + 4 + 5 = 15$$

The *order* in which the numbers appear does not matter because

$$2 + 1 + 5 + 3 + 4 = 15$$

This means that the first five integers have to be added together, using the notation for addition (+). In computers, integers are treated differently from the next type of numbers we will describe.

1.5 Real or decimal numbers

The idea of dividing numbers into decimal fractions allows us to deal with another type of number, the *real number*. For example 2.3 means an integer 2 plus three tenths of 1. Computers handle such numbers differently from integers. The number of decimal places handled depends on the computer and the number of places printed or displayed can often be chosen by the user.

Exercise 1 Divide 6 by 3. We write this as

$$6/3 \quad \text{or} \quad 6 \div 3 \quad \text{or} \quad \frac{6}{3} \quad \text{or} \quad 6 \times 3^{-1} = 2$$

all giving the integer 2 as answer.

Exercise 2 Divide 7 by 3.

$$7/3 = 2.333\ 333\ 33\ldots \text{ in decimals}$$
$$= 2 + 0.333\ 333\ 33.$$

The result can be expressed as an integer 2 plus a *fraction* $\frac{1}{3}$ that is as $2\frac{1}{3}$ .Computer calculations almost always express numbers in the decimal form.

1.6 Significant figures

If a real number is very large say

123 000 000

we may ask if the zeros are really *significant*, other than to indicate the very large number. For example a similar number might be

123 124 333

In terms of the problem does the last digit '3' have real meaning, or is it just thrown up by the calculation? It is important to ask which digits are actually significant. If we know from theory that a number is significant to four digits only, it is misleading to quote it to nine significant figures and it should be written as

123 100 000

It can be argued rightly that no harm is done if more figures are carried in a calculation than are significant. However, the question of significance should certainly be considered when giving the end result of a calculation. The problem of accuracy in calculations is quite complicated. Also, when we use arithmetic examples to explain mathematical theory there is sometimes a small discrepancy from theory due to rounding errors. The map maker also has to be careful to distinguish between numbers which are exact, such as coefficients given by theory, and numbers which arise from measurements, which can never be exact.

Exercise 1 Suppose the sides of a rectangle are measured on a map to be 23 mm and 45 mm. The area of the rectangle is then $23 \times 45 = 1035$ mm^2. If the measurements are uncertain to a millimetre and we had obtained 22 mm for the short side instead of 23, the result would have been $22 \times 45 = 990$ mm^2. Thus the result is only good to two significant figures if one side is also only good to two significant figures, and it might be safer to quote the answer as 1000 mm^2, or as 1035 ± 45 mm^2, or as a value lying in the range 990 to 1080 mm^2.

1.7 Truncated and rounded numbers

Numbers have to be cut short or *truncated* when quoting final results. For example a number such as 1.236 567 mm^2 for the area of a field measured from a map would be truncated to two decimal places or three significant figures as 1.23 or *rounded up* to 1.24. The figure 1.234 567 would be *rounded down* to 1.23. Computers do not automatically round numbers; they only truncate them. If rounding is required a short program segment, or algorithm, has to be used to do so.

Exercise 1 Devise a small *algorithm* to round up a number to two decimal places. An *algorithm* is the way a problem is set out for calculation. In this case, suppose the number is 1.236 567 which has to be rounded to two decimal places. If we add 0.005 to the number then truncate it we obtain the required result. The process works this way

$$1.236\,567 + 0.005 = 1.241\,567 \text{ which truncates to } 1.24$$

$$1.234\,567 + 0.005 = 1.239\,567 \text{ which truncates to } 1.23$$

Notice that the 5 is first added one digit to the right of the last place of decimals required, and then the number is truncated. This procedure is called a *rounding algorithm*. Normally this process is only used when printing out the final result of a computation.

1.8 Operators and stores

It has been assumed that the four basic operators of arithmetic are understood. The plus symbol '+' for addition, minus '–' for subtraction, the diagonal cross '×' for multiplication, and the quotient symbol '÷' for division.

In most computer languages not all of these are convenient. For multiplication a star or asterisk is used instead of the × which can easily be confused with the third last letter of the alphabet 'x'. For example 3 times 4 is written as

3 * 4

Also the division sign (÷) is often replaced by the slash / (or solidus) so we write

3 divided by 4 as 3/4

It is very useful for your calculator to have at least six stores. A number can be placed in each of these for further use. This saves keying in a number more than once (if the same number is going to be used in many calculations). Arrangements

vary from calculator to calculator. One common system is labelled *K in* and *K ou*
The six store locations are numbered 1 to 6. Thus if we key in

'123 *K in* 1' the number in store 1 is 123

To recall this number to the display we key

'K out 1'

The store locations can be used as a kind of program to carry out arithmet operations.

Exercise 1 Multiply 123 by 321 using stores 1 and 3.
The sequence of operations is as follows:

key in '123' → press 'K in 1' → key in '321' → press 'K in 3'
→ press 'K out 1 × K out 3 = '

and the result is displayed as 39 483. If now we want to add these numbers, all v need do is use the stores again and press

'K out 1 + K out 3 = '

giving the answer 444. Thus varied calculations can be carried out using tl stores without re-keying the numbers each time.

Exercise 2 Using stores 1 and 3 again, divide 123 by 321, add 123 to 321, mul ply 123 by itself and add 321.
(The last sequence is 'K out 1 × K out 1 + K out 3 =' 15 450.)

It is important to keep a record of which number is in which store for furth calculations. Also remember to check that the correct numbers have been key in by reading the displays back before any other operations are carried out.

1.9 Reciprocals

The number 1/30 is called the *reciprocal* of 30, 1/4 is the reciprocal of 4 and so o Thus $1/x$ is the reciprocal of any number called x. To calculate a result we key a number which becomes the x of the expression. Most calculators have a recipr cal key.

Exercise 1 Verify that the reciprocals of 30 and 4 are respectively 0.033 33 et and 0.25. Note these can be calculated directly using 1 ÷ 30 and 1 ÷ 4. The be efit of the reciprocal key is debatable.

1.10 Powers

When a number is multiplied by itself it is said to be *squared* or *raised to t power of two*. Thus two squared is written 2^2, two to the power of ten is 2^{10}. T power 10 etc. is called the *index*. (For more information about indices see Secti 1.12)

Most calculators have a key to square numbers and some have a key to rais number x to any power (The x^y key). However we shall do this calculation a different way to explain a general principle. To raise a number to any power we need to use the keys called *log* or *ln* and their reverse keys 10^x and e^x. These keys use *logarithms* and the *exponential* series which will be explained in Sections 1.14 and 1.15. However their use is quite easy. *(The reader may prefer to read Sections 1.11 to 1.14 before carrying out these simple exercises.)*

Exercise 1 Raise 4 to the power of 3. Answer $4^3 = 64$. There are three ways of working out the result:

(1) multiply directly $4 \times 4 \times 4 = 64$.

(2) use the *log* key (see 1.12)–the steps are

key 4 ; log ; × 3 = ; 10^x ; = 64

(3) use the *ln* key–the steps are

key 4 ; ln ; × 3 = ; e^x ; = 64

Note: some calculators will not give 64 exactly.

Exercise 2 Find the number whose cube is 64, i.e. find the *cube root* of 64. Put another way, if $x^3 = 64$ what is x? Although we could use the x^y key we will again use the *log* or *ln* keys. The steps with these keys are

key 64 ; log ; ÷ 3 = ; 10^x = 4

key 64 ; ln ; ÷ 3 = ; e^x = 4

This use of the *log* and 10^x or ln and e^x keys clearly gives a very general way of working out problems with powers.

Exercise 3 Find the *square root* of the number 123 201. This means we have to find the number which multiplied by itself equals 123 201. Using the *log* or *ln* keys as in the previous exercise we find

$$123\,201^{1/2} = \sqrt{123\,201} = 351$$

The check is that

$$351 \times 351 = 351^2 = 123\,201$$

Most calculators have a square root key '√' which can be used directly, and computer languages have SQRT function for the same purpose. However, because the *log* or *ln* keys can handle almost any problem with roots and powers, they are much more generally useful.

Exercise 4 Use the log key to find $123\,201^{-1/2}$.

The notation $^{-1/2}$ means that we wish to find $1/123\,201^{1/2}$ or $\frac{1}{123201^{1/2}}$

Using the *log* key the stages are

key 123 201 : log: change sign (+/–) : ÷ 2 = : 10^x : = $2.849\,003 \times 10^{-3}$.

The answer is the expected 1/351 in exponent form.

Thus the *log* key will deal with *negative indices*. It can also handle *fractional indices*.

Exercise 5 Find $2^{0.2345}$. It is difficult to think why such a sum would be required. What does the 0.2345 th root of 2 mean? Such expressions arise in some conical map projections. They are calculated in the same way as before. The stages are

$$\text{key in } 2 : \log : \times\ 0.2345 = : 10^x = 1.176\ 498\ 923$$

1.11 Factorials

Most calculators have a key marked '$x!$'. This is the factorial key which performs the following operation

$$x! = 1 \times 2 \times 3 \times 4 \times 5 \ldots x$$

Exercise 1 Calculate 'factorial 5' or 5!

$$5! = 1 \times 2 \times 3 \times 4 \times 5 = 120$$

1.12 Indices in multiplication

In 1.10 we showed how a calculator can be used to operate on expressions raised to powers, such as

$$4^3 = 64$$

We now discuss the matter of indices more generally. A simple index indicates how many times a number has to be multiplied by itself, for example

$$4^3 \text{ means } (4 \times 4 \times 4) = 64$$

The rule is that for *multiplication* we *add* the indices thus

$$4^1 \times 4^1 \times 4^1 \text{ means } 4^{1+1+1} = 4^3$$

The same rule applies to fractional indices such as

$$8 \times 8 = 64^{1/2} \times 64^{1/2} = 64^{1/2 + 1/2} = 64^1 = 64$$

For general indices n and m the same addition rule applies

$$4^n \times 4^m = 4^{n+m}$$

$$A^n \times A^m = A^{n+m} \qquad (1.2)$$

Exercise 1 Verify that

$$2^3 \times 2^9 = 4096$$

We have

$$2^3 \times 2^9 = 8 \times 512 = 4096$$

but using the index addition rule

$$2^3 \times 2^9 = 2^{3+9} = 2^{12} = 4096$$

1.13 Indices in division

If we define

$$\frac{1}{2^{12}} = 2^{-12}$$

or in general $$\frac{1}{A^{12}} = A^{-12}$$

we can treat division as part of multiplication. For example

$$\frac{A^8}{A^{12}} = A^8 A^{-12} = A^{8-12} = A^{-4} = \frac{1}{A^4}$$

Obviously

$$\frac{A^8}{A^{12}} = \frac{AAAAAAAA}{AAAAAAAAAAAA} = \frac{1}{AAAA} = A^{-4}$$

Exercise 1 Simplify the following expression and verify the result by calculation

$$16^2 \times 8^3.$$

Converting to powers of 2 we have

$$16^2 \times 8^3 = (2^4)^2 \times (2^3)^3 = 2^8 \times 2^9 = 2^{17}$$

To check by calculation consider the original expression $16^2 \times 8^3$.

This can be calculated from

$$(\exp(2 \times \ln 16)) \times (\exp(3 \times \ln 8)) = 256 \times 512 = 131\,072$$

The simplified expression 2^{17} can be calculated from

$$\exp(17 \times \ln 2) = 131\,072$$

We shall now explain the theory of logarithms.

1.14 Logarithms

Before the invention of *logarithms* by John Napier in the sixteenth century, all multiplications had to be carried out longhand. Logarithms were used by scientists and engineers for routine calculations for almost five centuries until mechanical, and later electronic, computers were invented. Logarithms are still important as mathematical operators in many branches of science, and in the derivation of mathematical functions themselves.

A logarithm is defined to enable the operations of multiplication and division to be made by the addition and subtraction of indices. We define:

The logarithm of a number *N* is the power to which the base *B* must be raised to give the number.

That is if

$$\log_B N = y \quad \text{then } B^y = N \quad B^{\log N} = N$$

This is really a very subtle idea. Its use is obvious from the following. Suppose we have to multiply two numbers N and M together. Then, using the addition of indices rule,

$$NM = B^{\log N} B^{\log M} = B^{(\log N + \log M)}$$

therefore

$$\log(NM) = \log N + \log M \qquad (1.3)$$

Thus we can carry out *multiplication* by the *addition* of logarithms. Of course the logarithms have first to be calculated:a very tedious task. However, once tables of logarithms had been devised, they could be used for all further calculations. Traditionally two bases were used:

a base of $B = 2.7$ (approximately) for *natural logarithms* denoted by *ln*

a base of $B = 10$ for *common logarithms* denoted by *log*

Today we are only concerned with natural logarithms although common logarithms are available on most calculators. See Section 1.15 for more information about Napier's base number $B = \mathrm{e}$.

Exercise 1 Using the common logarithms (log and 10^x keys of a calculator) show that $234 \times 567 = 132\,678$.

$$\log 234 = 2.369\,216$$

$$\log 567 = 2.753\,583$$

$$\log 234 + \log 567 = 5.122\,799\ (= x)$$

$$10^x\ (10^{5.122\,799}) = 132\,678$$

The number '2' before the decimals of the logarithm of 234 just indicates that

$$234 = 100 \times 2.34 = 10^2 \times 2.34$$

Similarly the '5' in the log of 132 678 indicates that

$$132\,678 = 10^5 \times 1.326\,78$$

Exercise 2 Using the natural logarithms (ln and e^x keys of a calculator) show that $234 \times 567 = 132\,678$.

$$\ln 234 = 5.455\,321$$

$$\ln 567 = 6.340\,359$$

$$\ln 234 + \ln 567 = 11.795\,680$$

$$\mathrm{e}^x = 132\,678$$

Thus the same answer is obtained with either base $B = \mathrm{e}$ or $B = 10$.

1.15 The base of natural logarithms e

Napier selected a special base, the *natural number e*, for his logarithms. This number is obtained by putting $x = 1$ in the *exponential series*

$$\mathrm{e}^x = 1 + x + \frac{x^2}{2!} + \frac{x^3}{3!} + \ldots + \frac{x^n}{n!} \qquad (1.4)$$

where $3! = 3 \times 2 \times 1$ is called 'factorial 3' (see Section 1.11).

Then

$$\mathrm{e} = \mathrm{e}^1 = 1 + 1 + \frac{1^2}{2!} + \frac{1^2}{3!} + \ldots + \frac{1^n}{n!} = 2.71828.$$

This may seem a strange choice of base. Its usefulness is apparent when the differentiation of $\mathbf{e}^x$ and $\mathbf{ln}\ x$ are considered in Section 9.13.

Exercise 1 Calculate e to four decimal places using the first five terms of the series. Then

$$e = 1 + 1 + \frac{1^2}{2!} + \frac{1^3}{3!} + \frac{1^4}{4!}$$

$$\begin{aligned} e &= 1 + 1 + \frac{1}{2} + \frac{1}{6} + \frac{1}{24} \\ &= 2.5 + 0.1666 + 0.0417 \\ &= 2.708 \end{aligned}$$

The error is 0.01.
Note: The error is the difference between the true value and the accepted value.

1.16 Division using logarithms

Suppose we have to divide number N by M. Then, using the subtraction of indices rule,

$$\frac{N}{M} = \frac{B^{\log N}}{B^{\log M}} = B^{(\log N - \log M)}$$

therefore

$$\log \frac{N}{M} = \log N - \log M \qquad (1.5)$$

Thus we can carry out *division* by the *subtraction* of logarithms.

Exercise 1 Using the natural logarithms show that $567/234 = 2.423\,077$

$$\begin{aligned} \ln 234 &= 5.455\,321 \\ \ln 567 &= 6.340\,359 \\ \ln 567 - \ln 234 &= 0.885\,038\ (= x) \\ e^x &= 2.423\,077 \end{aligned}$$

1.17 Powers by logarithms

The ability to raise numbers to powers by logarithms is most useful. Consider the number

$$N = P^k$$

By definition

$$\log_B P = y \quad \text{and} \quad B^y = P$$

then

$$N = P^k = (B^y)^k = B^{ky}$$

then

$$\log_B N = ky = k\log_B P$$

therefore

$$\log_B P^k = k\log_B P \qquad (1.6)$$

Exercise 1 Find the cube root of 2197. Let

$$R = 2197$$

then

$$R^{1/3} = 2197^{1/3}.$$

Taking logs

$$\ln R^{1/3} = \ln 2197^{1/3} = \tfrac{1}{3}\ln 2197 = \tfrac{1}{3}(7.694\,848) = 2.564\,949\ (= x)$$
$$e^x = 13$$

1.18 Arithmetic progression

An *arithmetic progression* (AP) is a series in which each successive term differs from adjacent terms by a fixed amount. For example

$$S = 1 + 2 + 3 + 4 + 5 \text{ to } n \text{ terms}$$

is an arithmetic progression because each term differs from adjacent ones by 1.

Exercise 1 Is the following series an arithmetic progression?

$$1 + 3 + 5 + 7 + \ldots$$

Test for the common difference (d)

$$3 - 1 = 2 \quad 5 - 3 = 2 \quad 7 - 5 = 2$$

Therefore it is an arithmetic progression (AP), $d = 2$.
We are generally interested in the sum to n terms of an AP such as

$$S = 1 + 2 + 3 + 4 + 5 \text{ to } n \text{ terms}$$

In general we can express the sum of an AP to n terms in the form

$$S = a + [a + d] + [a + 2d)] + [a + 3d] + \ldots + [a + (n-2)d] + [a + (n-1)d]$$

if we write this again backwards we have

$$S = [a + (n-1)d)] + [a + (n-2)d] + \ldots + [a + 2d] + [a + d] + a$$

then, summing the two series, we have

$$2S = [2a + (n-1)d] + [2a + (n-1)d] + [2a + (n-1)d] + \ldots \text{ to } n \text{ terms}$$
$$= n\,[2a + (n-1)d]$$

so

$$S = \frac{n}{2}\left[2a + (n-1)d\right] \qquad (1.7)$$

Then the sum to n terms of the series

$$S = 1 + 2 + 3 + 4 + 5 \text{ to } n \text{ terms}$$

where $a = 1$ and $d = 1$ is

$$S = \frac{n}{2}\left[2 + (n-1)\right]$$

$$= \frac{n}{2}(n+1) \qquad (1.8)$$

Exercise 2 Find the sum of the numbers from 1 to 10. Here $n = 10$ so

$$S = 5(11) = 55$$

which can be verified by direct addition.

1.19 Polynomials

A mathematical expression which contains several terms separated by the + or – operators is called a *polynomial*. For example

$$x + y + z$$

is a polynomial whilst

$$xyz$$

is a *monomial* expression because it has only one term. Usually a polynomial takes the form of increasing powers of a variable such as x, for example as

$$P = a + bx + cx^2 + dx^3 + \ldots$$

where a, b are *coefficients, or numbers supplied by theory.* For example P might be

$$P = 1 + 2x + 3x^2 + 4x^3 + \ldots$$

Exercise 1 Calculate P to four terms when $x = 2$ and

$$P = 1 + 2x + 3x^2 + 4x^3 + \ldots$$

We have

$$P = 1 + 2(2) + 3(4) + 4(8)$$

$$P = 1 + 4 + 12 + 32 = 49$$

A neater way to calculate P is to recast the polynomial as an algorithm in *nested form.* The expression

$$P = a + bx + cx^2 + dx^3 + ex^4 + \ldots$$

can be rearranged to

$$P = (((ex + d)x + c)x + b)x + a$$

which is much easier to calculate.

Exercise 2 Calculate P, by the nested method, to four terms when $x = 2$ and

$$P = 1 + 2x + 3x^2 + 4x^3 + \ldots$$

We have

$$P = ((4x + 3)x + 2)x + 1$$

$$P = ((4(2) + 3)(2) + 2)(2) + 1 = 49$$

1.20 Sum of the squares of the first *n* integers

Another important formula used in error theory is that for the sum of the squares of the first ***n*** whole numbers (integers). i.e.

$$S = 1^2 + 2^2 + 3^2 + 4^2 + 5^2 \text{ to } n \text{ terms}$$
$$= \frac{n}{6}(n + 1)(2n + 1) \qquad (1.9)$$

Exercise 1 Verify that

$$S = 1^2 + 2^2 + 3^2 + 4^2 = 30$$
$$S = \frac{n}{6}(n + 1)(2n + 1) = \frac{4}{6}(4 + 1)(8 + 1) = 30$$

and also

$$S = 1 + 4 + 9 + 16 = 30$$

Derivation of the formula

We derive this result by a general method using a *polynomial*. Let

$$S = a + bn + cn^2 + dn^3$$

We assume a third-order polynomial because we know that the sum of the integers is of second order. In any case if our assumption is wrong we will be unable to find the answer!

$$S = 1^2 + 2^2 + 3^2 + 4^2$$
$$= 1 + 4 + 9 + 16$$

when $n = 1$ $S = 1$ and $1 = a + b + c + d$ (A)
when $n = 2$ $S = 5$ and $5 = a + 2b + 4c + 8d$ (B)
when $n = 3$ $S = 14$ and $14 = a + 3b + 9c + 27d$ (C)
when $n = 4$ $S = 30$ and $30 = a + 4b + 16c + 64d$ (D)

The equations (A) to (D) may be solved by any elimination procedure giving the solutions

$$a = 0 \quad b = \frac{1}{6} \quad c = \frac{1}{2} \quad d = \frac{1}{3}$$

However it is instructive to employ a regular procedure as follows. Casting these equations in matrix form (see Chapter 7) gives

$$\begin{bmatrix} 1 & 1 & 1 & 1 \\ 1 & 2 & 4 & 8 \\ 1 & 3 & 9 & 27 \\ 1 & 4 & 16 & 64 \end{bmatrix} \begin{bmatrix} a \\ b \\ c \\ d \end{bmatrix} = \begin{bmatrix} 1 \\ 5 \\ 14 \\ 30 \end{bmatrix}$$

The solution may be obtained by elimination operating on rows. Subtracting row

1 from rows 1 and 2, row 2 from row 3, and row 3 from row 4, gives the new matrix

$$\begin{bmatrix} 0 & 0 & 0 & 0 \\ 0 & 1 & 3 & 7 \\ 0 & 1 & 5 & 19 \\ 0 & 1 & 7 & 37 \end{bmatrix} \begin{bmatrix} a \\ b \\ c \\ d \end{bmatrix} = \begin{bmatrix} 0 \\ 4 \\ 9 \\ 16 \end{bmatrix}$$

In the new matrix, subtracting row 2 from rows 2 and 3 and row 3 from row 4 gives another matrix

$$\begin{bmatrix} 0 & 0 & 0 & 0 \\ 0 & 0 & 0 & 0 \\ 0 & 0 & 2 & 12 \\ 0 & 0 & 2 & 18 \end{bmatrix} \begin{bmatrix} a \\ b \\ c \\ d \end{bmatrix} = \begin{bmatrix} 0 \\ 0 \\ 5 \\ 7 \end{bmatrix}$$

Finally subtracting row 3 from row 4 gives an equation in d only

$$\begin{bmatrix} 0 & 0 & 0 & 0 \\ 0 & 0 & 0 & 0 \\ 0 & 0 & 0 & 0 \\ 0 & 0 & 0 & 6 \end{bmatrix} \begin{bmatrix} a \\ b \\ c \\ d \end{bmatrix} = \begin{bmatrix} 0 \\ 0 \\ 0 \\ 2 \end{bmatrix}$$

$$\text{Hence } d = \frac{1}{3}$$

Substituting backwards and up through the decomposed matrices gives the complete solution

$$a = 0 \quad b = \frac{1}{6} \quad c = \frac{1}{2} \quad d = \frac{1}{3}$$

This solution can be verified by substitution. Finally we obtain the formula from

$$S = a + bn + cn^2 + dn^3$$

$$S = 0 + \frac{1}{6}n + \frac{1}{2}n^2 + \frac{1}{3}n^3$$

$$S = \frac{1}{6}n\left(1 + 3n + 2n^2\right)$$

$$S = \frac{n}{6}(n+1)(2n+1)$$

1.21 Ratios

In mathematics we often have to deal with *ratios*, for example the sides of *similar* triangles. Such a case might be

$$\frac{A}{B}=\frac{D}{E}$$

or in typical numbers

$$\frac{1}{5}=\frac{2}{10}$$

$A(1)$ is called the *numerator* and $B(5)$ the *denominator* of the first ratio or fraction. To deal with A, B, etc. separately we may put the ratio equal to a factor K such that

$$\frac{A}{B}=\frac{D}{E}=K$$

Then

$$A = BK \text{ and } D = EK$$

We use this to show that linear combinations of corresponding numerators and denominators are possible. For example

$$\text{If } \frac{A}{B}=\frac{D}{E} \text{ then } \frac{A+B}{B}=\frac{D+E}{E} \tag{1.10}$$

We see this is true because

$$\frac{A+B}{B}=\frac{BK+B}{B}=K+1$$

and

$$\frac{D+E}{E}=\frac{EK+E}{E}=K+1$$

Another useful device to equate numerators and denominator is to put

$$\frac{A}{B}=\frac{AP}{BP}$$

Then we can say that

$$AP = D \quad \text{and} \quad BP = E$$

For example we can see that if

$$\frac{1}{2}=\frac{4}{8}$$

then

$$P = 4 \quad \text{and} \quad 2P = 8$$

Exercise 1 (Miss this exercise until you have understood Chapter 3.) Suppose we have no table of tangents and only values of sin x are available, use the above device to evaluate x where

$$\tan x = 0.5 = \frac{1}{2}$$

We can put

$$\tan x = \frac{\sin x}{\cos x} = \frac{P\sin x}{P\cos x}$$

therefore

$$P\sin x = 1 \text{ and } P\cos x = 2$$

$$P^2\sin^2 x + P^2\cos^2 x = 1^2 + 2^2 = 5$$

$$P^2(\sin^2 x + \cos^2 x) = 5$$

But

$$\sin^2 x + \cos^2 x = 1$$

so

$$P^2 = 5 \qquad P = \pm\sqrt{5}$$

and

$$\sin x = \frac{1}{\pm\sqrt{5}} \qquad x = \pm\, 26.56°$$

1.22 Inequalities

In mathematical logic *inequalities* such as

'A is less than B' and 'B is greater than C'

are written in the form

$$A < B \quad B > C$$

It follows from these inequalities that

$$B > A \quad C < B$$

Again consider the fractions: if

$$\frac{A}{B} > \frac{D}{E}$$

then inverting both sides we have

$$\frac{B}{A} < \frac{E}{D}$$

Exercise 1 If $A = 2, B = 3, D = 3, E = 4$, clearly

$$\frac{2}{3} < \frac{3}{4} \quad \text{and} \quad \frac{3}{2} > \frac{4}{3}$$

Consider the inequality

$$\frac{2}{3} < \frac{3}{4}$$

Multiplying both sides by 12 we have

$$8 < 9$$

Inverting, we have

$$\frac{1}{8} > \frac{1}{9}$$

Multiplying again by 12 we have

$$\frac{12}{8} > \frac{12}{9}$$

therefore

$$\frac{3}{2} > \frac{4}{3}$$

1.23 The binomial theorem

One of the most important operations in ordinary arithmetic, and ordinary algebra, is the ability to expand a multiple expression in terms of its *factors*, for example, how to deal with an expression like

$$P = (a + b)(c + d) \tag{1.11}$$

$(a + b)$ and $(c + d)$ are said to be *factors* of P. For example, because

$$4 = 2 \times 2$$

the factors of 4 are 2 and 2. Other factors of 4 are 1 and 4. Consider the expression (1.11) again

$$P = (a + b)(c + d)$$

We can open out the brackets as follows

$$\begin{aligned} P = (a + b)(c + d) &= a(c + d) + b(c + d) \\ &= ac + ad + bc + bd \end{aligned}$$

The same rule applies to negative signs for

$$\begin{aligned} Q &= (a - b)(c + d) \\ &= ac + ad - bc - bd \end{aligned}$$

This expansion procedure can be generalised into a multiple product of a two termed, or *binomial*, expression such as $(a + b)^n$. The way this is done is by the *binomial theorem* which is explained below.

Exercise 1 Evaluate the following expressions by multiplying out their *factors*

$$P = (1 + 3)(6 + 2) \text{ and } Q = (5 - 1)(3 + 5)$$

Multiplying out gives

$$\begin{aligned} P &= 1 \times 6 + 1 \times 2 + 3 \times 6 + 3 \times 2 \\ &= 6 + 2 + 18 + 6 \\ &= 32 \end{aligned}$$

nd again

$$\begin{aligned} Q &= (5-1)(3+5) \\ &= 5 \times 3 + 5 \times 5 - 1 \times 3 - 1 \times 5 \\ &= 15 + 25 - 3 - 5 \\ &= 32 \end{aligned}$$

Ve see that these are correct because $4 \times 8 = 32$.

.24 Pascal's triangle

onsider the special case of (1.11) in which

$$P = (a+b)(a+b) = (a+b)^2 \qquad (1.12)$$

Iultiplying out gives

$$(a+b)^2 = a^2 + 2ab + b^2 \qquad (1.13)$$

nd similarly

$$Q = (a-b)(a-b) = (a-b)^2 \qquad (1.14)$$

$$(a-b)^2 = a^2 - 2ab + b^2 \qquad (1.15)$$

onsider also the triple product

$$R = (a+b)(a+b)(a+b) = (a+b)^3 \qquad (1.16)$$

Iultiplying out in two stages gives

$$\begin{aligned} R &= (a+b)(a+b)^2 = (a+b)(a^2 + 2ab + b^2) \\ &= a^3 + 2a^2b + ab^2 + ba^2 + 2ab^2 + b^3 \\ (a+b)^3 &= a^3 + 3a^2b + 3ab^2 + b^3 \qquad (1.17) \end{aligned}$$

xercise 1 It is left as an exercise for the reader to show that

$$(a+b)^4 = a^4 + 4a^3b + 6a^2b^2 + 4ab^3 + b^4 \qquad (1.18)$$

we examine the structure of the expressions (1.13), (1.17) and (1.18) we see ıat the coefficients follow the pattern produced by *Pascal's triangle,*

$$\begin{array}{c} 1 \\ 1\ 2\ 1 \\ 1\ 3\ 3\ 1 \\ 1\ 4\ 6\ 4\ 1 \end{array}$$

'he pattern can be seen best from line 4. An inner term is found by adding the wo terms immediately above and to its side: for example 4 = 1 + 3, 6 = 3 + 3, tc. Also the powers of a and b must always add up to the same figure: for example, in (1.17) the indices add to 3. Thus it would be easy to write down the result equested in Exercise 1 without the tedium of multiplying out.

1.25 General case of binomial theorem

It can be shown that the above process applies also to indices which are not positive whole numbers. The coefficients take the same pattern as in Pascal's triangle and the sum of the indices is always a fixed amount. The general expression is of the form

$$(a+b)^n = a^n + C_1 a^{n-1} b + C_2 a^{n-2} b^2 + \dots + C_r a^{n-r} b^r + \dots + C_{n-1} a b^{n-1} + b^n \qquad (1.19)$$

The coefficients follow the rule that

$$C_1 = n$$

$$C_2 = \frac{n(n-1)}{2!}$$

$$C_3 = \frac{n(n-1)(n-2)}{3!}$$

$$C_r = \frac{n(n-1)(n-2)\dots(n-r+1)}{r!} = \frac{n!}{r!(n-r)!} \qquad (1.20)$$

Remember that factorial 3 = 3! = 1 × 2 × 3 = 6, etc.

Exercise 1 Verify the coefficients of $(a+b)^4$ from the above formula (1.20) and Pascal's triangle. Here $n = 4$; $4! = 24$.

When $r = 1$ $\quad C_1 = \frac{4}{1!} = 4 \quad$ or $\quad \frac{24}{1 \times 6} = 4$

When $r = 2$ $\quad C_2 = \frac{4 \times 3}{2!} = 6 \quad$ or $\quad \frac{24}{2 \times 2} = 6$

When $r = 3$ $\quad C_3 = \frac{4 \times 3 \times 2}{3!} = 4 \quad$ or $\quad \frac{24}{6 \times 1} = 4$

1.26 Special case of binomial series

A very common use of the binomial theorem is as follows. Assuming $|a|$ is greater than $|b|$ (where $|a|$ is the absolute value of a, i.e. a if a is positive or negative) recast the expression as

$$\frac{S}{a^n} = \frac{1}{a^n}(a+b)^n = \left(1+\frac{b}{a}\right)^n$$

$$= (1+x)^n$$

where $|x| = \left|\frac{b}{a}\right| < 1$

Note: The symbol $|x|$ means that we take the positive value of x. This value is called the modulus of x. For example $|-2| = 2$.

Applying the binomial theorem gives

$$(1 + x)^n = 1^n + C_1 1^{n-1}x + C_2 1^{n-2}x^2 + \ldots + C_r 1^{n-r}x^r + \ldots + C_{n-1} 1x^{n-1} + x^n$$
$$= 1 + C_1 x + C_2 x^2 + \ldots + C_r x^r + \ldots + C_{n-1}x^{n-1} + x^n \qquad (1.21)$$

$$(1+x)^n = 1 + nx + \frac{n(n-1)}{2!}x^2 + \frac{n(n-1)(n-2)}{3!}x^3 + \ldots \qquad (1.22)$$

For many purposes, only the first few terms of this series are sufficient for approximate calculations. When x is small it is often sufficient to make the approximation to

$$\frac{1}{1+x} \approx 1 - x \qquad (1.23)$$

for

$$\frac{1}{1+x} = (1+x)^{-1} = 1 - x + x^2 - x^3 \text{ plus small terms}$$

Exercise 4 What is the error in using the approximation

$$\frac{1}{1.0006} = 1 - 0.0006 = 0.9994?$$

Answer: 0.0000004

SUMMARY OF KEY WORDS

normal, parallel, perpendicular, mathematical operators, braces {}, brackets [ ], parentheses ( ), addition +, multiplication ×, exponent form, integers, sequence, series, real number, significant figures, truncated and rounded numbers, algorithm, reciprocal, powers, index, log, ln, square root, cube root, negative indices, fractional indices, factorials, logarithms (common and natural), exponential series, arithmetic progression, polynomials, monomial, sum of squares of integers, ratios, numerator, denominator, inequalities, binomial theorem, Pascal's triangle, modulus

SUMMARY OF FORMULAE

$$A^n \times A^m = A^{n+m} \tag{1.2}$$

$$\log NM = \log N + \log M \tag{1.3}$$

$$e^x = 1 + x + \frac{x^2}{2!} + \frac{x^3}{3!} + \ldots + \frac{x^n}{n!} \tag{1.4}$$

$$\log \frac{N}{M} = \log N - \log M \tag{1.5}$$

$$\log_B P^k = k \log_B P \tag{1.6}$$

$$\text{Arithmetic progression } S = \frac{n}{2}\left[2a + (n-1)d\right] \tag{1.7}$$

$$\text{Sum of first } n \text{ integers} \quad S = \frac{n}{2}(n+1) \tag{1.8}$$

$$\text{Sum of squares of integers } S = \frac{n}{6}(n+1)(2n+1) \tag{1.9}$$

$$\text{If } \frac{A}{B} = \frac{D}{E} \text{ then } \frac{A+B}{B} = \frac{D+E}{E} \tag{1.10}$$

$$(a+b)^2 = a^2 + 2ab + b^2 \tag{1.13}$$

$$(a-b)^2 = a^2 - 2ab + b^2 \tag{1.15}$$

$$(a+b)^3 = a^3 + 3a^2b + 3ab^2 + b^3 \tag{1.17}$$

Binomial theorem

$$(a+b)^n = a^n + C_1 a^{n-1} b + C_2 a^{n-2} b^2 + \ldots + C_r a^{n-r} b^r + \ldots + C_{n-1} a b^{n-1} + b^n \tag{1.19}$$

$$\text{where } C_r = \frac{n(n-1)(n-2)\ldots(n-r+1)}{r!} = \frac{n!}{r!(n-r)!} \tag{1.20}$$

$$(1+x)^n = 1 + nx + \frac{n(n-1)}{2!}x^2 + \frac{n(n-1)(n-2)}{3!}x^3 + \ldots \tag{1.22}$$

Chapter 2
Plane Geometry

2.1 Planes and straight lines

In this chapter, the elementary properties of important geometrical entities are described and terminology is explained. The number of items has been kept to the very minimum, selection being restricted to those required in map making and for an understanding of later topics in this book.

We begin by looking at sheets of paper on which maps are drawn and printed. For this purpose a sheet of paper is laid flat on a table or drawing board. Thus we can say that the paper lies in a *plane*. A sheet of paper can be rolled into other shapes such as a *cylinder* and *cone,* folded into boxes and complex shapes in the art of origami. For mapping purposes we deal with the paper lying in a plane.

Formerly, many different sizes of paper were used for drawing and printing maps. Examples of these were *quarto, foolscap,* and *elephant.* One size bigger than *elephant* was called *double-elephant*; and, believe it or not, the *half double-elephant* size was not the same as an *elephant*! There has to be some agreement about paper sizes so that all maps fit together, so that printing machines are constructed to suit them, and so that map storage drawers are also made a convenient size. The accepted international paper sizes, called the *A series*, will now be used to explain some mathematical ideas.

Most people are familiar with the A4 size of paper used for everyday office work, student note pads, etc. (The size of paper used in this book is rather smaller than A4.) It will help your understanding of the following text if you have a sheet of A4 white paper ready to work some of the exercises as we go along.

The first thing to note about any paper sheet is that it forms a *rectangle.* A rectangle has its opposite sides parallel and equal, and its diagonals are also equal. When the sheet is held as in Figure 2.1 it is in a *portrait* position, and in Figure 2.2, it is *landscape*, named after the way artists use paper sheets for these two purposes.

Exercise 1 Measure the sides and diagonals of an A4 sheet of paper, called *ABCD*, as in Figure 2.1. The results, in millimetres (mm), may be something like the following

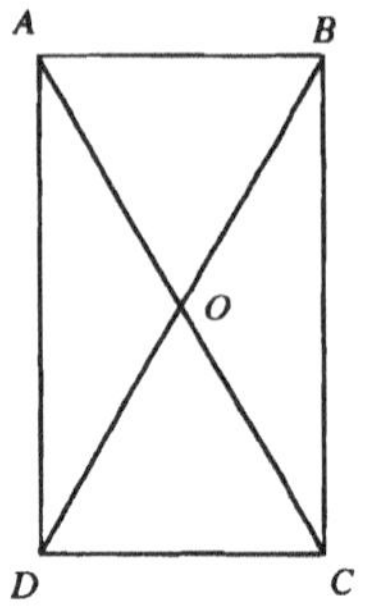

Figure 2.1

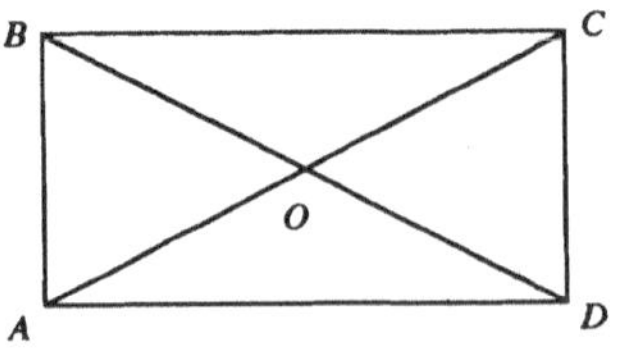

Figure 2.2

Short sides	Long sides	Diagonals
$AB = 211$	$BC = 296$	$AC = 362.8$
$CD = 210$	$AD = 297$	$BD = 363.5$

The reason that we do not have a perfect rectangle is due to errors in the measurements, but see below for the theoretically exact dimensions.

It is difficult to measure the diagonals directly with a 300 mm ruler. The first thing to do is to draw the diagonals in pencil by finding the middle of the paper, say point O. This can be done, without folding and creasing the paper, by stretching a piece of thread from corner to corner, marking a short pencil line section in each case. Once the centre has been found, the diagonals can be drawn in two parts, and the lines measured, also in two parts, in this case as $300 + 62.8 = 362.8$ and $300 + 63.5 = 363.5$.

It is particularly interesting to use a piece of thread for this exercise, because the name *straight line* means 'stretched linen thread'. It also shows that a *straight line* is the shortest distance in free space between two points. In practical work a point and a line are usually about 0.2 mm thick so that they can be seen. Sometimes it is necessary to think about the middle of these as having no thickness at all. In surveying and map construction, the setting-out or drawing of a precise grid is of fundamental importance.

Exercise 2 Show that the ratio $AB/BC = 0.713$ is approximately the same as the ratio $0.5 \times BC/AB = 0.701$.

This means that, if the paper is folded in two, the rectangle formed is almost the same shape as the original but half its size. The smaller paper size is A5. In fact it should be exactly the same shape. The way the A paper sizes are devised follows a pattern. If the paper proportions are chosen so that

$$\frac{AB}{BC} = \frac{0.5BC}{AB} \text{ or } AB^2 = \frac{1}{2}BC^2 \text{ or } AB = \frac{BC}{\sqrt{2}}$$

when folded over, the sides will be in the same ratios exactly. The A system of paper sizes begins with a sheet exactly one square metre in area with sides in these proportions, i.e. such that

$$AB \times BC = 1 \text{ and } AB = \frac{BC}{\sqrt{2}}$$

Thus

$$\frac{BC^2}{\sqrt{2}} = 1$$

therefore

$$BC = \sqrt{\sqrt{2}} = \sqrt{1.4142} = 1.1892$$

$$\text{and } AB = 0.8409$$

The sheet of paper thus created is called *AO* size. Halving and halving in sequence gives all the other A paper sizes: A4 is

$$1.1892/4 = 0.2973 \text{ m by } 0.8409/4 = 0.2102 \text{ m}$$

$$\text{or } 297.3 \text{ mm by } 210.2 \text{ mm}$$

2.2 Straight lines

Straight lines may be drawn mechanically by moving a pencil against the *straight edge* of a ruler. To test for straightness, the ruler can be reversed and another line drawn to coincide with the first. If the ruler is curved, a gap shows between the two lines which are also curved. It is important in map drawing to make this test, even if a computer plotter is used to draw the lines. In surveying, a straight line can be obtained by viewing objects in line, either with the unaided eye, or through a telescope. The finest line drawn on most maps is 0.2 mm thick.

As already explained above, yet another way to create a straight line is to pull a fine string tight. (**Note**: A light path may bend due to *refraction* and a string can sag appreciably due to *gravitational force*: effects which must be considered in precise work, but are ignored here.)

The length of a *segment* of a line may be measured by graduated scale or tape or laser beam or radio wave. For precise cartographic work a steel *straight edge* (manufactured for straightness) is used separately from the *scale* (for length). In ordinary work these two features are combined into a conventional ruler capable of measuring to 0.2 mm.

2.3 Intersecting, normal and parallel lines

Two lines drawn on a flat piece of paper (a *plane*) so that they cut each other are said to *intersect* at an angle, which is measured in the plane of the paper. Generally there are two such angles, an *acute* one such as B, and an *obtuse* one such as A (see Figure 2.3). Notice that in the sexagesimal system of angles (360° to the circle) we have

$$B = 180^\circ - A$$

If the angles A and B are equal (see Figure 2.4), they must both be right angles (or 90°). In this case the lines are said to be *orthogonal* to each other, or one line is *normal* to the other or at *right angles to* the other. If two lines in a plane do not intersect, they are said to be *parallel* to each other.

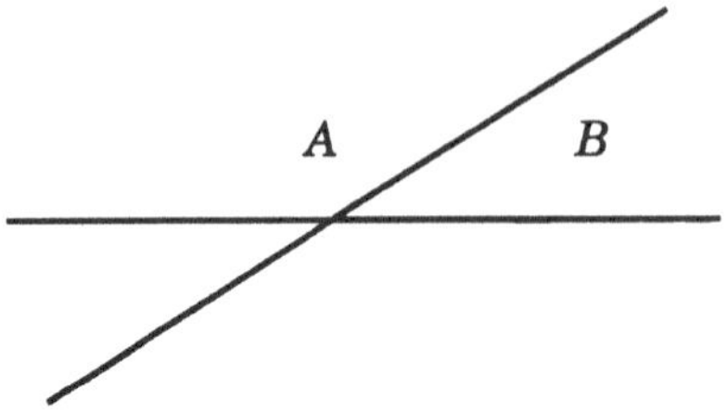

Figure 2.3

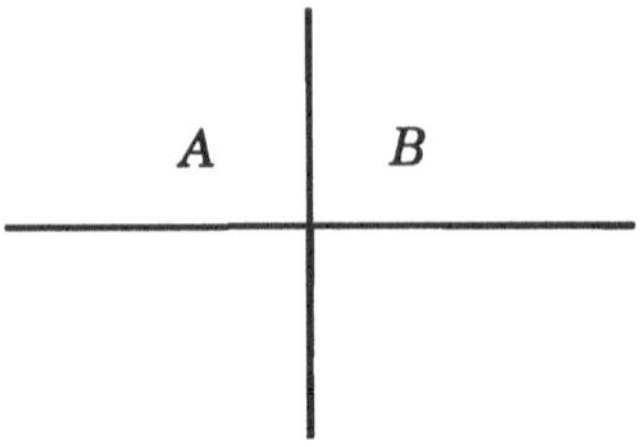

Figure 2.4

2.4 Reference grid

A network of regularly spaced parallel and orthogonal lines forms a *grid*. In surveying and mapping the accuracy of the grid is vital to most operations. It is therefore important to consider this in some detail both for its own sake and to introduce some basic geometrical concepts.

It has always been possible to create a right angle by simple methods and therefore to construct a grid. To capture the spirit of this truth, the reader is invited to perform the task of drawing a grid as outlined here using only a ruler, pencil and sheet of paper A4 size.

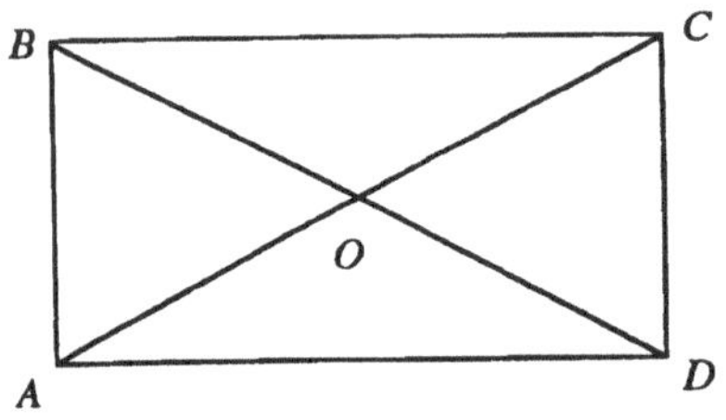

Figure 2.5

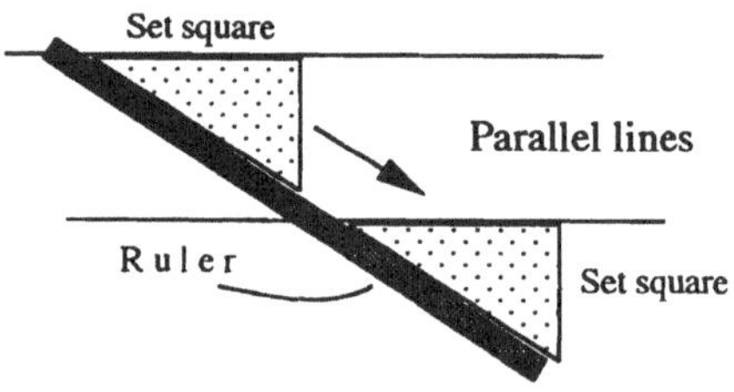

Figure 2.6

Start by drawing two straight lines roughly diagonally across the paper. Let them intersect at point O. From O mark off four equal lines OA, OB, OC and OD as shown in Figure 2.5, and join across the sides AB, BC etc. You have now drawn a rectangle.

Check that opposite sides are equal to drawing tolerance (0.2 mm) and that the diagonals AC and BD are also equal. If they are not, repeat the task and try to see why the error has crept in.

The opposite sides should also be parallel to each other. Test this by sliding a set square against a ruler, as shown in Figure 2.6.

In mathematics we can define that *ABCD* is a perfect rectangle, in reality it will only be a close approximation to one. We say that the perfect rectangle *ABCD* is a *mathematical model* of the real thing which you have just drawn. The differences between the two will be the subject of statistical and error analysis. In practical surveying, engineering, woodworking, etc. we have to decide on an acceptable tolerance for the practical creation of the rectangle. For example a wooden door is probably good to about 5 mm, a tennis court to about 10 mm, the grid of a map to about 0.2 mm.

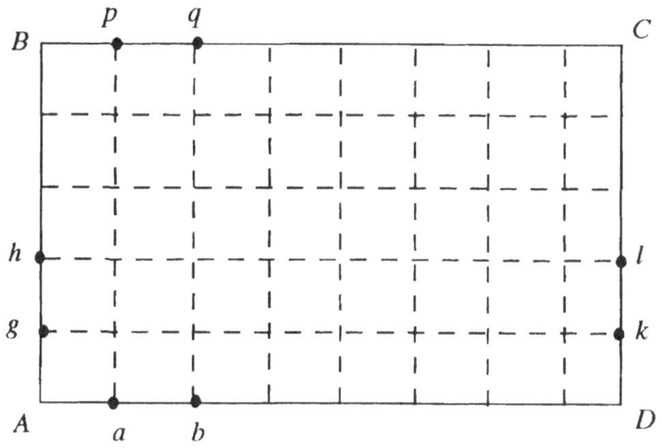

Figure 2.7

The basic rectangle *ABCD* is now used to construct a grid, as in Figure 2.7, say at 40 mm intervals. Point *A* is chosen as the *origin* of the grid and *AD* and *AB* its *axes*. Along *AD* and *BC* mark off points, *a*, *b*,... *p*, *q* at multiples of 40 mm apart. Do this first by stepping along at 40 mm intervals and see how the error accumulates! The correct way is to use the ruler to mark off all distances from the origin, i.e. at 40 mm, 80 mm and so on.

The lines *ap*, *bq*, etc. are all parallel and equally spaced. Again from the base line *AD* mark off the points *g*, *h*, ... and *k*, *l* on *AB* and *DC* and join the lines across to form the grid. Except by accident, or prior calculation, the original construction rectangle *ABCD* does not fit the edges of the grid exactly. Usually the original rectangle, draw in pencil, is erased and the final grid plotted in ink.

On a building or archaeological site, where the grid is used to set out or measure the footings of the walls, the grid intersections are marked by pegs or steel plates. Computer and cartographic plotters use a mechanical system of orthogonal rails to establish the grid quickly. Sometimes these rails are incorrectly aligned and need to be checked. Grids are used as the basis of a coordinate system. See Chapter 4.

2.5 Parallelogram, rectangle, square, and rhombus

Clearly knowledge of the geometry of the rectangle is essential for surveying and mapping. Consider the properties of the rectangle *ABCD*. Which of them are sufficient to define it ? Is it sufficient to say that 'opposite sides are parallel'?

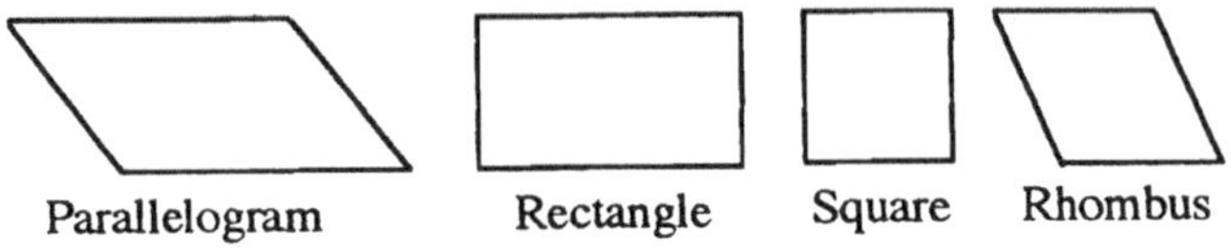

Figure 2.8

Figure 2.8 clearly shows that this is not so. Such a definition defines only a *parallelogram*. The method used to check the rectangle, by measuring its diagonals, gives a sufficient property to define it. *A rectangle is a parallelogram whose diagonals are equal.* A *square* is a special case of a rectangle *whose sides are equal.* A *rhombus* is a parallelogram whose sides are equal.

2.6 Pythagoras's theorem for a right angled triangle

The word 'theorem' is used in mathematics to describe some rule obeyed by a class of similar things, in this case the sides of a right angled triangle, such as ΔABC (Δ means 'triangle'). The long side *AC* is called the *hypotenuse*. The theorem of Pythagoras states that, *in a right angled triangle,*

the square on the hypotenuse = the sum of the squares on the other two sides

Applying the theorem to the Δ ABC this means that

$$AC^2 = AB^2 + BC^2$$

The notation used here for the *area* of a square is $AC^2 = AC$ times $AC = AC.\,AC$.

Exercise 1 Is the triangle whose sides are 3, 4, and 5 units long a right angled triangle? Answer: 'yes', because

$$5^2 = 3^2 + 4^2 \quad \text{i.e.} \quad 25 = 9 + 16$$

This gives an easy way to construct a right angle using a tape or ruler.

Exercise 2 Which of the following triangles are right angled? The three sides in each case are

(1) 6, 8, 10 (2) 5, 6, 7 (3) 2, 3, 9

The first is right angled, the second is not, and the third is not a triangle at all. For three sides to form a triangle, the sum of any two sides must be greater than the third which is not so in the third example.

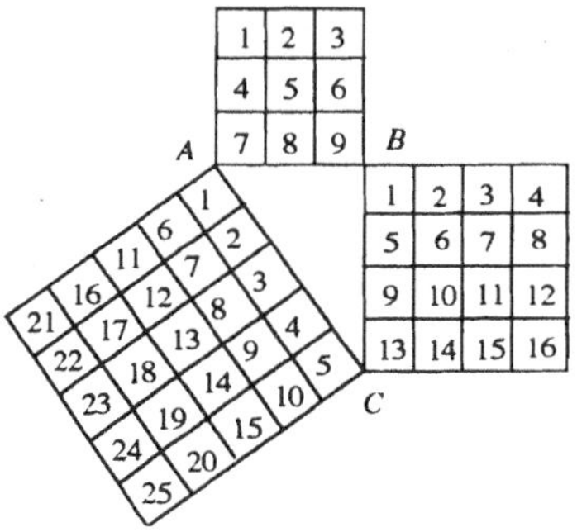

Figure 2.9

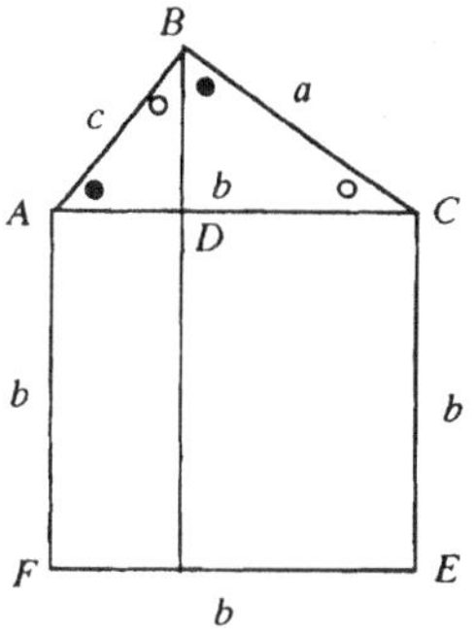

Figure 2.10

2.7 Proof of Pythagoras's theorem

That Pythagoras's theorem is true for the (3, 4, 5) triangle may be seen by inspection of Figure 2.9. It can be seen that AC^2 = the area of the square formed on the hypotenuse. If AC is measured in millimetres, the area is counted in units of *square millimetres*, written as mm^2. By inspection and counting of the squares

$$AC^2 = AB^2 + BC^2 = 25$$

Another notation is to name the sides after the angles opposite them, in small (lower case) letters thus

$$b^2 = c^2 + a^2$$

This *demonstrates* the theorem, but does not *prove* it. To do so, consider *any* right angled triangle ABC as in Figure 2.10. The angles of the triangle are written in capital letters (A, B, C), and the sides opposite them in lower case letters (a, b, c). Sometimes if there is some doubt as to which angle is meant, it is written in the full form such as

$$A\hat{B}C \text{ or } \angle ABC \text{ instead of the simpler } B.$$

The point D is the foot of the perpendicular from B to AC. By inspection we can see that

$$C\hat{B}D = 90 - C = A \text{ and } A\hat{B}D = 90 - A = C$$

Thus the two triangles BDC and ABC are the same shape but of different sizes. The two triangles are said to be *mathematically similar*. Notice we list the points in the correct order to indicate which angles in each triangle are equal. Since these triangles are similar, the corresponding sides differ by a constant *scale factor*, say k. Then we can say that

$$DC = kBC = ka \tag{2.1}$$

$$BC = kAC \text{ therefore } a = kb \tag{2.2}$$

If we wish to calculate k we do so from any of the ratios such as

$$k = \frac{a}{b} \tag{2.3}$$

Exercise 1 Show that, for the triangle whose sides are $c = 3, a = 4, b = 5,$

$$k = 0.8$$

Then

$$BD = kc = 0.8 \times 3 = 2.4 \quad \text{and} \quad DC = 0.8 \times 4 = 3.2$$

Verify these results by measurement.

In this theoretical discussion we do not need to know k. However we may p t

$$DC = ka = \frac{a}{b}a = \frac{a^2}{b}$$

In Figure 2.10, the large square $ACEF$ of side length b can be divided into tv : rectangles whose areas are $DC.b$ and $AD.b$, therefore

$$\text{area } DC.b = \frac{a^2}{b} \cdot b = a^2$$

In the same way by comparing similar triangles ADB and ABC we have

$$\text{area } AD.\ b = c^2$$

Finally by adding areas we have that

$$\text{square } ACEF = b^2 = \text{area } DC.b + \text{area } AD.b = c^2 + a^2$$

so

$$b^2 = c^2 + a^2 \tag{2.4}$$

This is Pythagoras's theorem proved for any triangle ABC, right angled at B.

Exercise 2 Write down the angles and sides of the triangle PQR right angled t Q. Also write out Pythagoras's theorem for this triangle. Sides are p opposite P ; opposite Q and r opposite R. Since Q is a right angle

$$q^2 = p^2 + r^2$$

2.8 The circle

The circle is surprisingly important and useful in practical life. Just look arou c your house to see how many circular objects there are in it: plates, knobs; wat ı pipes, bottles, tin cans with circular sections; buttons, wheels of all kinds, and c on. One reason for this popularity is that a circle is easy to make, say on a lath : or draw with a pair of compasses. In science too, the circle is important. It is t e basis of goniometers (devices for measuring angles), it is used to describe sectio s through a sphere, and is the basis for much theory about the shape of curves. r surveying and mapping the circle is second in importance only to the straig ı line.Therefore it is right to make some effort to understand the plane geomet y of the circle. A further treatment of the circle is given in section 10.3.

Exercise 1 Take a can from the kitchen, place its flat end on a sheet of pap r and draw round it to give a circle, as in Figure 2.11. Next, using your ruler, try (

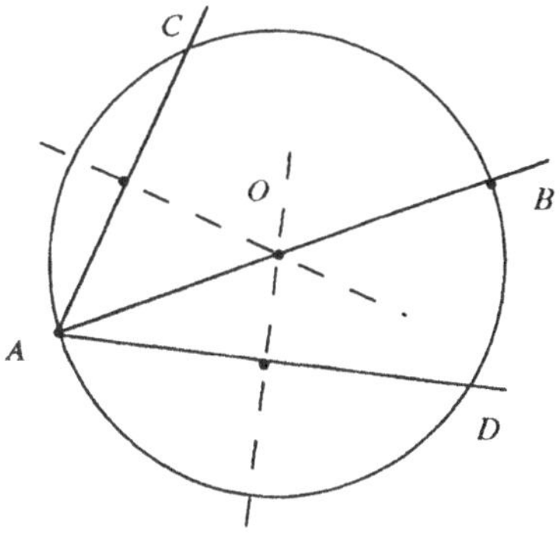

Figure 2.11

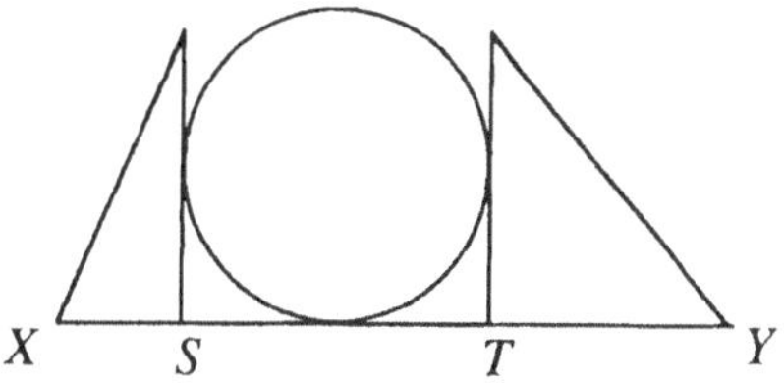

Figure 2.12

measure the greatest distance across the circle, the line *AB* of Figure 2.11. The line *AB* is a *diameter* of the circle, approximately 76 mm for a typical can of beans. Draw other lines through *A* to cut the circle in points *C* and *D* as shown. These lines *AC* and *AD* are called *chords* of the circle *ABC*. You can see by inspection or by measurement that the *diameter* of a circle is the greatest *chord* that can be drawn to a circle. Note that we are not really sure that *AB* is the longest chord, only nearly so.

Exercise 2 Try the following better way to measure the diameter of the circle or the can itself. Draw a line *XY* touching the circle as shown in Figure 2.12. Such a line is called a *tangent* to the circle. Using the set square, drop two other tangents perpendicular to *XY* cutting it at *S* and *T*. *ST* is of equal length to the diameter of the circle. In our example the result was 76 mm. This technique for measuring diameters is incorporated into surveying instruments and can be used optically.

It is often very important to find the centre of the circle. One way to do so is given in the next exercise.

Exercise 3 Find the mid points of the chords *AC* and *AD* of Figure 2.11 by measuring with the ruler. Using a set square draw other lines perpendicular to the original chords passing through their mid points. These are shown as broken lines in Figure 2.11. These broken lines meet at a point, the *centre O* of the circle *ABC*.

Exercise 4 Measure the lines *OA, OB, OC* and *OD*. Our results in millimetres were 38, 38, 37.5 and 37.5 respectively. Ideally they should all be equal to 38 mm, or half the length of *AB*. A line such as *OA* is called a *radius* of the circle. (The plural of radius is *radii*.) The radius is half the length of the diameter. If we call the length of the radius *r* and the diameter *d* we can express this connection as an *equation*

$$r = \frac{1}{2}d \tag{2.5}$$

Note that the equation (2.5) applies to all circles, not just this one. The quantities r and d are often called *parameters* of the circle because they are the key pieces of information needed to describe the circle. Note also that only one parameter, r or d, is needed because the other can be found from it using equation (2.5). The parameter d is said to be *dependent* on r, and vice versa.

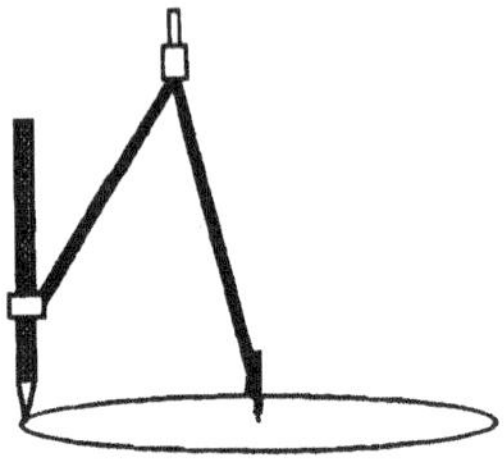

Figure 2.13

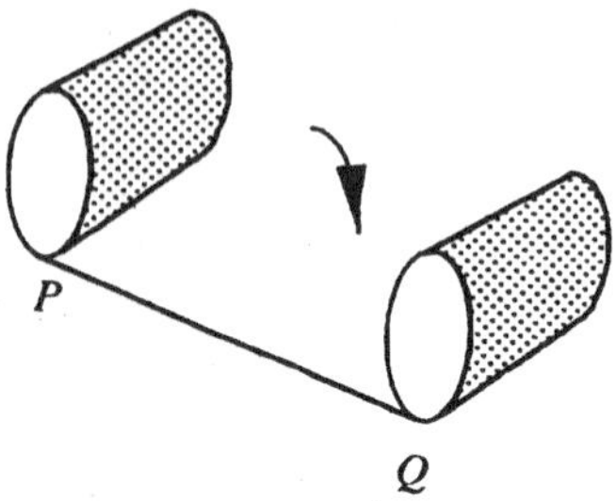

Figure 2.14

Another simple way to draw a circle from a known point as its centre is to use a pair of compasses (see Figure 2.13).

Exercise 5 With a pair of compasses, set the distance between the pencil and needle to be 38 mm and draw another circle. Clearly this way of drawing a circle is as easy as the first, but has the advantages that it gives the centre and radius straight away *by design*. Both these problems arise in surveying and mapping. Sometimes we need to find the parameters of a given circle such as an oil tank, or alternatively create a circle, such as a circular flower bed, from design parameters. The first method using chords gives a way to solve the oil tank problem, and the flower bed can be swung out using a piece of rope.

It should be noted that the above two ways of drawing a circle are *analogue* techniques. They depend on some mechanical device to do the drawing. The first uses another circle which is just copied, and the other a fixed dimension rotated about a point. The computer system, to be described later, uses *digital* techniques.

Exercise 6 Stand a tin can with its round side on a sheet of paper, mark a starting point, and roll it along a straight line until the starting point is reached again (see Figure 2.14). This transfers the distance round the circle (its *circumference*) into a straight line whose length can be measured. (**Note**: The rolling process needs care to avoid slipping.) Measure the length of the line PQ. Now divide the circumference c by the diameter d. *From a typical set of measurements the results were*

$$236/76 = 3.11$$

You should obtain a result of about 3.14. No matter how many circles we draw or of what size, this result is always the same, namely

$$\frac{c}{d} = \text{a constant, about } 3.14.$$

The Greek mathematicians, who first found this result, gave to the constant the symbol π (pronounced *pie* but spelled pi).Thus we have the important equation

$$\frac{c}{d} = \pi \tag{2.6}$$

Pi is so important that most calculators store its value permanently in a special memory accessed by a key labelled 'π'. Our calculator gives pi to nine decimal places as

$$3.141\,592\,654$$

Exercise 7 Show that pi is approximated by the following ratios:

22/7 to three decimal places, and

355/113 to six decimal places.

Exercise 8 The circumference of a tree was found by tape to be 3 metres. Verify that its radius is 0.477 m. (**Note**: This assumes that the tree has a circular cross-section.)

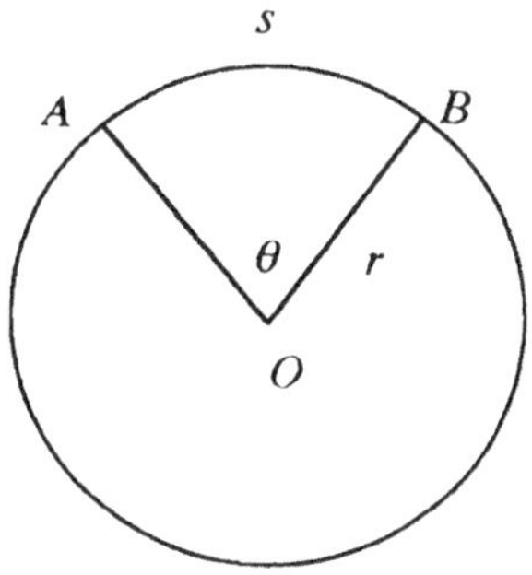

Figure 2.15

2.9 Angles

Figure 2.15 shows an angle *AOB* subtended, by an *arc AB* of length *s*, at the centre of a circle, radius *r*, whose centre is at *O*. It is common practice to label an angle in one of several ways.

It may be written as '*angle AOB*' placing the middle letter at the point from which the arms of the angle radiate. Angles are considered positive when measured in a clockwise manner. The *angle BOA*, on the other hand, is a negative acute angle. It may also be considered as a positive *exterior* angle. Thus it is important to label the points in a consistent manner, especially when calculating with a computer and to adopt a consistent convention. Other forms of writing an angle are

$$A\hat{O}B \quad \text{or} \quad \angle AOB$$

Where there is no ambiguity, a single letter may be used to represent an angle. Greek letters are often used (see page xiii). For example, at a point *A* an angle might be denoted by α (alpha), at B by β (beta) and at C by γ (gamma). The other Greek letters commonly used to denote angles are

theta θ, phi ϕ, and omega ω.

In Figure 2.15 we have denoted the angle at O by θ.

Angles are measured in a variety of units. First of all, one complete revolution of the radius may be called one *cycle*. Again, the whole circle may be divided equally into four parts, called right angles, which are themselves further divided in different ways.

2.10 Sexagesimal system

A common method is to divide a right angle into 90 parts or degrees, written 90°. Each degree is then further divided into 60 parts, or minutes of arc, written 60′; and finally each minute of arc is divided into 60 parts or seconds of arc, written as 60″.

Exercise 1 Verify that there are 324 000″ in a right angle and 1 296 000″ in a whole circle of 360°.

Exercise 2 Show that an angle of 47° 22′ 45″ is 47.379 167°. Most calculators have a special key to make this conversion. However the full calculation is written as

$$47 + 22/60 + 45/3600 = 47 + 0.366\,667 + 0.012\,500 = 47.379\,167$$

Exercise 3 Convert the angle 47.379 167° into degrees, minutes and seconds, as follows.

$$47.379\,167 - 47 = 0.379\,167, \quad 0.379\,167 \times 60' = 22.75\,002'$$

$$22.75\,002 - 22 = 0.75\,002, \quad 0.75\,002 \times 60'' = 45''$$ to the nearest second of arc.

2.11 Sexagesimal time system

You will have noticed that we used the words 'seconds of arc' in the above explanation. This is because of another way to divide up a whole circle: into units of time. The scientific clock face is divided into twenty four parts, one for each hour. Thus a right angle consists of six hours. If we divide each hour into 60 minutes of time, written 60^m, and each minute into 60 seconds of time written 60^s, these minutes and seconds are not the same size as the minutes and seconds of arc.

Exercise 1 Show that a minute of time is 15 times larger than a minute of arc. This follows from the fact that 360/24 = 15. Verify also that a second of time is numerically fifteen times larger than a second of arc. In surveying both units are used.

2.12 Centesimal system

There is another way to divide up a right angle, into 100 parts, called *gons*, written ^{g}, which is now the standard method on the continent of Europe. Each gon is divided into 100 centigons, written ^{c}, and each centigon is divided into 100 parts, written ^{cc}. This decimal system greatly simplifies the arithmetic. (**Note**: Another name for the 'gon' is the 'grad'.)

Exercise 1 An angle of $47.2245^{g} = 47^{g}, 22^{c}, 45^{cc}$. Thus no arithmetic is needed in a conversion.

2.13 Radian system

In mathematics however, instead of these arbitrary systems, the unit of angle employed is the *radian*. This quite simple concept can be explained as follows. Refer to Figure 2.15. If the angle θ is such that the *arc* of the circle $AB = s$ is equal in length to the radius r, then θ is defined to be one *radian*. Radians are sometimes called 'circular measures'. The reason for adopting this system of angular units is to allow us to relate the angle at the centre of a circle to the length of arc it subtends in a simple way.

Exercise 1 How many radians are there in a full circle? Answer: two pi radians, or 2π radians, or approximately 6.28 rad. The size of an angle in radians is given by the length of the circular arc divided by the radius, thus in this case

$$\text{angle} = c/r = 2\pi r/r = 2\pi$$

where c is the circumference. It might be thought that this is a complicated way of dealing with angles, especially as there is no exact number in a complete cycle. In mathematics it is the simplest system, but in practical measurement of angle it is not. Thus both methods are needed. The degree and gon systems are used for instruments, such as theodolites or protractors, and the radian system in mathematics.

When angles are mentioned in mathematical formulae, the unit of measurement is always the radian unless otherwise stated.

Exercise 2 Show that there are approximately 57.3° in a radian. This is so because

$$2\pi\ \text{rad} = 360°, \text{ therefore one rad} = 360°/2\pi = 360°/6.28 \approx 57.3°$$

where the symbol $\approx$ means approximately equal to. To convert any angle $\alpha°$ to radians we use the formula

$$\alpha\ \text{rad} = \alpha° \times \pi/180° \qquad (2.7)$$

Exercise 3 Convert the sexagesimal angle 47° 22′ 45″ to its radian equivalent. First convert this angle to decimal degrees as above to obtain 47.379 167°, and convert this to radians using equation (2.7). (**Note**: Use the value of pi from your calculator.)

$$\text{angle in radians} = 47.379\,167° \times \pi/180° = 0.826\,922$$

The converse calculation from radians to degrees is carried by the formula

$$\alpha^\circ = \alpha \,\text{rad} \times 180^\circ/\pi \qquad (2.8)$$

Exercise 4 Convert the angle 0.826 922 rad to sexagesimal degrees using equation (2.8)

$$\text{angle}^\circ = 0.826\,922 \times 180^\circ/\pi = 47.379\,167^\circ$$

Exercise 5 Find the length of the arc of a circle of radius 300 mm subtended by an angle of 45° at its centre. The arc s is given by

$$s = r\theta \quad \text{where } \theta \text{ is in radians.}$$

Thus $s = r\theta^\circ \times \pi/180^\circ = 235.62$ mm

This is one of the most important calculations in mapping.

Exercise 6 What is the length of arc subtended by an angle of one sexagesimal second of arc (1″) on the surface of the Earth whose radius is 6 378 140 m?

$$s = r\theta = \frac{6378140 \times 1 \times \pi}{60 \times 60 \times 180} = 30.922\text{m (approx. 100 ft)}$$

Note: $1'' = \left(\frac{1}{60 \times 60}\right)^\circ$

Exercise 7 In seconds of arc, what angle is subtended at the centre of the Earth by a distance of 3 mm?

We will use the result in Exercise 6 to arrive at an approximate answer. The angle subtended by 30.922 m (= 30 922 mm) is 1". Therefore 3 mm subtends an angle of 3/30 922" = 0.0001" approximately. The implication of this for the cartographer is that very small angles are involved in maps and map projections. These need special care in calculations.

SUMMARY OF KEY WORDS

plane, cylinder, cone, A series, rectangle, portrait, landscape, straight line, segment, straight edge, scale, plane, orthogonal, normal, right angle, parallel, grid, origin, parallelogram, rhombus, hypotenuse, area, diameter = d, chord, tangent, radius = r, equation, centre, acute, obtuse, parameters, analogue, digital, circumference = c, $\pi = \text{pi} = \frac{c}{d}$, *angle, cycle, sexagesimal, degrees, minutes, seconds, time system, centesimal system, gons, radian system*

SUMMARY OF FORMULAE

Pythagoras's theorem: right angle at B $\quad b^2 = c^2 + a^2$ (2.4)

radius of circle $\quad r = \frac{1}{2}d\quad$ diameter (2.5)

circumference / diameter $\quad \frac{c}{d} = \pi \quad$ constant pi (2.6)

$$\alpha \text{ rad} = \alpha^\circ \times \pi/180^\circ \qquad (2.7)$$

$$\alpha^\circ = \alpha \text{ rad} \times 180^\circ/\pi \qquad (2.8)$$

Chapter 3 Trigonometry

3.1 Introduction

Within the subject of *trigonometry*, many of the ideas and theorems of geometry are converted into a language, or algebra, which can be used to develop the subject without the need for complicated diagrams. In turn, these algebraic expressions are used to make calculations for all manner of practical purposes in engineering, science and map making.

3.2 Functions of angles

In surveying and mapping, angles by themselves are not much use. To be really useful they are modified into *functions*. We will explain this concept with an example.

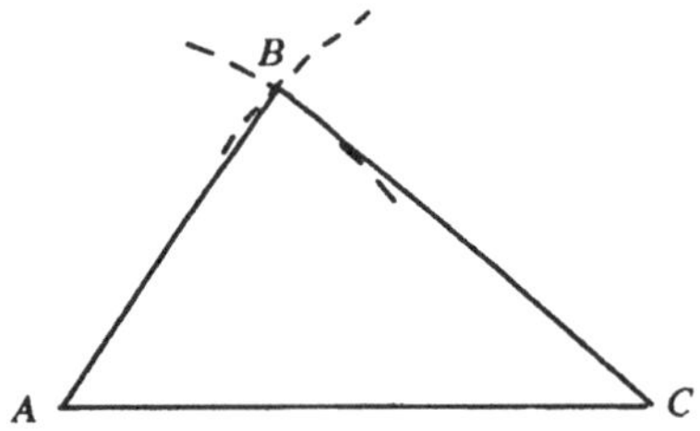

Figure 3.1

Exercise 1 Draw a triangle ABC whose sides are 7, 9 and 10 cm long. This is best done by drawing the longest side first: call it AC. Then using the compasses describe arcs $AB = 7$ cm and $CB = 9$ cm as indicated in Figure 3.1. This completes the triangle ABC.

Exercise 2 With a protractor, measure the angles A, B and C. You should find them to be about the following values in sexagesimal degrees

$$A = 61.5 \quad B = 75.5 \quad C = 43.5$$

Their sum is

$$A + B + C = 180.5°$$

Theoretically this sum should be two right angles, or 180°, or π radians. In practice there is usually an error of measurement. How can we find what the exact values of these angles should be, given the lengths of the sides? The answer is to use *functions of the angles.* We can calculate the angle A using the cosine formula (3.6) given later and find it to be 60.94°. Then using the same or another formula (3.9) we can calculate angle B. Once angles A and B are known, we find C from the equation

$$C° = 180° - A° - B°$$

Once the angles are known, several other important calculations can be made, for example, to enable the triangle to be plotted on a map, or for other purposes. Most of the mathematics involved in cartography concerns functions of angles and sides: this is the branch of mathematics called *trigonometry.*

3.3 The cosine and secant function

We start by considering the circle of Figure 3.2 in which two diameters AB and CD, meeting at O, are inclined to each other by an angle of 56°. Lines CE and DF are perpendicular to AB. Thus CE is parallel to FD. Angles ECD and FDC are both 34°.

Exercise 1 Draw a copy of Figure 3.2, making AB = 84 mm i.e. radius r = 42 mm. Measure OF, OE, FD, and EC. They should be approximately 23.5, 23.5, 34.5 and 34.5 mm respectively.

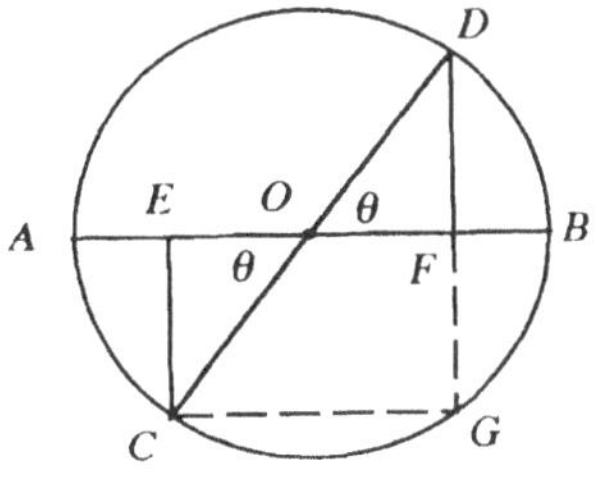

Figure 3.2

Now calculate the ratio OF/OD = 23.5/42 = 0.5595. Clearly this ratio has something to do with the angle DOF = 56°. No matter what size we make the circle, this ratio will be the same, subject to drawing and measuring tolerances.

Exercise 2 Show that the ratio EF/CD = 47/84 = 0.5595. The broken lines indicate why this is so.

Exercise 3 Now consider the problem around the other way. If we know that OD = 42 mm and that the ratio is 0.5595, we can calculate

$$OF = OD \times \text{ratio} = 42 \times 0.5595 = 23.5 \text{ and also}$$

$$EF = CG = CD \times \text{ratio} = 84 \times 0.5595 = 47$$

This ratio is called the *cosine of 56°* or *cos 56°* for short. Remember cos 56° is just one number. Thus cos θ is also a number. The function name is 'cos θ'. The 'cos' cannot be separated from 'θ'. This *cos* function is to be found in most calculators, which also allow for the angle to be input in 'deg' (sexagesimal degrees), or 'grad' (gons), or 'rad' (radians).

Exercise 4 Verify with your calculator that cos 56° = 0.5592 to four decimal places. The result we obtained by drawing is close enough. The cosine function is *the basic function of trigonometry*. It can be calculated from the value of the angle itself (see 9.13) without drawing. Normally a calculator will give the result to nine decimal places, which is adequate for most surveying work.

Exercise 5 A sloping line is measured to be 123.456 m long. If the slope of the line is 2°, verify that its horizontal equivalent is 123.381 m.

This is a common calculation, because maps show only horizontal distances. If the sloping line is called 's' and the horizontal equivalent 'd' they are related by the equation

$$d = s\cos\theta \tag{3.1}$$

Exercise 6 A horizontal line is 123.381 m long. If its slope is 2° show that its slant length is 123.456 m. The result is obtained from

$$123.381/\cos 2° = 123.381/0.999\ 390 = 123.456$$

For convenience in writing, 1/cosine is called the *secant*. Thus 1/cos 2° = sec 2°. However the secant function is not usually given in a calculator because it is not really needed. If we look again at Figure 3.2 we can see that sec $\theta = OD/OF$. Thus another equation relating s and d is

$$s = d\sec\theta \tag{3.2}$$

Which is neater than writing $s = d/\cos\theta$, even though this formula is used in calculations.

Exercise 7 Using a calculator verify that sec 2° = 1.000 609 544

Notation for cosine squared

Because cos θ is just a number, we can square it, or take its square root and so on. It would be cumbersome to write 'the square of cos θ' or to write it as

$$(\cos\theta)^2$$

Instead, it is written

$$\cos^2\theta$$

Remember this just means cos $\theta \times$ cos θ.

Exercise 8 Show that $\cos^2 56° = 0.3127$ to four decimals.

$$\cos 56° = 0.559\ 19 \quad \text{and}\ 0.559\ 19 \times 0.559\ 19 = 0.3127$$

3.4 The cosine formula for a plane triangle

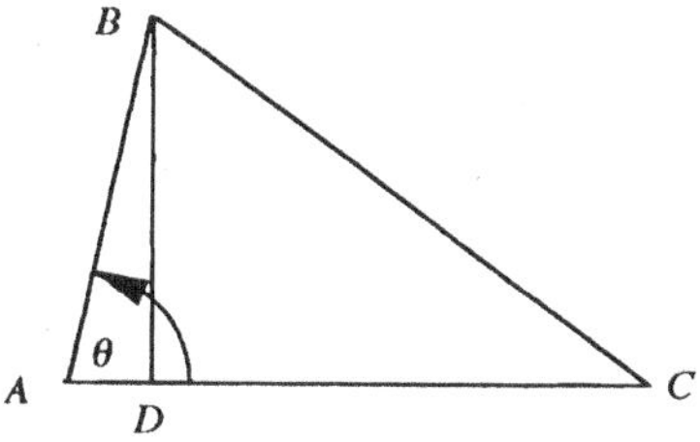

Figure 3.3

Consider the triangle ABC of Figure 3.3. BD is perpendicular to AC. Using Pythagoras's theorem 2.7, in triangles ABD and BCD we have

$$AB^2 = AD^2 + BD^2 \quad \text{and} \quad BC^2 = DC^2 + BD^2$$

therefore

$$AB^2 - BC^2 = AD^2 + BD^2 - (DC^2 + BD^2) = AD^2 - DC^2 \qquad (3.3)$$

But

$$DC = AC - AD$$

therefore (1.15)

$$DC^2 = (AC - AD)^2$$

$$= AC^2 - 2AC\,AD + AD^2$$

Substituting in (3.3) for DC^2 gives

$$AB^2 - BC^2 = AD^2 - (AC^2 - 2AC\,AD + AD^2) = -AC^2 + 2AC\,AD$$

therefore

$$BC^2 = AB^2 + AC^2 - 2AC\,AD$$

But

$$AD = AB \cos\theta$$

therefore

$$BC^2 = AB^2 + AC^2 - 2\text{AC AB} \cos \theta \qquad (3.4)$$

If we put $BC = a$, because it is the side opposite the angle A, and $AB = c$ and $AC = b$, and also angle $\theta = A$ this formula (3.4) becomes

$$a^2 = b^2 + c^2 - 2\,bc \cos A \qquad (3.5)$$

This is the *cosine formula* for the plane triangle ABC. Its form should be studied and memorised. You will notice that a is the side opposite the angle A, and b and c are the other two sides of the triangle. By the same argument we could have shown that

$$b^2 = a^2 + c^2 - 2\,ac \cos B$$

$$c^2 = a^2 + b^2 - 2\,ab \cos C$$

Exercise 1 Verify that the cosine formula for the triangle PQR is

$$p^2 = q^2 + r^2 - 2\,qr \cos P$$

Exercise 2 Recast formula (3.5) to calculate the angle A of triangle ABC given its three sides a, b, and c. Proceed as follows

$$2bc \cos A = b^2 + c^2 - a^2$$

therefore

$$\cos A = \frac{b^2 + c^2 - a^2}{2bc} \tag{3.6}$$

Exercise 3 Show that if two sides of the triangle *ABC* are equal, the ang opposite them are equal. Although this might appear obvious by inspection formal proof is needed. Suppose the sides are a and c, then from (3.6) we hav

$$\cos A = \frac{b^2 + c^2 - a^2}{2bc} \quad \text{and} \quad \cos C = \frac{a^2 + b^2 - c^2}{2ab}$$

If $a = c$ these expressions become equal, for

$$\cos A = \frac{b^2}{2bc} \quad \text{and} \quad \cos C = \frac{b^2}{2cb}$$

therefore $\cos A = \cos C$

and therefore $A = C$

Note: The sum of any two sides of a triangle must be greater than the third fo triangle to be formed.

Exercise 4 If a = 11.358 cm, b = 10 cm and c = 7 cm calculate the angle at from equation (3.6). We have

$$\cos A = \frac{100 + 49 - 129.004}{2 \times 10 \times 7} = \frac{19.996}{140} = 0.1428$$

This is the cosine of the angle 81.79°. The angle is found using the *inverse cosi* key of a calculator. It is usually written in red or brown as $\cos^{-1}$.

Note: It has to be pointed out that this notation for the reverse functional pr ess is quite inconsistent with most other branches of mathematics. For exam it is usual to write the reciprocal of x or $1/x$ as x^{-1}.

If this notation had been used for the cosine function then

$$\cos^{-1} x$$

would be secant x or sec x.

For trigonometric functions it is used in the following sense: if

$$x = \cos A$$

then $A = \cos^{-1} x$.

Summarising the forward and inverse operations of the calculator keys we ha

$$\cos 81.79° = 0.1428 \quad \text{and} \quad \cos^{-1} 0.1428 = 81.79°$$

Exercise 5 Using the cosine formula, calculate the angles B and C of the same triangle as in the last exercise, and verify that all three angles add up to 180°. Table 3.1 shows a tabular layout of the results of the calculations for all three angles.

Table 3.1 Solution of plane triangle ABC

side	a	b	c
value	11.358	10	7
side squared	129.004	100	49

Angle	cos	angle deg
A	0.1428	81.79
B	0.4906	60.62
C	0.7924	37.59
		sum 180.00

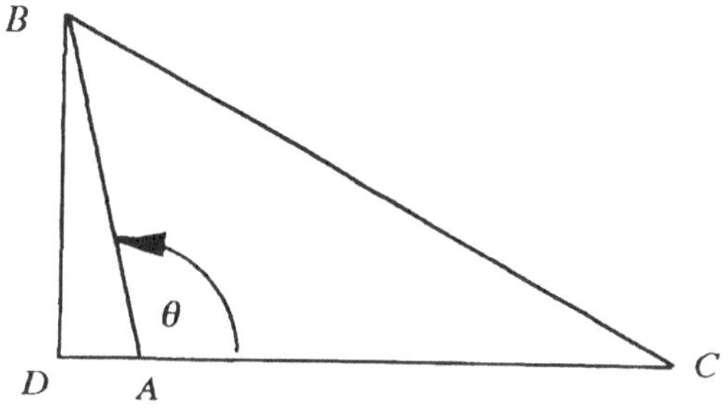

Figure 3.4

3.5 Angles greater than a right angle

Figure 3.4 shows a triangle ABC in which one of the angles is greater than 90° and less than 180°. Let's see what happens to the cosine formula in this case. Let the sides of the triangle be a = 13 cm, b = 10 cm and c = 7 cm. Drawing the triangle enables us to measure the angle A to be 98°. From formula (3.6) we have

$$\cos A = \frac{100 + 49 - 169}{2 \times 10 \times 7} = \frac{-20}{140} = -0.142857$$

The calculator gives the correct answer as

$$\cos^{-1} - 0.142\,857 = 98.21°$$

The negative sign for cos A has to be interpreted in the following manner. From Figure 3.4 you will see that the cosine of A is AD/AB. Because D is on the opposite side of A from C, *the line AD is considered to be negative*. Putting this another way, if the line A to C, i.e. from left to right, is *positive*, a line AD from right to left is considered *negative*. This is an example of using the negative sign as an *operator* which tells which direction along a line we are considering.

Exercise 1 If AC is a positive line, what is the sign of CA? (Answer: negative)

The negative cosine indicates an angle greater than 90°, but (as we shall see later) less than 270°. It is vital to deal with signs correctly to obtain correct results from a formula such as (3.6).

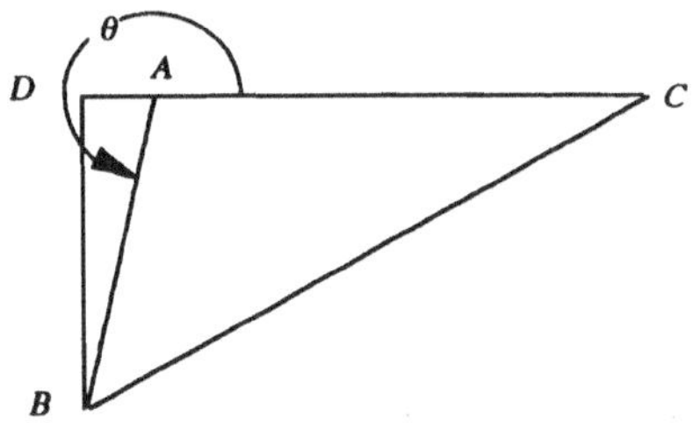

Figure 3.5

Now let us see what happens when the angle θ is more than 180°. In Figure 3.5 the sides of the triangle are unchanged but B now lies below A, and θ is the *exterior angle* of the triangle at point A. By definition,

$$\cos\theta = \frac{AD}{AB} = -0.142857$$

This is the same as cos ϕ where $\phi = 360° - \theta$ is the *interior angle* at A of the triangle ABC. Thus we have the rule that

$$\cos A = \cos(360° - A)$$

Exercise 2 If cos A is – 0.142 857 what is the angle A?

The calculator gives the result as 98.21°, but we have just shown that the result could equally be

$$360° - 98.21° = 261.79°$$

Exercise 3 Demonstrate by calculator that

$$\cos 261.79° = -0.142\,802$$

(The slight difference is due to rounding errors in the arithmetic. If the whole calculation is carried out without writing down intermediate figures the results for cos A and cos (360 – A) are identical.)

It is clearly very important in mapping to be able to tell which result is correct when calculating the positions of points such as B. In short, given the three sides of a triangle alone we cannot tell whether Figure 3.4 or 3.5 is correct. Some other information is needed, such as 'B lies above A'. Using a computer to solve problems we cannot easily deal with written rules. The maths alone must supply the complete answer. The way this problem is solved will be discussed in section 3.6. We will return to this problem after considering the last possible case, in which B is below A but to the right of it, as in Figure 3.6.

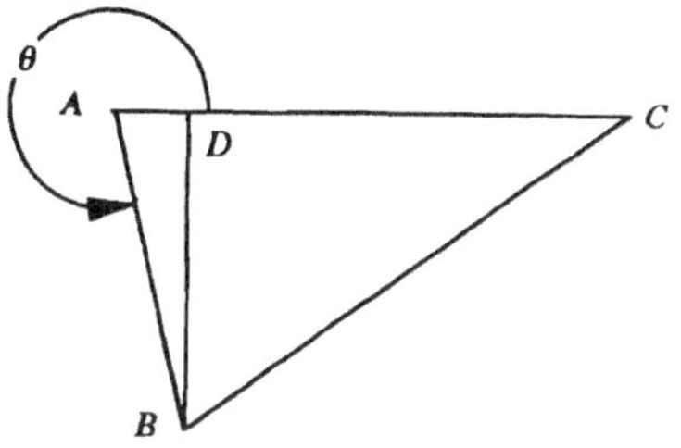

Figure 3.6

In this case the angle θ is greater than 270°. Here it is 180° + 98.21° = 278.21°.

Exercise 4 Show that the cosine of 278.21° is 0.142 802, i.e. positive, and that the cosine of (278.21° – 90°) = cos 188.21° = – 0.989751, i.e. negative.

Exercise 5 To complete the picture, show that the cosine of 81.79° is 0.142 802, i.e. positive, and of 81.79° – 90 = – 8.21° is 0.989 751, i.e. also positive.

It is instructive to summarise all four cases of the triangle from Figures 3.3 to

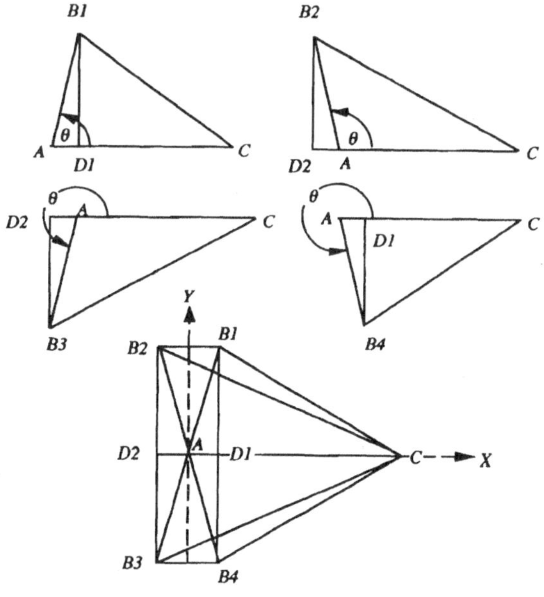

Figure 3.7

3.6, with B in four positions $B1$ to $B4$, round the circle as in the combined figures of Figure 3.7, and D in positions $D1$ and $D2$.

With reference to Figure 3.7, we recall the following conventions:

(1) The angle θ, which shows the direction of AB, is considered positive anticlockwise from the starting line AC.
(2) A line *directed* from left to right is positive, and from right to left is negative.
(3) The line AB is considered to be a fixed positive constant. Only the angle θ is varied.
(4) The cosine of θ is always given by the ratio AD/AB. It therefore takes the sign of AD.
(5) Also, the *projection* of the line AB on to the line AC is given by $AB \cos \theta$. This has size and sign.

From Figure 3.7 it can be seen that the sign of AD, and therefore of the cosine, follows the scheme shown in Figure 3.8.

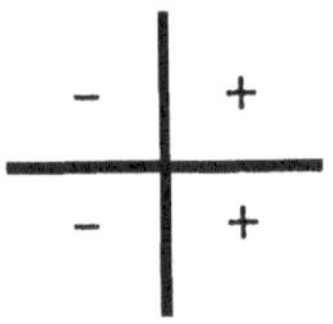

Figure 3.8

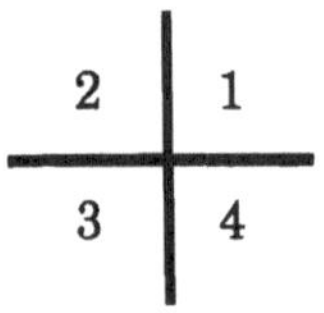

Figure 3.9

Each space of the figure is called a *quadrant*. The quadrants are numbered anticlockwise in positive order from the chosen starting line AC, as in Figure 3.9.

Putting the information of Figures 3.8 and 3.9 into words, 'the cosine is positive in the first and fourth *quadrants*, and negative in the second and third quadrants'.

Exercise 6 Draw a scale figure for all four cases of the triangle ABC similar to Figure 3.7, but to your own dimensions, and show by measurement and using your calculator that the signs of the cosines follow the pattern of Figures 3.8 and 3.9.

3.6 Coordinate axes and bearings

We have shown that the angles of a triangle can be calculated using the cosine formula, but that there is ambiguity as to which way up or which way round the triangle may be. To overcome this problem, a system of *axes* is introduced at the point A round which the line AB is rotated through an angle which can be up to 360°. It is usual to call such a whole circle angle a *bearing*. Point A becomes the *origin O* and the starting line *AC* becomes the *OX axis*. The first axis is called the *primary axis* of the system. If we consider *anticlockwise bearings to be positive*, the axis at a bearing of +90° from the *OX* axis, or *OY* axis, is called the *secondary axis* of the system. Usually we just refer to these axes as the '*X*' and '*Y*' axes for short.

Note: In surveying, the axis system is usually chosen differently. Positive bearings are reckoned clockwise from north with the *OX* axis (the first axis) pointing North and the second axis, the *Y* axis, to the east. This should not cause problems with formulae provided the sign rules are strictly adhered to. We will look at this difference later when we come to three-dimensional problems.

In Section 3.4 we came across the problem that when calculating an angle from its cosine two results are possible. To resolve the matter we also use the cosine of a second angle commencing at the secondary axis of the system. If the first angle is *A* the second angle will be *A* – 90°. If both cosines, i.e. cos *A* and cos (*A* – 90°), are available we can resolve any ambiguity. The signs of the cosines for these two systems will follow the schemes

First angle

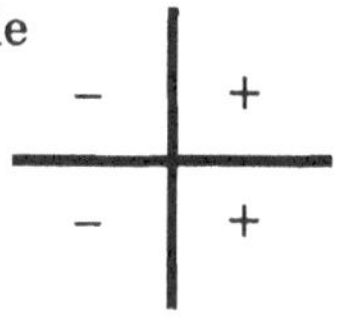

Second angle

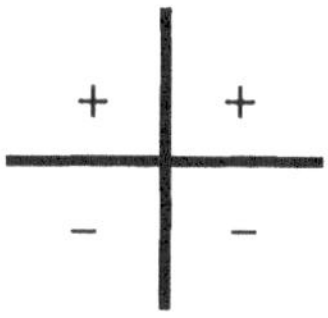

Combining these schemes in order gives the scheme for the four quadrants.

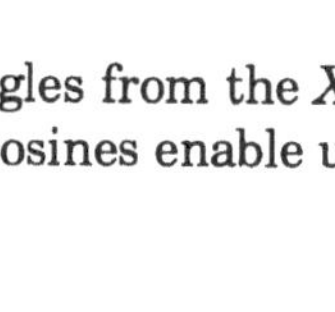

Figure 3.10

Exercise 1 Show that, if the angles from the *X* and *Y* axes to the line *OB* are 150° and 60° respectively, their cosines enable us to say in which quadrant the point *B* lies.

We have cos 150° = – 0.8660 i.e. negative, cos 60° = 0.5 i.e. positive. Inspecting the scheme of Figure 3.10, we see that *B* is in the second quadrant (as expected from drawing either angle). Using a procedure similar to this, a computer is able to avoid ambiguity in mapping points. (See section 4.4 dealing with the ATAN2 function)

If a point B lies in the first quadrant, say at an angle of 56°, $A - 90°$ is negative, i.e. $56° - 90° = -34°$. Its cosine is positive = 0.829. This is because an angle of $-A$ means a clockwise movement from AC to a position which would also be achieved by an anticlockwise turn of $360° - A$.

Exercise 2 Verify that $\cos(360° - 34°) = \cos(326°) = 0.829 = \cos(-34°)$. Also verify that $\cos(145°) = \cos(360° - 145°) = -0.819$.

3.7 Direction cosines

Where the cosines of angles are referred to the primary and secondary axes in this way they are called the *direction cosines* of the line AB with respect to the OX and OY axes. For brevity they are given single letters, usually L and M or l and m. Thus

$$\cos\theta = L \quad \text{and} \quad \cos(\theta - 90°) = M$$

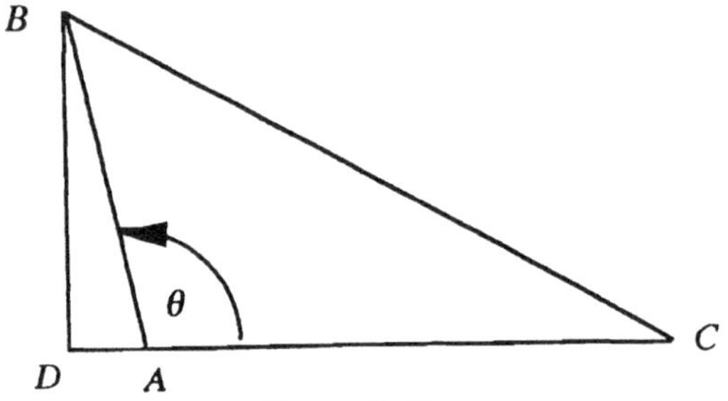

Figure 3.11

Consider Figure 3.11, we may put $L = AD/AB$ and $M = DB/AB$ where L and M are the direction cosines of AB. From the theorem of Pythagoras we have

$$AB^2 = AD^2 + DB^2$$

Dividing throughout by AB^2 we obtain

$$\frac{AB^2}{AB^2} = \frac{AD^2}{AB^2} + \frac{DB^2}{AB^2}$$

Therefore

$$1 = L^2 + M^2 \tag{3.7}$$

3.8 Other trigonometrical functions

Although the direction cosines are much used in surveying and mapping, conventional trigonometry defines other functions for brevity and convenience.

The sine of angle θ or sin θ

From Figure 3.11, the ratio $M = DB/AB = \cos(\theta - 90°)$ is also called the *sine of the angle θ (sin θ* for short).

Exercise 1 Using a calculator, verify that cos (78° – 90°) = sin 78° = 0.9781.

Exercise 2 Show that the signs of sine θ follow the scheme shown in Figure 3.12

Figure 3.12

As expected, these signs are the same as the cosines of the second angles referred to the *OY* axis, The sine is positive in the first two quadrants and negative in the last two.

Exercise 3 Show that cos 45° = sin 45° = 0.7071.

Because $L = \cos\theta$ and $M = \sin\theta$ we have from equation (3.7)

$$\cos^2\theta + \sin^2\theta = 1 \tag{3.8}$$

Exercise 4 Verify by calculator that equation (3.8) is valid for the following angles

56°, 146°, 236°, 326°

3.9 The sine rule in triangle *ABC*

Another very useful formula connecting parts of a plane triangle is the *sine rule*. In triangle *ABC*, if the sides opposite the angles *A*, *B*, and *C* are respectively *a*, *b*, and *c*
we have the rule

$$\frac{a}{\sin A} = \frac{b}{\sin B} = \frac{c}{\sin C} = 2R \tag{3.9}$$

Here *R* is the radius of the circle which passes through *ABC*, called its *circumscribing circle.*

Exercise 1 Calculate the length of the side *a* given that *b* = 10 cm, *A* = 60.94° and *B* = 76.23°.

$$a = \frac{b \sin A}{\sin B} = \frac{10 \times 0.8741}{0.9713} = 9\,\text{cm}$$

Verify that the third side is 7 cm. (**Hint:** $C = 180° - (A + B)$) and that *R* = 5.15 cm. It is a good idea to draw this triangle to scale to check these results.

Proof of the sine rule

To prove this very important rule we first need to prove other useful properties of a triangle and a circle.

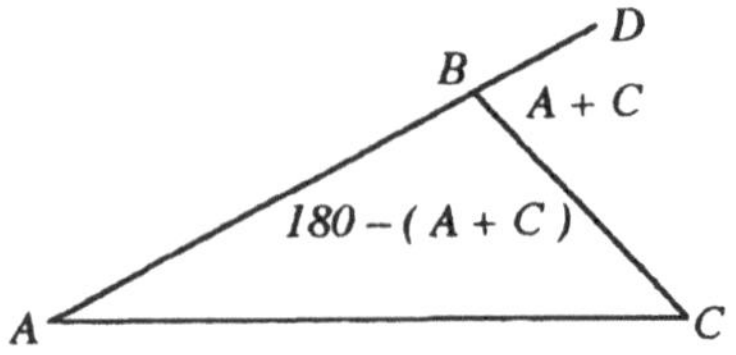

Figure 3.13

Consider Figure 3.13

$$\text{angle } CBA = B = 180° - (A + C)$$

Since ABD is a straight line

$$\angle DBC = 180° - B = A + C$$

In words this property is usually written 'The exterior angle of a triangle is equal to the sum of its interior opposites.' We now use this property applied to Figure 3.14.

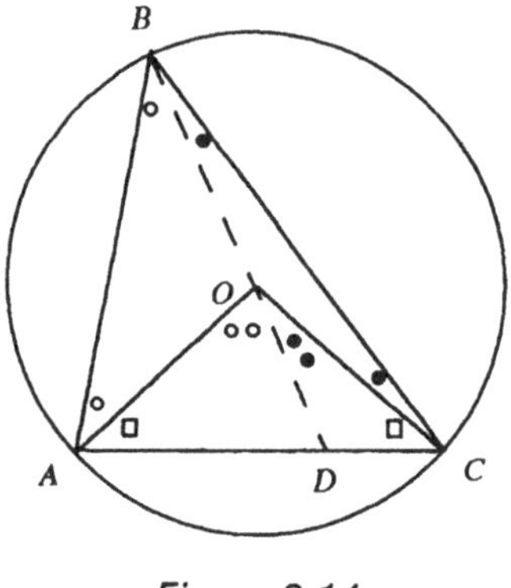

Figure 3.14

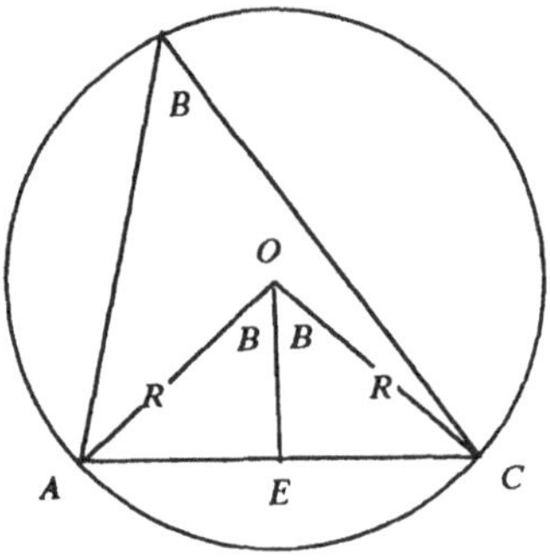

Figure 3.15

The sides OA, OB, and OC are all radii of the circle and equal to R. Therefore triangles ABO, ACO and CBO are *isosceles* (they have two equal sides). Therefore angles opposite equal sides in these three triangles are equal. They are marked by dot and square symbols. By the rule just proved

$$\angle AOD = 2\angle ABO \text{ and } \angle COD = 2\angle CBO$$

Thus

$$\angle AOC = \angle AOD + \angle COD = 2\angle ABO + 2\angle CBO = 2\angle ABC \quad (3.10)$$

This important result is much used in surveying especially in setting out engineering curves.

Now consider Figure 3.15 in which OE is perpendicular to AC and, using the last result, bisects $\angle AOC = 2B$. From triangles AOE and COE we have

$$AE = R \sin B \quad \text{and} \quad CE = R \sin B$$

therefore

$$b = AC = AE + CE = 2R \sin B$$

Thus

$$\frac{b}{\sin B} = 2R$$

In an identical way we can show that

$$\frac{a}{\sin A} = 2R \quad \text{and} \quad \frac{c}{\sin C} = 2R$$

These three expression are combined into the sine rule as

$$\frac{a}{\sin A} = \frac{b}{\sin B} = \frac{c}{\sin C} = 2R \tag{3.9}$$

Exercise 1 In triangle ABC, $a = 9$ cm, $c = 7$ cm and $C = 40°$, calculate side b.

First calculate angle A from

$$\sin A = \frac{a \sin C}{c} = \frac{9 \sin 40°}{7} = 0.8264$$

Therefore

$$A = \sin^{-1} 0.8264 = 55.73°$$

But the sine of $180° - 55.73° = 124.27°$ is also 0.8264, so A could also be 124.27°. That this is possible can be seen from Figure 3.16 in which the two positions of A are shown at A and A'.

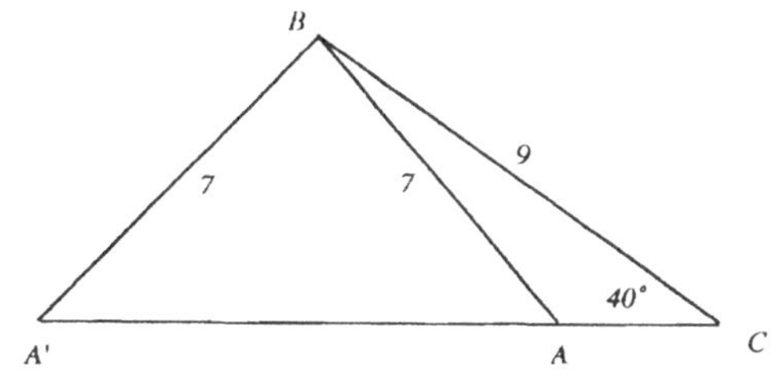

Figure 3.16

This exercise warns against using the sine rule without thinking. A correct result will be obtained if two sides and the *angle included between them* is given, but not in the example set. Of course there may be other evidence to say which result is correct, but the maths alone is unclear.

3.10 The tangent of the angle θ or tan θ

In Figures 3.7, it is useful to call the ratio *DB*/*AD* the *tangent of the angle* θ or *tan* θ for short.
It follows that

$$\tan\theta = \frac{\sin\theta}{\cos\theta} \qquad (3.11)$$

because

$$\tan\theta = \frac{DB}{AD} = \frac{DB}{AB}\frac{AB}{AD} = \frac{\sin\theta}{\cos\theta}$$

Exercise 5 Show that

$$\frac{\sin 56^\circ}{\cos 56^\circ} = \frac{0.8290}{0.5592} = 1.4825 = \tan 56^\circ$$

Exercise 6 Show that the signs of the tangents follow the scheme of Figure 3.17.

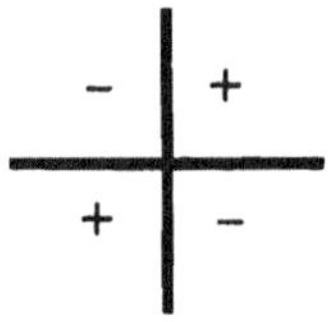

Figure 3.17

The tangent is positive in the first and third quadrants. Verify these signs using the tangents of the following angles

$$56^\circ, 146^\circ, 236^\circ, 326^\circ$$

3.11 The inverse functions $\sin^{-1}$ and $\tan^{-1}$

As with the cosine, the notation for the inverse function of sin and tan are something of an anomaly. They mean respectively

$$\text{If } \sin 56^\circ = 0.8290 \text{ then } \sin^{-1} 0.8290 = 56^\circ$$
$$\text{and if } \tan 56^\circ = 1.4826 \text{ then } \tan^{-1} 1.4826 = 56^\circ$$

3.12 Cosecant and cotangent

The reciprocal of the sine is called the *cosecant* and the reciprocal of the tangent is called the *cotangent*. These are abbreviated to cosec and cot respectively. Thus we have

$$\frac{1}{\tan A} = \cot A \quad \text{and} \quad \frac{1}{\sin A} = \operatorname{cosec} A$$

The cosecant should not be confused with the secant which is the reciprocal of the cosine.

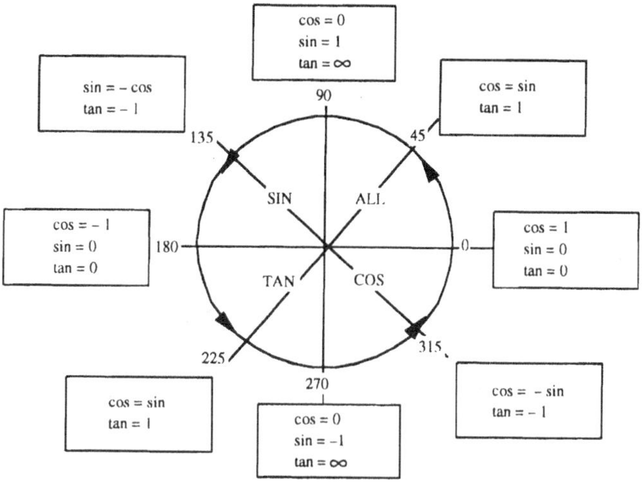

Figure 3.18

3.13 Summary of the trigonometrical functions

The signs and key values of the three main trigonometrical functions, cosine, sine and tangent, are summarised in Figure 3.18. The centre scheme shows which functions *are positive* in the four quadrants, with the convention that an anti-clockwise angle is positive. Notice that when

$$\theta = 45° = 50 \text{ gons} = \frac{\pi}{4} \text{ radians}$$

the *sine and cosine are numerically* equal to $\frac{1}{\sqrt{2}} = 0.7071$

Exercise 1 Using your calculator, verify the signs and values of the trigonometrical functions shown in Figure 3.18. Having done so, use your powers of reason to deduce the same answers with reference to basic definitions.

Exercise 2 Sketch the graph $y = \sin\theta$ within the range 0° to 360°. (The characteristic wave shape is used to model many physical phenomena.) The sketch for the key values of Figure 3.18 is shown in Figure 3.19.

3.14 The cosine and sine of the compound angle $(A + B)$

One of the most important problems in surveying and mapping concerns coordinate systems and their relationships. A geodetic satellite operates on a world

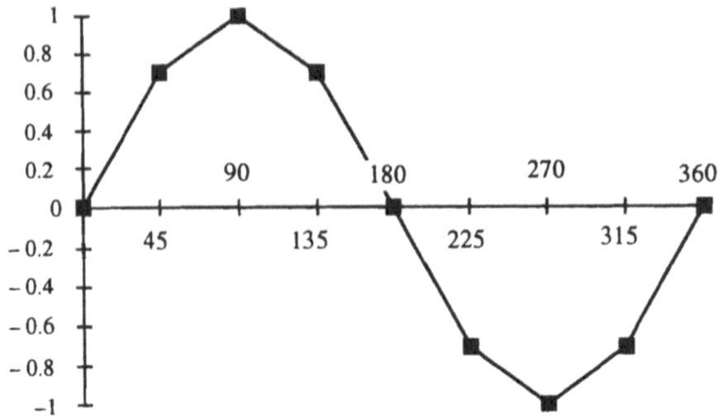

Figure 3.19

wide system of coordinates: in contrast, local maps are usually based on a national or more local system. When you are out walking in the countryside, map reading is an important way of locating your position. Usually you have to turn the map round to the correct orientation before finding your way. In computer mapping systems, the equivalent of rotating a map is also needed. To understand how this is done involves the cosine and sine of *compound angles* such as $(A + B)$. The following are fundamental

$$\sin(A + B) = \sin A \cos B + \cos A \sin B \quad (3.12)$$

$$\cos(A + B) = \cos A \cos B - \sin A \sin B \quad (3.13)$$

We shall derive these results for angles in the first quadrant, although they are valid for all quadrants. Consider Figure 3.20. The following pairs of lines are orthogonal

QP and *OP*   *QS* and *OR*   *PR* and *OR*

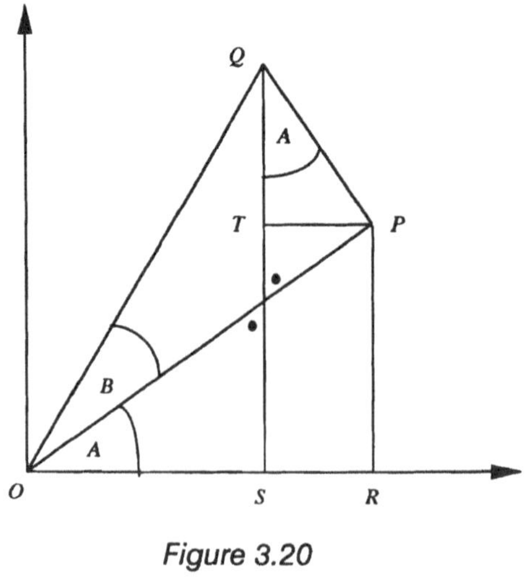

Figure 3.20

In triangle PQT the angle at $Q = A$, because the angles QPO and QSO are equal (both 90°) and the angles marked with dots are also equal.

Hence

$$\sin(A+B) = \frac{QS}{OQ} = \frac{QT+TS}{OQ} = \frac{QT+PR}{OQ}$$
$$= \frac{QP\cos A + OP\sin A}{OQ}$$
$$= \frac{QP}{OQ}\cos A + \frac{OP}{OQ}\sin A$$

therefore

$$\sin(A+B) = \cos A \sin B + \sin A \cos B \tag{3.12}$$

Exercise 1 Show by calculation that

$$\sin 70° = \cos 20° \sin 50° + \sin 20° \cos 50°$$

we have

$$\sin 20° = 0.342\,020 \quad \cos 20° = 0.939\,693$$
$$\sin 50° = 0.766\,044 \quad \cos 50° = 0.642\,788$$

therefore by multiplication

$$\sin 70° = (0.939\,693 \times 0.766\,044) + (0.342\,020 \times 0.642\,788)$$
$$\sin 70° = 0.939\,693$$

which can be verified by direct calculation.

Exercise 2 Verify that

$$\cos 20° = \sin 70° = 0.939\,693$$

3.15 Expression for cos (*A* + *B*)

We derive the expression for cos $(A + B)$ in a similar manner. Refer to Figure 3.20.

$$\cos(A+B) = \frac{OS}{OQ} = \frac{OR-SR}{OQ} = \frac{OR-TP}{OQ}$$
$$= \frac{OP\cos A - QP\sin A}{OQ}$$
$$= \frac{OP}{OQ}\cos A - \frac{QP}{OQ}\sin A$$

therefore

$$\cos(A + B) = \cos A \cos B - \sin A \sin B \tag{3.13}$$

Note the change of sign.

Exercise 2 By calculation derive cos 300° by putting 300° = 270° + 30°

$$\cos 270° = 0 \quad \sin 270° = -1$$

$$\cos 30° = 0.866\ 025 \quad \sin 30° = 0.5$$

$$\begin{aligned}\cos\,(270° + 30°) &= \cos 270° \times \cos 30° - \sin 270° \times \sin 30° \\ &= (0 \times 0.866\ 025) + (1 \times 0.5) \\ &= 0.500\ 000\end{aligned}$$

This is the result expected because 300° = 360° – 60°:

$$\cos\,(360° - A) = \cos A = \cos 60° = 0.5$$

3.16 Expressions for sin (*A* – *B*) and cos (*A* – *B*)

If we put $B = -B$ in the above expressions for sin $(A + B)$ and cos $(A + B)$ and remember that

$$\sin\,(-B) = -\sin B \quad \text{and} \quad \cos(-B) = \cos B$$

we have at once the expressions

$$\sin\,(A - B) = \sin A \cos B - \cos A \sin B \tag{3.14}$$

$$\cos\,(A - B) = \cos A \cos B + \sin A \sin B \tag{3.15}$$

3.17 Expression for tan (*A* + *B*)

The result for tan $(A + B)$ is simply obtained from

$$\tan(A + B) = \frac{\sin(A + B)}{\cos(A + B)} = \frac{\cos A \sin B + \sin A \cos B}{\cos A \cos B - \sin A \sin B}$$

Division of numerator and denominator by cos A cos B gives

$$\tan(A + B) = \frac{\tan A + \tan B}{1 - \tan A \tan B} \tag{3.16}$$

And because tan $(-B) = -$ tan B we have at once

$$\tan(A - B) = \frac{\tan A - \tan B}{1 + \tan A \tan B} \tag{3.17}$$

3.18 Expressions for sin 2*A* and cos 2*A*

If we put $B = A$ in the above expressions (3.12) and (3.13) we obtain the special cases

$$\sin 2A = 2 \sin A \cos A \tag{3.18}$$

$$\cos 2A = \cos^2 A - \sin^2 A \tag{3.19}$$

Because

$$\cos^2 A + \sin^2 A = 1$$

(3.19) may also be cast in the useful forms

$$\cos 2A = 2 \cos^2 A - 1 \tag{3.20}$$

$$\cos 2A = 1 - 2 \sin^2 A \tag{3.21}$$

3.19 Expressions for sin 3A and cos 3A

If we put $B = 2A$ in the above expression (3.13) and use appropriate results from (3.19) and (3.21) we obtain the special cases.

$$\cos 3A = 4 \cos^3 A - 3 \cos A \tag{3.22}$$

and

$$\sin 3A = 3 \sin A - 4 \sin^3 A \tag{3.23}$$

3.20 Expressions for sums and differences in terms of half angles

The following identities are used in surveying and cartography to derive important formulae such as the mid latitude formulae for computing geographical coordinates.

$$\sin A + \sin B = 2 \sin \tfrac{1}{2}(A + B) \cos \tfrac{1}{2}(A - B) \tag{3.24}$$

$$\sin A - \sin B = 2 \cos \tfrac{1}{2}(A + B) \sin \tfrac{1}{2}(A - B) \tag{3.25}$$

$$\cos A + \cos B = 2 \cos \tfrac{1}{2}(A + B) \cos \tfrac{1}{2}(A - B) \tag{3.26}$$

$$\cos A - \cos B = -2 \sin \tfrac{1}{2}(A + B) \sin \tfrac{1}{2}(A - B) \tag{3.27}$$

We prove the first of these formulae to indicate how all can be derived, and because the proof uses an important mathematical procedure. We can express two quantities A and B as follows

$$A = \tfrac{1}{2}(A + B) + \tfrac{1}{2}(A - B)$$

$$B = \tfrac{1}{2}(A + B) - \tfrac{1}{2}(A - B)$$

Then using (3.12) and (3.14)

$$\begin{aligned} \sin A + \sin B &= \sin \tfrac{1}{2}(A + B) \cos \tfrac{1}{2}(A - B) + \cos \tfrac{1}{2}(A + B) \sin \tfrac{1}{2}(A - B) \\ &\quad + \sin \tfrac{1}{2}(A + B) \cos \tfrac{1}{2}(A - B) - \cos \tfrac{1}{2}(A + B) \sin \tfrac{1}{2}(A - B) \\ &= 2 \sin \tfrac{1}{2}(A + B) \cos \tfrac{1}{2}(A - B) \end{aligned}$$

which is formula (3.24). The other formulae (3.25), (3.26) and (3.27) are proved in identical manner.

Note the sign change in (3.27) .

3.21 Half angle formulae

Formulae which express functions of the angles of a plane triangle in terms of its sides are also important. These are

$$\sin^2 \tfrac{1}{2}A = \frac{(s-b)(s-c)}{bc} \tag{3.28}$$

$$\cos^2 \tfrac{1}{2}A = \frac{s(s-a)}{bc} \tag{3.29}$$

$$\tan^2 \tfrac{1}{2}A = \frac{(s-b)(s-c)}{s(s-a)} \tag{3.30}$$

where the sum of the sides

$$a + b + c = 2s$$

In triangle ABC we have from (3.5)

$$a^2 = b^2 + c^2 - 2\,bc \cos A$$

but from (3.21)

$$\cos A = 1 - 2\sin^2 \tfrac{1}{2}A$$

so

$$a^2 = b^2 + c^2 - 2\,bc(1 - 2\sin^2 \tfrac{1}{2}A)$$
$$a^2 = b^2 + c^2 - 2\,bc + 4bc \sin^2 \tfrac{1}{2}A$$
$$4bc \sin^2 \tfrac{1}{2}A = a^2 - (b-c)^2$$

Remembering that

$$x^2 - y^2 = (x + y)(x - y)$$

$$\begin{aligned} 4bc\sin^2 \tfrac{1}{2}A &= a^2 - (b-c)^2 \\ &= [a+(b-c)][a-(b-c)] \\ &= (a+b-c)(a-b+c) \end{aligned}$$

But if

$$a + b + c = 2s$$
$$a + b - c = 2(s-c)$$
$$a - b + c = 2(s-b)$$

therefore

$$4bc \sin^2 \tfrac{1}{2}A = 4(s-b)(s-c)$$

$$\sin^2 \tfrac{1}{2}A = \frac{(s-b)(s-c)}{bc} \tag{3.28}$$

which is formula (3.28).

Exercise 1 Prove the formula

$$\cos^2 \tfrac{1}{2}A = \frac{s(s-a)}{bc} \tag{3.29}$$

Hint: Use the general expression from (3.8).

$$\cos^2 A = 1 - \sin^2 A$$

3.22 Hero's formula for the area of a triangle

The *area of the triangle ABC* (6.2) is given by

$$\Delta = \tfrac{1}{2}bc\sin A = \tfrac{1}{2}bc.2\sin\tfrac{1}{2}A\cos\tfrac{1}{2}A$$
$$= bc\sin\tfrac{1}{2}A\cos\tfrac{1}{2}A$$

$$\Delta^2 = b^2c^2\sin^2\tfrac{1}{2}A\cos^2\tfrac{1}{2}A$$

$$\Delta^2 = b^2c^2\frac{(s-b)(s-c)}{bc}\frac{s(s-a)}{bc}$$
$$= s(s-a)(s-b)(s-c)$$

$$\Delta = \sqrt{[s(s-a)(s-b)(s-c)]} \tag{3.31}$$

See Section 6.2 for an exercise using hero's formula.

SUMMARY OF KEY WORDS

trigonometry, functions, cosine, secant, cosine formula, inverse cosine, exterior angle, interior angle, quadrants, direction cosines, sine, isosceles, tangent, bearing, primary axis, secondary axis, sine rule, circumscribing circle, cosecant, cotangent, compound angles, area of a triangle (Hero's formula)

SUMMARY OF FORMULAE

$$a^2 = b^2 + c^2 - 2\,bc\cos A \qquad (3.5)$$

$$L^2 + M^2 = 1 \qquad (3.7)$$

$$\cos^2\theta + \sin^2\theta = 1 \qquad (3.8)$$

$$\frac{a}{\sin A} = \frac{b}{\sin B} = \frac{c}{\sin C} = 2R \qquad (3.9)$$

$$\tan\theta = \frac{\sin\theta}{\cos\theta} \qquad (3.11)$$

$$\sin(A + B) = \sin A\cos B + \cos A\sin B \qquad (3.12)$$

$$\cos(A + B) = \cos A\cos B - \sin A\sin B \qquad (3.13)$$

$$\sin(A - B) = \sin A\cos B - \cos A\sin B \qquad (3.14)$$

$$\cos(A - B) = \cos A\cos B + \sin A\sin B \qquad (3.15)$$

$$\tan(A+B) = \frac{\tan A + \tan B}{1 - \tan A\tan B} \qquad (3.16)$$

$$\tan(A-B) = \frac{\tan A - \tan B}{1 + \tan A\tan B} \qquad (3.17)$$

$$\sin 2A = 2\sin A\cos A \qquad (3.18)$$

$$\cos 2A = \cos^2 A - \sin^2 A \qquad (3.19)$$

$$\cos 2A = 2\cos^2 A - 1 \qquad (3.20)$$

$$\cos 2A = 1 - 2\sin^2 A \qquad (3.21)$$

$$\cos 3A = 4\cos^3 A - 3\cos A \qquad (3.22)$$

$$\sin 3A = 3\sin A - 4\sin^3 A \qquad (3.23)$$

$$\sin A + \sin B = 2\sin\tfrac{1}{2}(A+B)\cos\tfrac{1}{2}(A-B) \qquad (3.24)$$

$$\sin A - \sin B = 2\cos\tfrac{1}{2}(A+B)\sin\tfrac{1}{2}(A-B) \qquad (3.25)$$

$$\cos A + \cos B = 2\cos\tfrac{1}{2}(A+B)\cos\tfrac{1}{2}(A-B) \qquad (3.26)$$

$$\cos A - \cos B = -2\sin\tfrac{1}{2}(A+B)\sin\tfrac{1}{2}(A-B) \qquad (3.27)$$

$$\sin^2\tfrac{1}{2}A = \frac{(s-b)(s-c)}{bc} \qquad (3.28)$$

$$\cos^2\tfrac{1}{2}A = \frac{s(s-a)}{bc} \qquad (3.29)$$

$$\tan^2\tfrac{1}{2}A = \frac{(s-b)(s-c)}{s(s-a)} \qquad (3.30)$$

$$\Delta = \sqrt{[s(s-a)(s-b)(s-c)]} \qquad (3.31)$$

Chapter 4
Plane Coordinates

4.1 Plane coordinates in map making

Fundamental to map making are *plane cartesian coordinates*, either in two or three-dimensions. Problems arise with the conventions in use.

In a two dimensional system (*X*, *Y*) or (N, E), it is customary to direct the *X* axis to the north (N) and the *Y* axis to the east (E), because a bearing *T* is reckoned *clockwise* from north in surveying. This convention accords with mathematics in which the *x* axis points to the east, the *y* axis to the north, but angles are reckoned positive in an *anticlockwise* sense. The mathematical convention is used in this book.

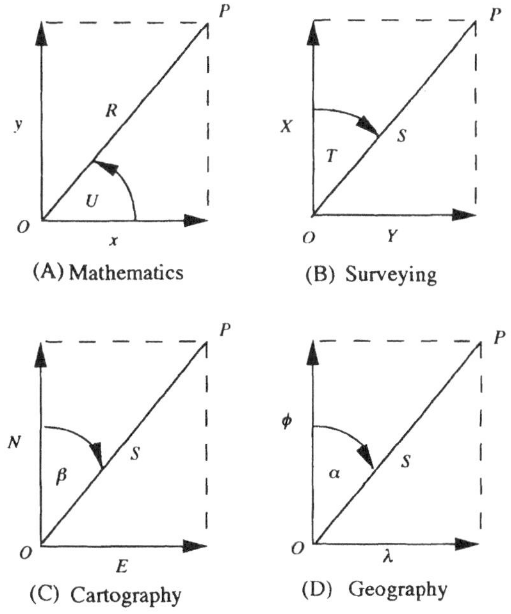

Figure 4.1

The mathematical convention for angles is in direct contrast with the conventions of map reading and cartography, where although it is customary to quote easting (E) before northing (N), angles (bearings and azimuths) are reckoned positive *clockwise from north, not anticlockwise from east.*

Again, the other convention is to refer to *geographical coordinates* in the order as latitude (ϕ) before longitude (λ). To avoid potential confusion, all listed coordinates should clearly indicate which definition is being used, as well as the units of measurement.

In practice, a point on a map has to be sufficiently large for it to be seen. Usually this is a dot of 0.2 mm. However in strict mathematics we sometimes consider the centre of this dot to be the actual point, and say that it has no size but only position relative to other points.

4.2 Cartesian coordinates (*x*, *y*)

This system uses a network of squares to locate points in a plane and, as has already been pointed out in Section 2.4, is a very practical system both on a map for plotting points and on the ground for surveying. Such a system of cartesian coordinates (x, y) is illustrated in Figure 4.2.

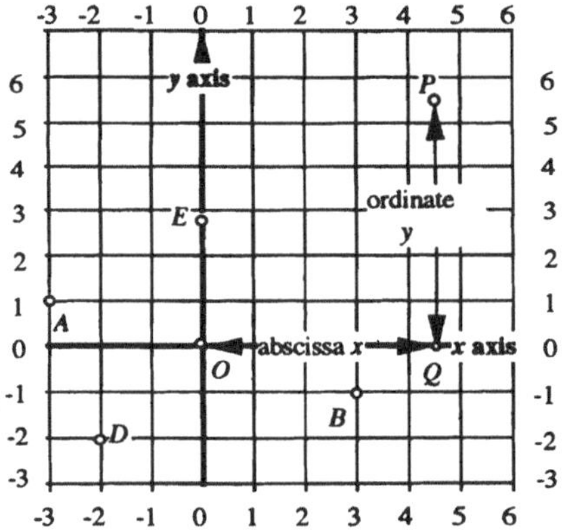

Figure 4.2

A point such as P is located on the sheet of paper by a distance OQ (called the *abscissa* of P) measured parallel to the Ox axis, and another distance QP (called the *ordinate* of P) parallel to the Oy axis. The abscissa and ordinate form the cartesian coordinates of a point which are written in the forms

$$P(x, y) \quad \text{or as} \quad (x_P, y_P)$$

Points to the left of the Oy axis and/or below the Ox axis have negative values.

Exercise 1 From Figure 4.2 verify that the coordinates of the points are as follows

$A(-3, 1)$ $B(3, -1)$ $D(-2, -2)$ $E(0, 2.8)$ $P(4.5, 5.5)$

A computer screen can be thought of as a very fine grid in which cartesian coordinates (x, y) can be read or plotted to an accuracy of at least 0.1 mm. When the cursor, driven by a mouse, rests at a point on the screen, the coordinates of this point are registered in the computer for further use. The reverse of this process is also used in computer plotting. Input coordinates of a point are used to drive a pen or cursor to its position in the cartesian grid, and then marked by some symbol chosen by the operator. The first process is called *digitising* and the second is called *plotting*.

4.3 Plane polar coordinates (r, U)

In this system we use *radial lines* to locate points in a plane. Such a system of polar coordinates (r, U) is illustrated in Figure 4.3. A point such as P is located on the sheet of paper by a radial distance OP and an angle U (called the *bearing* of P). The distance and bearing form the *polar coordinates* of a point which are written in the forms

$$P(r, U) \quad \text{or as} \quad (r_P, U_P)$$

The distance (r) is always positive.

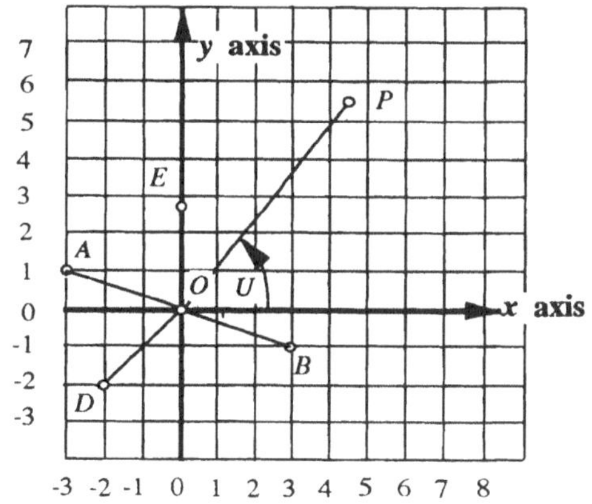

Figure 4.3

Exercise 1 From Figure 4.3 verify by measurement, using the figure scale for distance, that the polar coordinates of the points are as follows

$$A(3.2, 161.6°) \quad B(3.2, -18.4°) \quad D(2.8, -135.0°)$$

$$E(2.8, 90.0°) \quad P(7.11, 50.7°)$$

Note: After studying Section 4.4, you may verify these results by calculation.

4.4 Conversion of polar and cartesian coordinates

Consider Figure 4.3. The position of the point P with respect to the origin O and orthogonal axes Ox and Oy can be defined in two ways:

(1) by its *rectangular cartesian coordinates* (x, y)

(2) by its *polar coordinates* (r, U)

U is reckoned positive in an anticlockwise direction from the Ox axis. To convert from one system to the other the following relationships are used.

(a) Given (r, U), we find x and y from

$$x = r \cos U \quad y = r \sin U \tag{4.1}$$

(b) Given (x, y), we find U and r from

$$\tan U = \frac{y}{x} \tag{4.2}$$

and

$$r = \sqrt{\left(x^2 + y^2\right)} \tag{4.3}$$

A useful check is

$$r = x \cos U + y \sin U \tag{4.4}$$

Exercise 1 The coordinates of P in mm are (25, 55). Calculate r and U.

$$\tan U = \frac{55}{25} = 2.2$$

$$U = 65.556°$$

$$r = \sqrt{(25^2 + 55^2)} = 60.42 \text{ mm}$$

Check

$$r = 25 \cos 65.556° + 55 \sin 65.556° = 60.42 \text{ mm}$$

Exercise 2 The coordinates of P in mm are (–25, –55). Calculate r and U.

$$\tan U = \frac{-55}{-25} = +2.2$$

$$U = 65.556° \,?$$

How can U be the same as in Exercise 1? We have already discussed this problem in Section 3.5. More information than the tangent alone is needed to place P correctly in the third quadrant. Two pieces of information are required.

$$r = \sqrt{(25^2 + 55^2)} = 60.42 \text{ mm}$$

Check

$$r = 25 \cos 65.556° + 55 \sin 65.556° = 60.42 \text{ mm}$$

If we use the *rectangular-to-polar* function key of a calculator we obtain the following values

$$r = 60.42 \quad U = -114.443°$$

This correctly places P in the third quadrant. The calculator algorithm has solved the ambiguity. Before explaining how, check that tan (–114.443°) = 2.2. The calculator has examined the signs of x and y separately before calculating tan U and then U. In most systems the function used is called

ATAN 2

The function has taken note of the negative signs of both x and y, thus ensuring that an erroneous first quadrant angle is not returned, because

$$-5.5 / -2.5 = 2.2$$

Because not all algorithms to compute the reverse trigonometrical functions deal properly with the various quadrants, a check should be made when using a calculator or computer algorithm for the first time. Also, a check should be made to see if a correct result is given for the limiting cases of points on the coordinate axes.

Exercise 3 Use your calculator to obtain *the bearings* U from the origin O to the four axis points A, B, C and D, of Table 3.2.

Table 3.2

Point	A	B	C	D
x	6	0	– 6	0
y	0	6	0	– 6
U	0	90	180	– 90

The last result (– 90°) is usually added to 360° to give the whole circle bearing of 270°.

4.5 Coordinate differences

In practice, coordinates are seldom referred directly to the origin O, but in terms of *coordinate differences*. Usually connections between two points, such as A and B, are required. In this case the above formulae (4.1) to (4.4) need only a little modification to substitute the following

$$\Delta x = x_B - x_A \text{ for } x \quad \text{and} \quad \Delta y = y_B - y_A \text{ for } y \tag{4.5}$$

the Greek letter Δ indicating a difference. The transformation equations become

$$\Delta x = r \cos U \quad \Delta y = r \sin U \tag{4.6}$$

Given (Δx, Δy), we find U and r from

$$\tan U = \frac{\Delta y}{\Delta x} \tag{4.7}$$

and

$$r = \sqrt{(\Delta x^2 + \Delta y^2)} \tag{4.8}$$

A useful check is

$$r = \Delta x \cos U + \Delta y \sin U \tag{4.9}$$

Exercise 1 The coordinates of A and B in millimetres are A (50,70) and B (75 125). Calculate the distance (r), and the direction (U) from A to B.

$$\tan U = \frac{125-70}{75-50} = \frac{55}{25} = 2.2$$

$$U = 65.556°$$

$$r = \sqrt{(25^2 + 55^2)} = 60.42 \text{ mm}$$

Check

$$r = 25 \cos 65.556° + 55 \sin 65.556° = 60.42 \text{ mm}$$

Note: The direction from B to A is 245.556° = (180° + U).

4.6 Mathematical expressions for a straight line

To deal with the drawing of straight lines by computer, and many other problem in map making, we have to describe a line in a mathematical way. In a two dimensional rectangular coordinate system with axes Ox and Oy, all points on a line are usually expressed in the form of an equation. The first form of such an equation is the *gradient form*

$$y = mx + c \tag{4.10}$$

In this equation (m) and (c) are constants which are particular to this line, while (x) and (y) are the coordinates of any point (such as P) lying on the line. Refer to the Figure 4.4.

Exercise 1 Draw a copy of Figure 4.4 at any scale using graph paper. To draw the line AEP, plot two points A (0,4) and E (3, 5.8) and draw a line through them with a ruler. Measure the angle that the line makes with the Ox axis (it should be just over 30°). To plot the point P we have to extend the line a little. Mark the fixed points A and E with a black dot and P with a circle. This indicates that P can be anywhere on the line. To form the equation of the line we express the (x) and (y) values of P in terms of things we know. First we see that the gradient (m) of the line is obtained from

$$m = \tan U = \frac{y_E - y_A}{x_E - x_A} = \frac{5.8-4}{3-0} = 0.6$$

Next we express the unknown coordinates of P in terms of something we know a

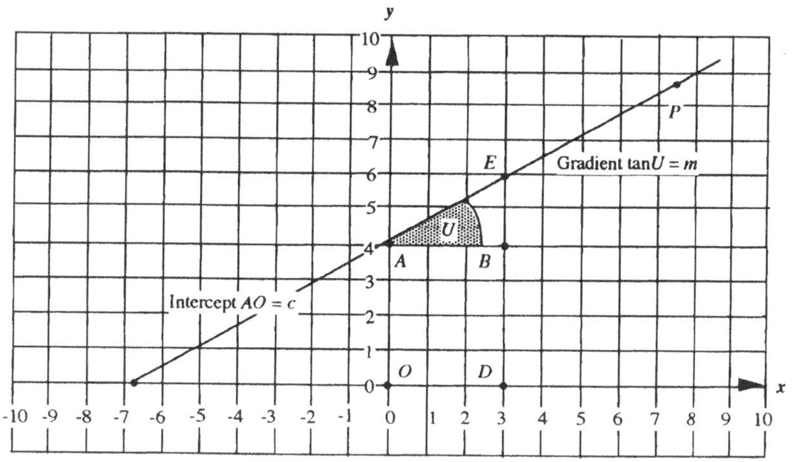

Figure 4.4

follows

$$\frac{y_P - y_A}{x_P - x_A} = \tan U = m$$

Rearranging gives

$$y_P - y_A = m(x_P - x_A)$$
$$= mx_P$$

because $x_A = 0$.
Then

$$y_P = mx_P + y_A$$

Because A is the point on the Oy axis cut by the line we put $y_A = c$, called the *intercept of the line on the Oy axis,* and obtain finally the equation of the line

$$y_P = mx_P + c \qquad (4.11)$$

This is the general form of the equation. The constant m is the *slope* or *gradient* of the line relative to the Ox axis. (The notation is copied from an uphill or downhill slope relative to the horizontal.) Substituting the known values in this case gives the equation of this particular line to be

$$y_P = 0.6x_P + 4$$

Since the general point P could have any name we drop the suffixes and write

$$y = 0.6x + 4$$

for this particular line, and for the general case

$$y = mx + c \qquad (4.10)$$

It is important to grasp how to set up such an equation. The process of drawing the line shows how this is done. We plotted the point A, thus giving c, then obtained the gradient m from two points A and E, and finally expressed these known elements in terms of the unknown coordinates. If the gradient had been given as data, there would be no need to find it from two points. One fixed point and a gradient defines the line. So too do two fixed points. It is always a good idea to try to draw the line to see if enough information is given before working out its equation.

Exercise 2 Use the equation of the line AE to find where it cuts the Ox axis. When the line cuts this axis $y = 0$ therefore

$$0 = 0.6x + 4$$

$$x = -4/0.6 = -6.6$$

Exercise 3 Verify that the point P (7.5, 8.5) lies on the line

$$y = 0.6x + 4$$

Substitution for $x = 7.5$ in the right hand side of the equation gives

$$y = 0.6 \times 7.5 + 4 = 4.5 + 4 = 8.5$$

Exercise 4 Calculate the values of all ordinates (y) of points on the line corresponding to the abscissae

$$x = 1, 2, 3, 4, 5$$

When several repetitive calculations have to be performed by hand, it is best to make a table of the various steps, as follows:

x	1	2	3	4	5
$0.6x$	0.6	1.2	1.8	2.4	3.0
c	4	4	4	4	4
y	4.6	5.2	5.8	6.4	7.0

When a line is drawn by computer from, for example, a given starting point and gradient, a vast number of close points are plotted using the equation of the line just as in this exercise. These closely stepped points give the effect on the eye of a continuous line. With the aid of a magnifying glass, look at a slanting line on a computer screen to verify this statement.

Exercise 5 See what happens to the equation when a line through A is
(i) parallel to the Ox axis and,
(ii) parallel to the Oy axis.

We can still draw both these lines with a ruler so the problem is not impossible. We are asked to see what happens to the equation of the line in the form $y = mx + c$.

In the first case $m = 0$ and the equation is $y = c$. In the second case $m = \tan 90°$ and the line does not cut the Oy axis. The equation has no meaning. Thus if this equation is used in a computer program it can fail. It is not a *robust equation*, because it does not work in all cases.

In the second case, when the equation fails ($m = \infty$), we can write another similar one in terms of the intercept of the line on the Ox axis namely $x = d$. Thus all lines parallel to the Ox axis have equations of the form $y = c$, and all lines parallel to the Oy axis have equations of the form $x = d$. We can use these equations to tell a computer to plot a grid on a map. In the next section a more robust equation for a straight line is considered.

4.7 Polar equation of a line in a plane

In the above section it was demonstrated that the gradient form of the equation for a line fails when the line runs parallel to the Oy axis. The equation (4.10) can be recast as follows

$$y = x \tan U + c$$

$$y = x\frac{\sin U}{\cos U} + c$$

therefore

$$y \cos U = x \sin U + c \cos U$$

$$y \cos U - x \sin U = c \cos U \tag{4.12}$$

but

$$c \cos U = OG = p \text{ (say)}$$

Also, the angle that OG makes with the positive direction of the Oy axis is $+U$ (see Figure 4.5), and the angle it makes with the positive direction of the Ox axis is

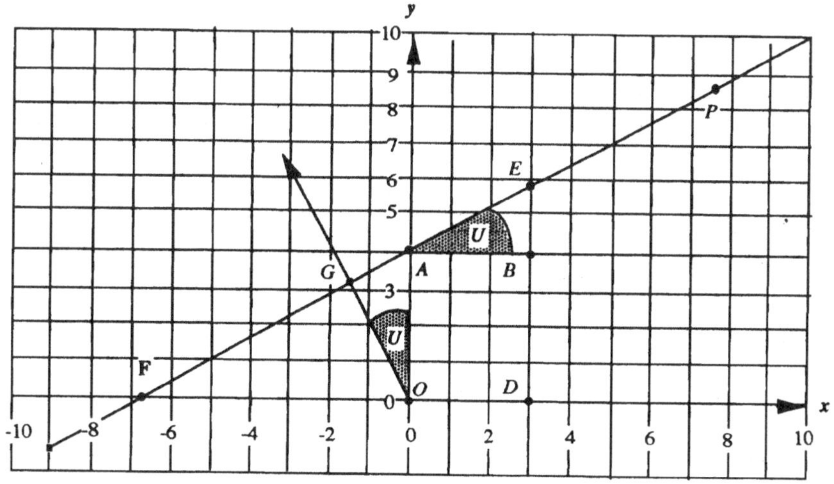

Figure 4.5

(90° + *U*). Thus the direction cosines of *OG* are respectively

$$L' = \cos(90° + U) = -\sin U$$

and

$$M' = \cos(U) = \cos U$$

and (4.12)

$$-x \sin U + y \cos U = p$$

becomes

$$L'x + M'y = p \qquad (4.13)$$

This is a robust form of the equation of a line in a plane which has no failure case. When

$$U = 90° \quad M' = 0 \quad L' = -1$$

equation (4.13) becomes

$$x = -p$$

When

$$U = 0° \quad M' = 1 \quad L' = 0$$

equation (4.13) becomes

$$y = p$$

Thus the *polar equation of a line* in the form (4.13) is universally applicable.

4.8 Parametric equations of a line in a plane

If *E* is any point on the line with coordinates (*a*, *b*) and the distance *EP* along the line to any other point (say *P*) is *t* then the coordinates of *P* are given by

$$x = a + t\cos U \quad y = b + t\sin U \qquad (4.14)$$

Remembering that cos*U* and sin*U* are the direction cosines of the line *EP* we see that these equations (4.14) can also be written

$$x = a + tL \quad y = b + tM \qquad (4.15)$$

Note in the polar form of the equation of the line (4.12), the direction cosines of the ray from the origin perpendicular to the line, i.e. of *OG* are used. The unknown variable distance *t* is called a *parameter*. Given the values of *a*, *b* and *U* this parameter defines all other points on the line.

In geodetic surveying, the distance *t* is measured by tape, by optical tachymetry or by an electromagnetic system, and the angle *U* by theodolite circle. These equations (4.14) or (4.15) are used to calculate the coordinates of points, from measurements taken at a known point such as (a,b).

Exercise 1 Refer to Figure 4.6. Two points *A* (20, 40) and *B* (80, 80) lie on the centre line of a road. Find the equation of the line *AB* in its three forms. From (4.7)

$$m = \tan U = \frac{\Delta y}{\Delta x} = \frac{40}{60} = 0.6667$$

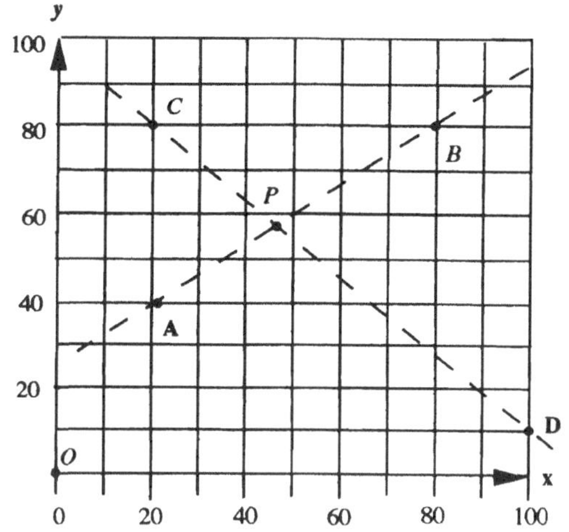

Figure 4.6

therefore $$U = 33.69°$$

From (4.11) and the coordinates of A

$$c = y - m\,x = 40 - 0.6667\ (20) = 26.67$$

The equation of the line in the gradient form is

$$y = 0.6667x + 26.67 \tag{4.16}$$

Also

$$L' = \cos\ (90 + U) = -\sin U = -0.5547$$

and

$$M' = \cos\ (U) = \cos U = 0.8320$$

From (4.13) and the coordinates of A (20, 40)

$$-x \sin U + y \cos U = p$$

$$p = 40 \cos U - 20 \sin U = 22.19$$

Therefore the equation of the line in the polar form is

$$-\sin U\ (x) + \cos U\ (y) = p$$

$$-0.5547x + 0.8320y = 22.19 \tag{4.17}$$

Using point A as reference the equations of the line in the *parametric form* are

$$x = 20 + 0.8320t$$

$$y = 40 + 0.5547t \tag{4.18}$$

The unknown parameter is t. All points on the line are defined by its values. For example, point B is defined when

$$t = AB = \sqrt{(60^2 + 40^2)} = 72.111$$

and from (4.18)

$$x_B = 20 + 0.8320 \times 72.111 = 20 + 60 = 80$$
$$y_B = 40 + 0.5547 \times 72.111 = 40 + 40 = 80$$

Exercise 2 Two points C (20, 80) and D (100, 10) lie on the centre line of another road. See Figure 4.6 again. Find the equations of the line CD in its three forms. This is a repeat of the calculations for Exercise 1 with different figures. The summary of results is

$$m = \frac{-70}{80} = -0.875 \qquad U = -41.186°$$

$$c = y - mx = 80 + 0.875\ (20) = 97.5$$

The equation of the line in the gradient form is

$$y = -0.875x + 97.5 \tag{4.19}$$

Also

$$L' = \cos\ (90 + U) = -\sin U = +\ 0.6585$$

and

$$M' = \cos\ (U) = \cos U = 0.7526$$

From (4.12) and the coordinates of C (20, 80)

$$p = 80 \cos U - 20 \sin U = 80 \times 0.7526 + 20 \times 0.6585 = 73.378$$

Therefore the equation of the line in the polar form is

$$0.6585x + 0.7526y = 73.378 \tag{4.20}$$

Using point C as reference the equations of the line in the parametric form are

$$x = 20 + 0.7526t$$
$$y = 80 - 0.6585t \tag{4.21}$$

To check these equations, we fix D from C. The parameter t is given by

$$t = \sqrt{(80^2 + 70^2)} = 106.3$$

Exercise 3 Find the coordinates of P at the intersection of AB with CD. To obtain this result we solve the simultaneous equations (4.16) and (4.19)

$$y = 0.6667x + 26.67 \tag{4.16}$$
$$y = -0.8750x + 97.50 \tag{4.19}$$

$$0.6667x + 26.67 = -0.8750x + 97.50$$
$$1.5417x = 70.83$$
$$x = 45.94$$

From (4.16)

$$y = 0.6667\ (45.94) + 26.67 = 57.3$$

Therefore the lines intersect at the point P (45.9, 57.3).

Exercise 4 Find the angle at which the lines *CD* and *AB* intersect at *P*. The anticlockwise angle *DPB* is given directly from the two gradients

$$\text{angle} = \text{gradient of } AB - \text{gradient of } CD = 33.69° - (-41.19°) = 74.88°$$

Exercise 5 The road *AB* of Exercise 1 is to be plotted 10 mm wide on the map. What are the equations of its edges? The polar form is most appropriate here, since the edges of the road are formed by shifting the centre line by +5 mm and – 5 mm. This is achieved by altering *p* of (4.17) by ± 5 mm to give the equations for the road edges to be

$$-0.5547x + 0.8320y = 27.19$$
$$-0.5547x + 0.8320y = 17.19$$

4.9 Angle between two lines

It was shown in Exercise 4 of Section 4.8 that the angle between two lines can be obtained from the two gradients of the lines. A neater way is to use direction cosines. Either the direction cosines of the lines themselves or of the normals to these lines can be used. Let the *angle between two lines* whose gradients are U_1 and U_2 be W, then we have

$$W = U_2 - U_1$$
$$\cos W = \cos (U_2 - U_1)$$

From equation (3.15)

$$\cos (U_2 - U_1) = \cos U_1 \cos U_2 + \sin U_1 \sin U_2$$

therefore, if we put $\cos U_1 = L_1$, $\cos U_2 = L_2$, $\sin U_1 = M_1$, and $\sin U_2 = M_2$, we have

$$\cos W = L_1L_2 + M_1M_2 \tag{4.22}$$

This a most important result. For similar results in three dimensions see Section 5.10. It follows, because $\cos 90° = 0$, that two lines are perpendicular if

$$L_1L_2 + M_1M_2 = 0 \tag{4.23}$$

and, because $\cos 0° = 1$, that two lines are parallel if

$$L_1L_2 + M_1M_2 = 1 \tag{4.24}$$

In this simple case of two dimensions, we have the further simplifications that for perpendicular lines

$$L_2 = -M_1 \quad \text{and} \quad M_2 = L_1$$

and for parallel lines

$$L_1 = L_2 \quad \text{and} \quad M_1 = M_2$$

It should also be remembered from (3.7) that

$$L^2 + M^2 = 1$$

Exercise 1 Find the angle between the lines *CD* and *AB* of Exercise 4 of Section 4.8. From previous exercises we have

$$L_1 = 0.7526 \quad M_1 = -0.6585$$
$$L_2 = 0.8320 \quad M_2 = 0.5547$$
$$\cos W = 0.7526 \times 0.8320 - 0.6585 \times 0.5547 = 0.2609$$
$$W = \cos^{-1}(0.2609) = 74.88°$$

Exercise 2 Show that the lines whose equations are given below are perpendicular to each other.

$$0.6585x + 0.7526y = 73$$
$$-0.7526x + 0.6585y = 12$$

By inspection,

$$L_1L_2 + M_1M_2 = 0$$

therefore the cosine of the angle between the lines is zero, and therefore the lines are orthogonal.

4.10 The equation of a circle

The circle features widely in surveying and cartography. A circle is a curve lying in a plane. It can be completely described in a two-dimensional coordinate system in this plane. (See also Sections 5.20 and 10.3.) To define a circle we need three pieces of essential information. Most commonly there are three cases to be considered.

(1) the coordinates (a, b) of the centre and the radius r, or
(2) the coordinates (x, y) of three points on the circle, or
(3) three tangents of the circle from known points.

If the centre of a circle of radius r lies at the origin of coordinates, the position of any point P (x, y) on its circumference is given by

$$x^2 + y^2 = r^2$$

This is the equation of a circle in a two-dimensional coordinate system. Refer to Figure 4.7. If the centre of the circle is moved to the point Q (a, b), the equation of the circle becomes

$$(x - a)^2 + (y - b)^2 = r^2$$

Expanding these expressions gives

$$x^2 - 2ax + a^2 + y^2 - 2by + b^2 - r^2 = 0$$
$$x^2 + y^2 - 2ax - 2by + a^2 + b^2 - r^2 = 0$$

which can be recast as

$$x^2 + y^2 + 2gx + 2fy + c = 0 \qquad (4.25)$$

where

$$g = -a \quad f = -b \quad c = a^2 + b^2 - r^2$$

Equation (4.25) is the usual form of the equation of a circle in two dimensions.

Exercise 1 The centre of a circle, shown in Figure 4.7, of radius 5 units lies at the point (5, 2). Find its equation in the standard form (4.25). Here

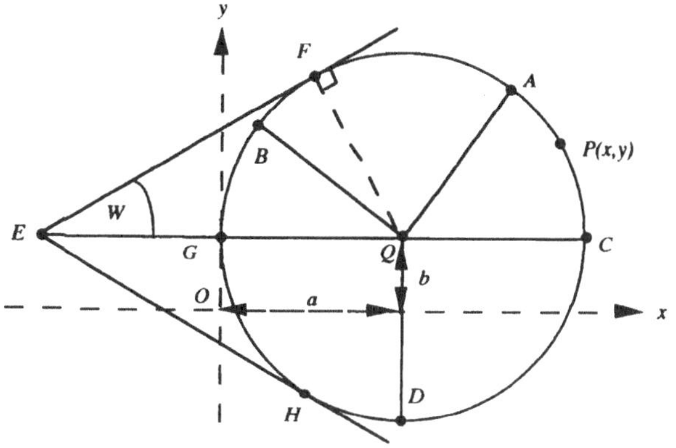

Figure 4.7

$$g = -5 \quad f = -2 \quad c = 25 + 4 - 25 = 4$$

The equation of the circle is

$$x^2 + y^2 - 10x - 4y + 4 = 0 \tag{4.26}$$

Exercise 2 Do the points D (5,–3) and K (8, 5) lie on this circle? Substituting the coordinates of D in (4.26) gives

$$5^2 + 3^2 - (10 \times 5) + (4 \times 3) + 4 = 25 + 9 - 50 + 12 + 4 = 0$$

Therefore D lies on the circle. Substituting the coordinates of K we find that it does not lie on the circle.

Exercise 3 A circle passes through three points A (8, 6), B(1, 5) and C (10, 2). Find its equation in the standard form. The procedure is to substitute the coordinates of each point in equation (4.25) and solve for g, e, and c. Thus we have

$$64 + 36 + 16g + 12f + c = 0$$
$$1 + 25 + 2g + 10f + c = 0$$
$$100 + 4 + 20g + 4f + c = 0$$

or

$$16g + 12f + c = -100$$
$$2g + 10f + c = -26$$
$$20g + 4f + c = -104$$

Giving the solution

$$g = -5 \quad f = -2 \quad c = 4$$

Therefore the equation of the circle is

$$x^2 + y^2 - 10x - 4y + 4 = 0 \tag{4.26}$$

Its centre is (5, 2) and its radius is given by

$$r = \sqrt{(25 + 4 - 4)} = 5$$

4.11 Tangent to a circle

Tangents to a circle are of interest in surveying because a line of sight can be taken to a circular object such as a pond or pipe, or tower. A common situation is to measure the angle $FEH = 2W$ say (see Figure 4.7) and the distance $EG = s$ say. Then

$$r = FQ = (EG + GQ)\sin W = (s + r)\sin W$$

$$r = \frac{s \sin W}{1 - \sin W} \qquad (4.27)$$

Thus we have the distance EQ to the centre of the circle $= r + s$.

Exercise 1 Find the centre and radius of the circle of Figure 4.7 using the following information.

EQ is parallel to the x axis, the coordinates of E are (–5, 2), $EG = s = 5$, and angle $W = 30°$.

$$r = \frac{5 \times 0.5}{1 - 0.5} = 5$$

so the coordinates of Q are (5, 2) because EQ is parallel to the x axis.

4.12 Length of a tangent to a circle from a given external point

The length of the tangent t from an external point such as E (x, y) to the circle is given very neatly by the expression

$$x_E^2 + y_E^2 + 2gx_E + 2fy_E + c = t^2 \qquad (4.28)$$

This result follows from the fact that

$$EQ^2 = (x_E - a)^2 + (y_E - b)^2 = t^2 + r^2$$

which on rearranging and remembering that

$$g = -a \quad f = -b \quad c = a^2 + b^2 - r^2$$

results in expression (4.28).

Exercise 1 Find the length of the tangent EF to the circle of Figure 4.7. We have from (4.28)

$$t^2 = 25 + 4 - 10(-5) - 4(2) + 4 = 33 + 50 - 8 = 75$$

$$t = 8.66$$

This result can be verified from

$$t = EQ \cos W = (s + R)\cos W = 10 \cos 30° = 8.66.$$

Exercise 2 Find the coordinates of point F. The gradient of EF is 30° therefore

$$x_F = -5 + 8.66 \cos 30° = 2.5$$
$$y_F = 2 + 8.66 \sin 30° = 6.33$$

4.13 Equation of a tangent to a point on a circle

The equation of the tangent to a circle at a point F *on the circle* is given by

$$x\,x_F + y\,y_F + g(x + x_F) + f(y + y_F) + c = 0 \qquad (4.29)$$

or

$$(x_F + g)\,x + (y_F + f)\,y + (gx_F + fy_F + c) = 0 \qquad (4.30)$$

The proof requires an understanding of differentiation (see Chapter 9). Consider the equation of the circle (4.25)

$$x^2 + y^2 + 2gx + 2fy + c = 0 \qquad (4.25)$$

Differentiating partially with respect to x and y we have

$$2x\,\delta x + 2y\,\delta y + 2g\,\delta x + 2f\,\delta y = 0$$
$$\delta x\,(x + g) + \delta y\,(y + f) = 0$$

$$\frac{dy}{dx} = \lim = \frac{\delta y}{\delta x} = -\frac{g + x}{f + y} \qquad (4.31)$$

Thus (4.31) is the gradient of the *tangent* at any point on the circle. The gradient at point F is therefore

$$-\frac{g + x_F}{f + y_F}$$

(**Note**: The gradient of the *normal* to the circle at this point is $\frac{f + y}{g + x}$.)

But the gradient by definition through any other point (x, y) on the line is given by

$$\frac{y - y_F}{x - x_F}$$

Therefore

$$\frac{y - y_F}{x - x_F} = -\frac{g + x_F}{f + y_F}$$

or

$$(y - y_F)(f + y_F) = -\,(x - x_F)(g + x_F)$$

so

$$-y_F^2 + fy - fy_F + yy_F = +x_F^2 - gx + gx_F - xx_F$$
$$x_F^2 + y_F^2 - gx + gx_F - fy + fy_F - yy_F - xx_F = 0$$

Since the point F lies on the circle

$$x_F^2 + y_F^2 = -\,(2gx_F + 2fy_F + c)$$

we have

$$-\,(2gx_F + 2fy_F + c) - gx + gx_F - fy + fy_F - yy_F - xx_F = 0$$

therefore

$$-gx - gx_F - fy - fy_F - yy_F - xx_F - c = 0$$

Changing all signs and rearranging we have

$$(x_F + g)x + (y_F + f)y + (gx_F + fy_F + c) = 0 \qquad (4.30)$$

Exercise 1 Find the equation of the tangent through $F(2.5, 6.33)$ to the circle

$$x^2 + y^2 - 10x - 4y + 4 = 0$$

The equation is

$$(2.5 - 5)x + (6.33 - 2)y + 2.5(-5) + 6.33(-2) + 4 = 0$$

$$-2.5x + 4.33y - 21.16 = 0$$

As a check, we verify that E (–5, 2) lies on this tangent

$$-2.5(-5) + 4.33(2) - 21.16 = 0$$

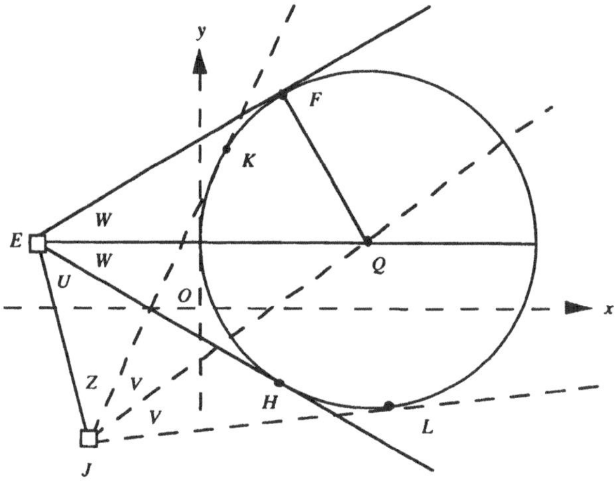

Figure 4.8

4.14 Fitting a circle to three tangents

A common case in surveying is to observe only tangents to a circle, from at least two points whose coordinates relative to each other are known. Usually four tangents are obtained in such a case. The arrangement is illustrated in Figure 4.8. The coordinates of Q the centre of the circle and its radius r have to be found. The ray EQ bisects the measured angle FEH, and the ray JQ bisects the angle KJL. The angles U and Z are also known. Thus the angle EQJ can be found from

$$\text{angle } (EQJ) = 180° - U - W - V - Z$$

The simplest solution is to calculate EQ from the triangle EQJ using the sine rule (3.9)

$$EQ = \frac{EJ\sin(Z+V)}{\sin(EQJ)}$$

and r from

$$r = EQ \sin W$$

To keep Figure 4.8 simple, in this example the derivation of the coordinates of Q is simple because EQ is parallel to the Ox axis. In a more general case the bearing of EQ would be found from the known data and Q calculated from (4.6).

Exercise 1 Find the coordinates of the centre and the radius of the circle fitting the tangents observed from E and J of Figure 4.8, and the following data.

$$E\ (-5, 2) \quad J\ (-4, -7)$$

$$W = 30° \quad U = 53.66° \quad Z = 28.21° \quad V = 23.13°$$

From the coordinates of E and J we find

$$EJ = 9.055$$

$$EQ = \frac{EJ\sin(Z+V)}{\sin(EQJ)} = \frac{9.055\sin 51.34°}{\sin 45°} = 10$$

The coordinates of the centre are Q (5, 2). and the radius is given by

$$r = 10 \sin 30° = 5$$

SUMMARY OF KEY WORDS

plane cartesian coordinates, geographical coordinates, abscissa, ordinate, bearing, polar coordinates, rectangular-to-polar key, coordinate differences, equation of a line in gradient form, polar equation of a line, parameter, parametric form of equation of a line, angle between two lines, equation of a circle.

SUMMARY OF FORMULAE

CARTESIAN AND POLAR COORDINATES

Given (r, U), find x and y from

$$x = r\cos U \quad y = r\sin U \tag{4.1}$$

Given (x, y), find U and r from

$$\tan U = \frac{y}{x} \tag{4.2}$$

and

$$r = \sqrt{\left(x^2 + y^2\right)} \tag{4.3}$$

A useful check is

$$r = x\cos U + y\sin U \tag{4.4}$$

COORDINATE DIFFERENCES

$$\Delta x = r\cos U \quad \Delta y = r\sin U \tag{4.6}$$

$$\tan U = \frac{\Delta y}{\Delta x} \tag{4.7}$$

$$r = \sqrt{(\Delta x^2 + \Delta y^2)} \tag{4.8}$$

$$r = \Delta x\cos U + \Delta y\sin U \tag{4.9}$$

EQUATIONS OF A STRAIGHT LINE

Gradient form $\quad y = mx + c \quad$ (4.10)

Polar form $\quad L'x + M'y = p \quad$ (4.13)

Parametric forms $\quad x = a + t\cos U \quad y = b + t\sin U \quad$ (4.14)

$\quad x = a + tL \quad y = b + tM \quad$ (4.15)

Angle between lines $\quad \cos W = L_1L_2 + M_1M_2 \quad$ (4.22)

Orthogonal lines $\quad L_1L_2 + M_1M_2 = 0 \quad$ (4.23)

Parallel lines $\quad L_1L_2 + M_1M_2 = 1 \quad$ (4.24)

THE CIRCLE

Equation of a circle $\quad x^2 + y^2 + 2gx + 2fy + c = 0 \quad$ (4.25)

Length of tangent $\quad {x_E}^2 + {y_E}^2 + 2gx_E + 2fy_E + c = t^2 \quad$ (4.28)

Eqn of tangent at F $\quad (x_F + g)\,x + (y_F + f)\,y + (gx_F + fy_F + c) = 0 \quad$ (4.30)

Chapter 5
Problems in Three Dimensions

5.1 Reference systems

Consider a solid, such as a tin can from the kitchen. Its geometrical properties, size, shape, etc., are unaltered if it is moved about in three-dimensional space, say if it is taken from a shelf in position B and placed on a stool in position A of Figure 5.1. What will change is its *position* and possibly also its *attitude*. In this case it was originally lying on its side on the shelf and has been moved to a vertical attitude on top of the stool. The unchanged properties can be measured anywhere, but the spacial information is very special to each position of the can. To describe the position and attitude of the solid a coordinate reference system is needed. The commonest system is the three dimensional grid, based on three axes (OX, OY, OZ) perpendicular to each other (mutually orthogonal). Figure 5.1 shows a room with three orthogonal axes with the origin O located at one corner of the room. Two positions, A and B, of the centre of the top of a tin can are also shown. The origin is chosen at one corner to avoid negative values for coordinates. This choice of origin is convenient but not essential. It makes the arithmetic easier.

Figure 5.1 is drawn as an *isometric view* of the room. The three axes are drawn at angles of 120° to each other and the scales along each axis are made the same. Each floor tile is 0.1 m square. This produces the effect that all parallel lines in three dimensions remain as parallel lines in the two dimensional drawing.

Exercise 1 By measurement on Figure 5.1 show that the point P has coordinates

$$X_P = 0.7 \quad Y_P = 0.5 \quad Z_P = 0.7$$

Exercise 2 Verify that the stool is approximately 0.22 m high, 0.18 m wide and 0.4 m long.

Because only lines parallel to the axes are true to scale in isometric drawings, we cannot *measure* the horizontal distance from P to the OZ axes. This distance may

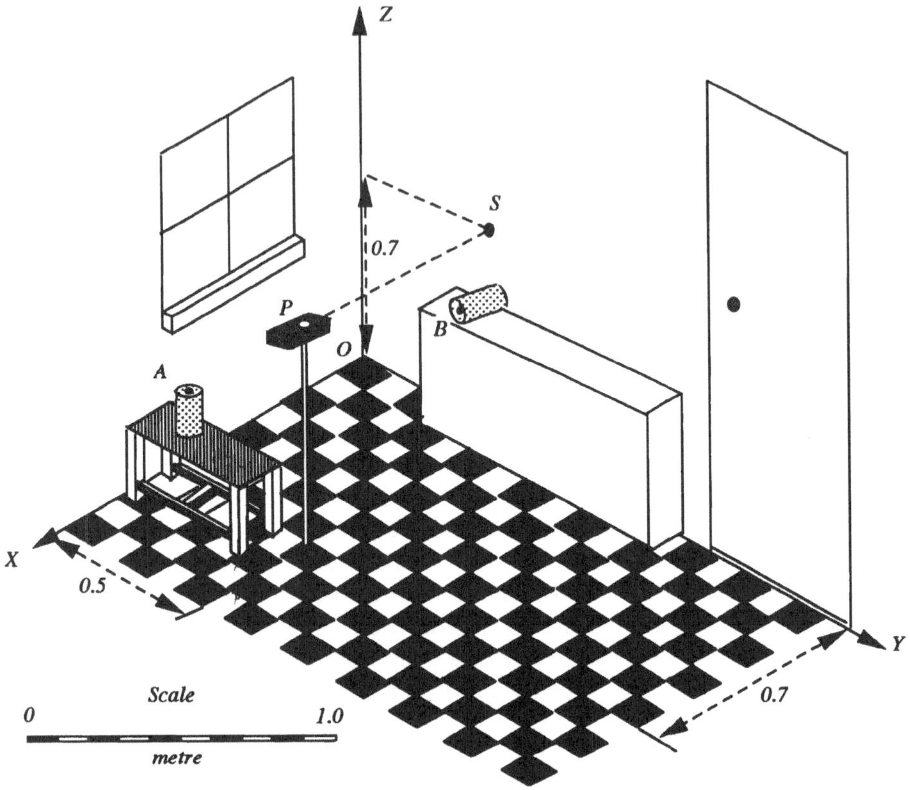

Figure 5.1

however be calculated from the measured coordinates using Pythagoras's theorem. Verify that this distance is 0.86 m.

Exercise 3 Check that the dimensions of the door and window are approximately (0.5 × 1.5) and (0.5 × 0.5) metres respectively.

In surveying and cartography we do not normally deal with tin cans but with artificial satellites or pipework in a factory. However the mathematical principles used to deal with these problems are much the same in all cases. Isometric drawings are easy to construct from dimensions and are useful for taking off measurements. However they are disturbing to the eye which is more used to seeing objects in *perspective views*, in which parallel lines are depicted converging to the horizon. Perspective views are much less convenient for making measurements.

Exercise 4 Imagine a grid system set up in one corner of the room in which you are sitting. Estimate your position in the system by estimating your distance from the two walls and floor used as the reference planes. Suppose your estimations are 2, 3 and 1 metre respectively. Thus your position P in the room is uniquely described in metres by the coordinates

$$OX = 2 \quad OY = 3 \quad OZ = 1$$

A shorter way to describe this position is to write

$$P(2, 3, 1)$$

Exercise 5 Suppose you now move to a new position Q whose estimated coordinates are

$$Q(3, 3, 1)$$

Clearly you have only moved by 1 metre parallel to one wall.

Exercise 6 Where are you if your coordinates are (3, 3, 0)? (Lying on the floor of course!)

For the survey of the interior of buildings, a laser electronic distance meter is available to measure distances from walls as depicted at P in Figure 5.1. This device is accurate to about 3 mm.

5.2 A three-dimensional model

Before reading further, the reader should refer to Section 4.4 where polar and rectangular coordinates in two dimensions are introduced. These basic ideas are now developed further. To assist the understanding of three-dimensional geometry the reader is strongly advised to construct a three-dimensional model according to the following instructions.

(1) Copy Figure 5.2 on a sheet of paper at an enlarged scale if possible.
(2) Cut out the shape and also make a cut along the OZ axis as shown.
(3) Fold along the axes and, with a paper clip, hold the tab in position behind the ZY plane to form a three dimensional model which is represented in Figure 5.3 by the isometric view.

It is worthwhile comparing the two Figures 5.2 and 5.3, and the three dimensional model, in some detail, and noting the following.

(1) All the plotted points lie in axes planes for ease in constructing the three dimensional model.
(2) In Figure 5.3 the point O lies inside a box open towards the reader. Unfortunately the human vision system will not always see it this way. The impression changes suddenly so that that O appears towards the reader *on the outside of the box.*

Exercise 1 Verify that this visual change from one aspect to another does not occur with Figure 5.1.

The reason for this stability is probably due to the fact that the brain is presented with other information, such as the stool and the table, which it knows to have only one acceptable aspect. If we redraw Figure 5.3 with a man standing inside the box, the unstable tendency is reduced. However we again encourage the reader

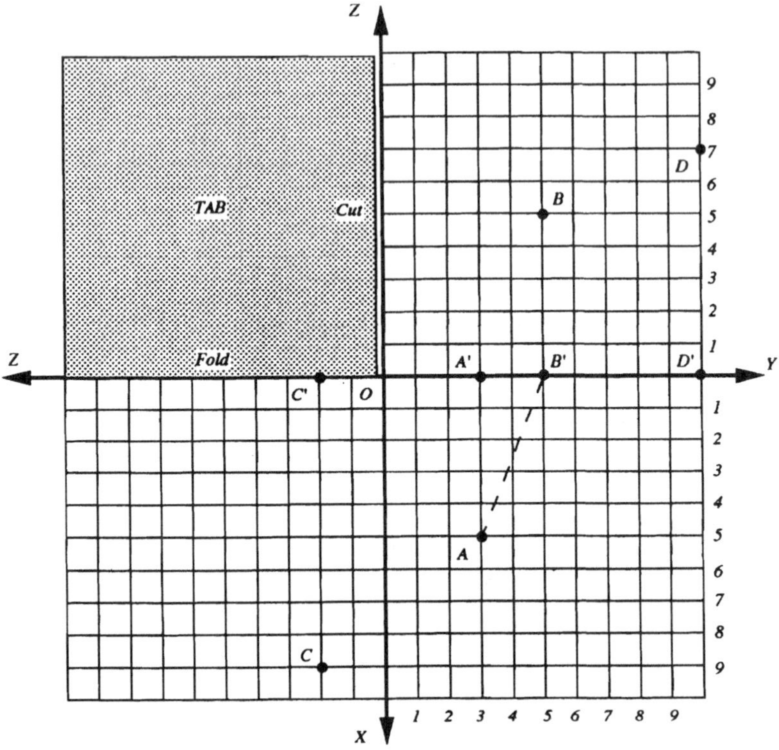

Figure 5.2

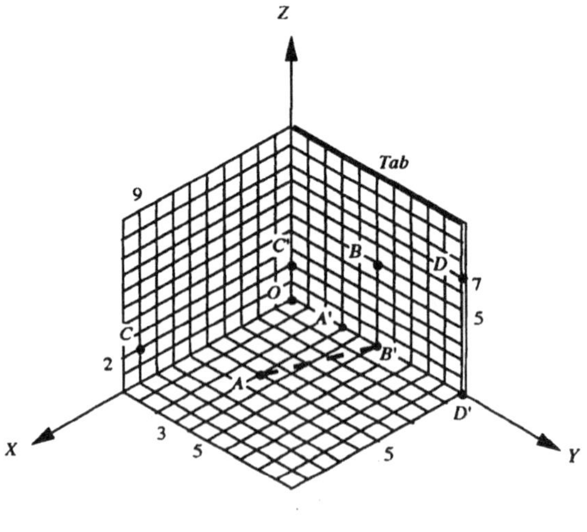

Figure 5.3

to construct the three dimensional model so that this visual problem is removed, and for other reasons which will become apparent as this chapter proceeds.

Exercise 2 Verify from both Figures 5.2 and 5.3 that the coordinates of the points are

Point	A	A'	B	B'	C	C'	D	D'	O
X	5	0	0	0	9	0	0	0	0
Y	3	3	5	5	0	0	10	10	0
Z	0	0	5	0	2	2	7	0	0

5.3 Signs and coordinate differences

The three axes OX, OY and OZ of Figure 5.3 form a *right handed* set, so called for the following reason. Consider the right hand, as in Figure 5.4, with the thumb pointing upwards along the OZ axis, the index finger along the OX axis, and the second finger along the OY axis. This system is called 'right handed'. The positive directions are given by the pointing direction of fingers and thumb. Negative directions are opposite to the finger pointings. The similar *left handed* system is less frequently used.

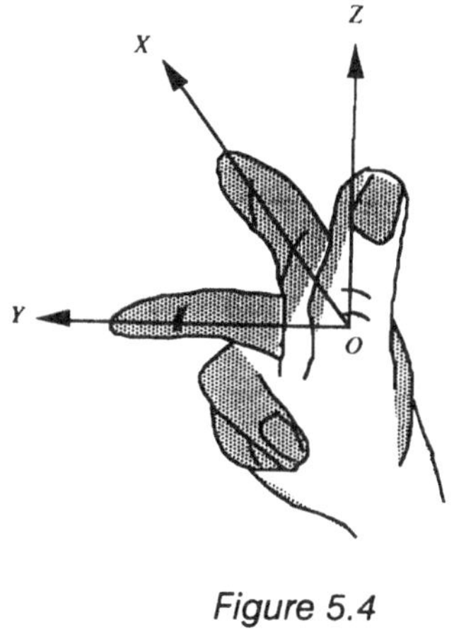

Figure 5.4

In coordinate geometry it is most important to stick strictly to sign conventions. When we refer to the steps AA', $A'B'$ and $B'B$ we mean

$$X_{A'} - X_A \quad Y_{B'} - Y_{A'} \quad Z_B - Z_{B'}$$

Notice that the coordinate of the second point of a step is written before the first point to ensure that the correct sign is obtained.

Exercise 1 Verify that the step $B'B$ is positive (+5) and that BB' is negative (–5).

Exercise 2 Verify that the steps $A'A$, $B'A'$ and BB' are

$$-(X_{A'} - X_A) \quad -(Y_{B'} - Y_{A'}) \quad -(Z_B - Z_{B'})$$

or

$$X_A - X_{A'} \quad Y_{A'} - Y_{B'} \quad Z_{B'} - Z_B$$

Note that because A', B' and B all lie in the YZ plane

$$X_{A'} = X_{B'} = X_B = 0$$

therefore in a similar way

$$AA' = X_{A'} - X_A = X_B - X_A$$

and

$$A'B' = Y_B - Y_A \quad \text{and} \quad B'B = Z_B - Z_A$$

5.4 Directed steps

The distance AA' is called a *directed step*. It takes the sign of the direction of an axis. In moving from A to B for instance we are often forced to move in steps parallel to the three axes.

Exercise 1 Verify that it does not matter in which order we make the steps AA', $A'B'$ and $B'B$ in moving from A to B. The six possibilities are illustrated in Figure 5.5.

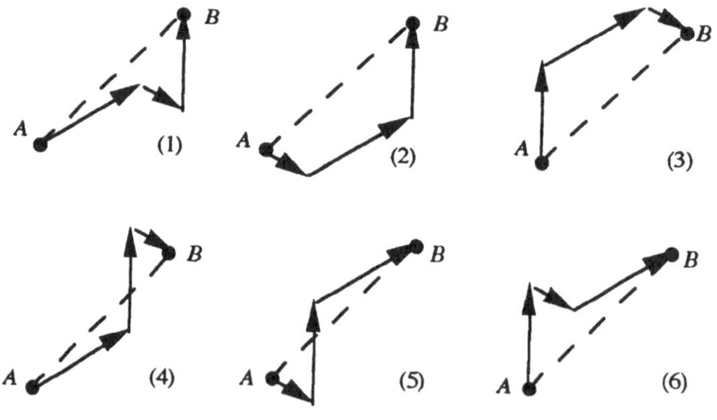

Figure 5.5

That there are six possible ways of moving from A to B can be reasoned as follows. There are three possible *first* moves: parallel to the X axis (1), parallel to the Y axis (2), parallel to the Z axis (3). After the first move there are *two* possible *second* moves, shown in (4), (5) and (6). Thus we have 3×2 possibilities.

Mathematics also uses these three steps for numerical operations. We can think of the movement from A to B as a *vector AB*, but when calculations are required, this movement has to be reduced to its three steps or *components* parallel to the axes. For more information about vectors see Chapter 8.

A flatbed plotter is a mechanical device which draws maps by moving a pen in steps along two orthogonal axes. Industrial three-axes measuring machines move tools and measuring heads along three orthogonal axes. Robots however generally use the alternative of polar movements involving rotations about axes, which are discussed later in Section 5.6 (see also 7.18).

Other notation for directed steps

Because the notation used for the step

$$AB = X_B - X_A$$

is quite cumbersome, other ways of writing it are employed, for example we write

$$\Delta X = X_B - X_A$$

when the step AB is large and

$$\delta X = X_B - X_A$$

when it is small. This notation uses the Greek characters 'delta' for 'D' (Δ) and 'd' (δ), indicating a difference.

Sometimes the steps are abbreviated even further using the lower case letters x, y, and z as follows

$$x = \Delta X \quad y = \Delta Y \quad z = \Delta Z$$

These various forms of notation representing the same thing can be confusing to a beginner. When dealing with any branch of mathematics it is vital to understand the notation being used, before trying to use the notation to develop further ideas. Notation is the vocabulary of mathematics. If we don't know the meaning of a word, we cannot communicate.

5.5 Length of a line

The length of the line AB is related to the coordinates of A and B by the expression

$$AB^2 = (X_B - X_A)^2 + (Y_B - Y_A)^2 + (Z_B - Z_A)^2 \tag{5.1}$$

This is obtained by the successive application of Pythagoras's theorem, first to triangle ABB' giving

$$AB^2 = AB'^2 + BB'^2$$

and then to triangle $AA'B'$ giving

$$AB'^2 = AA'^2 + A'B'^2$$

Therefore

$$AB^2 = AA'^2 + A'B'^2 + BB'^2$$
$$= (X_B - X_A)^2 + (Y_B - Y_A)^2 + (Z_B - Z_A)^2$$

The length $AB = +\sqrt{AB^2}$.

Exercise 1 Show that the lengths of the lines *AB* and *CD* are respectively

$$\sqrt{((0-5)^2 + (5-3)^2 + (5-0)^2)} = \sqrt{(25 + 4 + 25)} = \sqrt{54} = 7.35$$

and

$$\sqrt{((0-9)^2 + (10-0)^2 + (7-2)^2)} = \sqrt{(81 + 100 + 25)} = \sqrt{206} = 14.35$$

Exercise 2 Check these results by measurement on the three-dimensional model.

If elastic bands are fed through small holes in the paper model to join *AB* and *CD* a better impression of the space is created. Figure 5.7 is a picture of this model with the elastic bands in position.

5.6 Polar to rectangular coordinates in three dimensions

The extension to a three-dimensional system is illustrated in Figure 5.6. The axis system is right handed, with *U* positive anticlockwise and *V* positive upwards, $OP = r$. The transformation equations are

$$x = r \cos V \cos U \quad y = r \cos V \sin U \quad z = r \sin V \tag{5.2}$$

$$r = \sqrt{(x^2 + y^2 + z^2)} \tag{5.3}$$

$$\tan U = \frac{y}{x} \tag{5.4}$$

$$\tan V = \frac{z}{\sqrt{(x^2 + y^2)}} \tag{5.5}$$

A useful check is

$$r = (x \cos U + y \sin U) \cos V + z \sin V \tag{5.6}$$

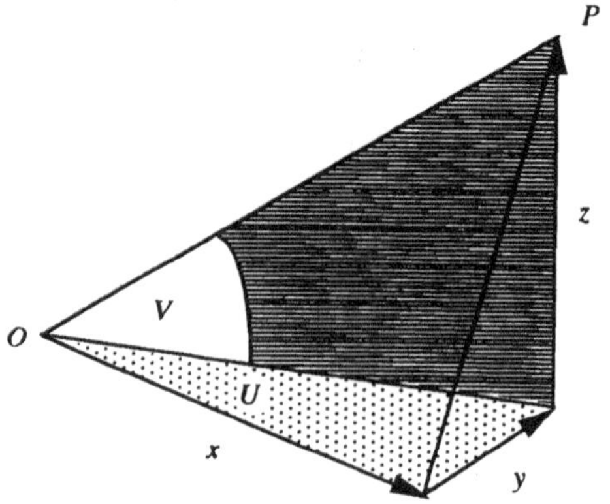

Figure 5.6

Exercise 1 The coordinates of P in millimetres are (25, 55, 20). Calculate r, U and V.

$$\tan U = \frac{55}{25} = 2.2$$

$$U = 65.556°$$

$$\tan V = \frac{20}{60.45} = 0.3310$$

$$V = 18.317°$$

$$r = \sqrt{(25^2 + 55^2 + 20^2)} = 63.640 \text{ mm}$$

Check

$$r = (25 \cos 65.556° + 55 \sin 65.556°) \cos 18.317° + 20 \sin 18.317°$$
$$= 63.640 \text{ mm}$$

Exercise 2 Given that $r = 63.64$ mm, $U = 65.556°$, and $V = 18.317°$ calculate the cartesian coordinates (x, y, z) using equations (5.2).

$$x = 63.640 \times \cos 18.317° \times \cos 65.556° = 25.00 \text{ mm}$$
$$y = 63.640 \times \cos 18.317° \times \sin 65.556° = 55.00 \text{ mm}$$
$$z = 63.640 \times \sin 18.317° = 20.00 \text{ mm}$$

5.7 Coordinate differences

In practice coordinates are seldom referred directly to the origin O. Usually connections between two points, such as A and B, are required. In this case the above formulae (5.1) to (5.5) need only a little modification to put

$$\Delta x = x_B - x_A \text{ for } x \quad \Delta y = y_B - y_A \text{ for } y \quad \Delta z = z_B - z_A \text{ for } z \tag{5.7}$$

the Greek letter Δ indicating a difference. The transformation equations become

$$\Delta x = r \cos V \cos U \quad \Delta y = r \cos V \sin U \quad \Delta z = r \sin V \tag{5.8}$$

$$r = \sqrt{(\Delta x^2 + \Delta y^2 + \Delta z^2)} \tag{5.9}$$

$$\tan U = \frac{\Delta y}{\Delta x} \tag{5.10}$$

$$\tan V = \frac{\Delta z}{\sqrt{\left(\Delta x^2 + \Delta y^2\right)}} \tag{5.11}$$

A useful check is

$$r = (\Delta x \cos U + \Delta y \sin U) \cos V + \Delta z \sin V \tag{5.12}$$

Exercise 1 The coordinates of A and B in millimetres are

A (50, 70, 20) B (75, 125, 40).

Calculate r, U and V.

$$\Delta x = 75 - 50 = 25 \quad \Delta y = 125 - 70 = 55 \quad \Delta z = 40 - 20 = 20$$

$$\tan U = \frac{55}{25} = 2.2$$

$$U = 65.556°$$

$$\tan V = \frac{20}{\sqrt{(25^2 + 55^2)}} = \frac{20}{60.415} = 0.3310$$

$$V = 18.317°$$
$$r = \sqrt{(25^2 + 55^2 + 20^2)} = 63.640 \text{ mm}$$

Check

$$r = (25 \cos 65.556° + 55 \sin 65.556°) \cos 18.317° + 20 \sin 18.317°$$
$$= 63.640 \text{ mm}$$

5.8 Direction cosines in three dimensions

The directed steps of *AB* and the line *AB* itself are related by the expressions

$$\frac{AA'}{AB} = \cos\alpha \qquad \frac{A'B'}{AB} = \cos\beta \qquad \frac{B'B}{AB} = \cos\gamma \tag{5.13}$$

where the angles α, β and γ are the angles that *AB* makes with the positive directions of the *OX*, *OY* and *OZ* axes respectively. To visualise these angles properly, the reader is advised to refer to the three-dimensional paper model already constructed which is illustrated in Figure 5.7, and more specifically in Figure 5.8.

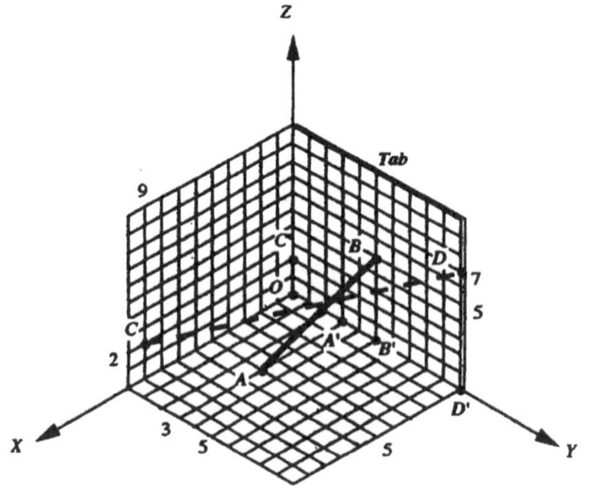

Figure 5.7

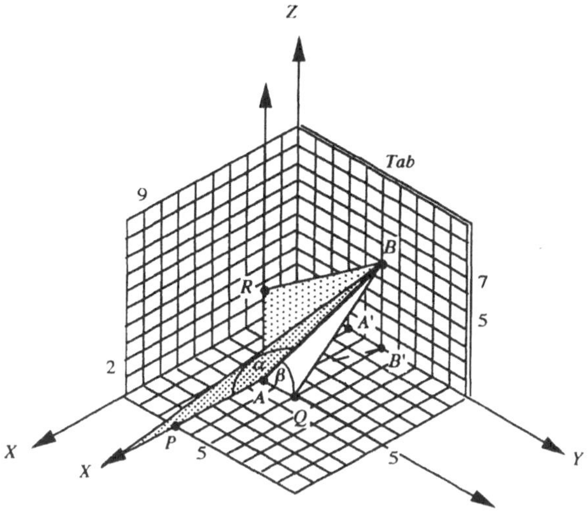

Figure 5.8

n this case, the angle α which *AB* makes with the *OX* axis is greater than 90° and ess than 180°. Therefore its cosine is negative. This can also be seen from the efinition

$$\cos\alpha = \frac{X_B - X_A}{AB} = \frac{0-5}{7.35} = -0.680272$$

'he two other cosines are

$$\cos\beta = \frac{Y_B - Y_A}{AB} = \frac{5-3}{7.35} = 0.272108$$

$$\cos\gamma = \frac{Z_B - Z_A}{AB} = \frac{5-0}{7.35} = 0.680272$$

'he three cosines cos α, cos β and cos γ are called the *direction cosines* of the line ιB since they tell in which direction *AB* is pointing relative to the axes. A very ommon notation for these direction cosines is

$$L = \cos\alpha \quad M = \cos\beta \quad N = \cos\gamma$$

'o determine the angles from the direction cosines, we need to consider the signs ›f all three *L*, *M* and *N*. Since each can be positive or negative, there are eight ›ossible combinations, according to the scheme for quadrants in the *OXY* plane.

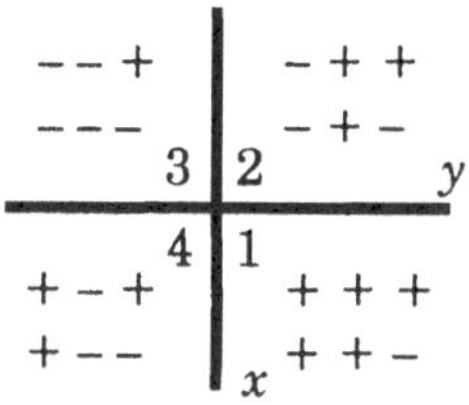

If AB lies in the first quadrant and is pointing upwards, the signs of L, M and N are all positive (shown as + + +). If AB lies in the first quadrant and is pointing downwards, L and M are positive and N is negative (shown as + + –), and so on for all eight possibilities. In the exercise above, the respective signs are – + +, therefore AB is in the second quadrant pointing upwards. Hence the respective angles are

$$\alpha = 132.865° \quad \beta = 74.210° \quad \gamma = 47.135°$$

Exercise 1 Verify that the direction cosines of the line DC are $L = +0.62718$, $M = -0.69686$ and $N = -0.34843$. Therefore DC is in the fourth quadrant pointing downwards and the correct angles are

$$\alpha = 51.16° \quad \beta = 225.82° \quad \gamma = 110.39°$$

These results should be tested against the paper model.

Exercise 2 Show that the direction cosines L, M and N are related by the expression

$$L^2 + M^2 + N^2 = 1 \tag{5.14}$$

From (5.12) we have

$$AB^2 = (X_B - X_A)^2 + (Y_B - Y_A)^2 + (Z_B - Z_A)^2$$

and since

$$L = \cos\alpha = \frac{X_B - X_A}{AB}$$

$$M = \cos\beta = \frac{Y_B - Y_A}{AB}$$

$$N = \cos\gamma = \frac{Z_B - Z_A}{AB} \tag{5.15}$$

we have

$$AB^2 = (L.AB)^2 + (M.AB)^2 + (N.AB)^2$$
$$= L^2 AB^2 + M^2 AB^2 + N^2 AB^2$$

therefore

$$L^2 + M^2 + N^2 = 1$$

Note: It might be thought that only two direction cosines need be known, leaving the third to be calculated from (5.14). However, because the sign of the square root cannot be resolved, no unique solution is possible with only two direction cosines.

5.9 Equations of a line in three dimensions

If two points on a line AB are known, its direction cosines, L, M and N, can be calculated from equations (5.15). Reversing the argument, the coordinates of B, a distance AB along AB from A, will be given by the formulae

$$X_B = X_A + L.\,AB$$
$$Y_B = Y_A + M.\,AB$$
$$Z_B = Z_A + N.\,AB \qquad (5.16)$$

Thus any point P a distance AP along AB from A will be given by the formulae

$$X_P = X_A + L.\,AP$$
$$Y_P = Y_A + M.\,AP$$
$$Z_P = Z_A + N.\,AP \qquad (5.17)$$

The distance AP is usually denoted by 't' and formulae (5.17) become

$$X_P = X_A + t.L$$
$$Y_P = Y_A + t.M$$
$$Z_P = Z_A + t.N \qquad (5.18)$$

The coordinates of all points on the line, corresponding to all values of t can then be found. Therefore these are the equations of the line AB and its extensions beyond the points A and B.

Exercise 1 If A and B are the points A (5, 3, 0) and B (0, 5, 5), show that the coordinates of E, the mid point of AB, are E (2.5, 4.0, 2.5). Refer to Figure 5.9. Also find the coordinates of F and G where $GA = AF = 2AB$.

It is wise to lay out the calculations in a tabular form, as in Table 5.1. All columns of the table are listed in alphabetical order from left to right, and the rows are numbered from the top downwards. Thus we can give all cells a reference. For example, the coordinates of point A are located in cells $B2$, $C2$, and $D2$. The differences of coordinates between A and B are in cells $B4$, $C4$, and $D4$, their squares in $B5$, $C5$, and $D5$, and the square root of the sum of these squares in $B6$. This is the distance AB obtained from

$$\sqrt{((X_B - X_A)^2 + (Y_B - Y_A)^2 + (Z_B - Z_A)^2)}$$

or in terms of the cells as

$$\text{SQRT}(B5 + C5 + D5) = \sqrt{(25 + 4 + 25)} = \sqrt{54} = 7.348\,469\,23$$

where the computer code for square root is 'SQRT'. A note of this calculation is placed at some convenient spot in the table to remind the operator what has been done to obtain the number in cell $B6$.

Note: Some readers will see that this method of working is that used by *spreadsheet computer systems*. To the reader who only uses a hand calculator, this orderly layout is still advisable.

Table 5.1

	A	B	C	D
1	Point	*X*	*Y*	*Z*
2	*A*	5	3	0
3	*B*	0	5	5
4	Delta	–5	2	5
5	Delta Squared	25	4	25
6	Dist	7.34846923	SQRT(*B*5+*C*5+*D*5)	
7		*L*	*M*	*N*
8		–0.6804138	0.27216553	0.68041382
9	*t*, *E*	3.67423461	0.5*B7	
10	*B*	0	5	5
11	*E*	2.5	4	2.5
12	*t*, *F*	14.6969385		
13	*F*	–5	7	10
14	*t*, *G*	–14.696938		
15	*G*	15	–1	–10

Having obtained the length of *AB*, the next stage is to calculate the direction cosines *L*, *M* and *N*. These are obtained from the operations

B4/B6, C4/B6, and D4/B6 abbreviated to B4:D4 /B6

and the results placed in cells B8, C8, and D8.

To test the calculation method we compute the coordinates of *B* from the formulae (5.18), placing the results in cells B10:D10. To calculate the coordinates of *E* the mid point of *AB*, we put $t = \frac{1}{2}AB = 3.674$, and again use formulae (5.18). The points *F* and *G* are obtained by making $t = 2AB$ and $t = -2AB$ respectively. The minus sign places the point *G* along *AB* in the opposite direction to *B*. Once again we encourage the reader to check the results for *F* and *G* by using the paper model.

Exercise 2 Verify that the coordinates of *E*, *F* and *G* could have been obtained by direct scaling of the directed steps of *AB* without the need to find the direction cosines at all.

If we call the directed steps of *AB* respectively *x*, *y*, and *z* that is

$$x = X_B - X_A = -5 \qquad y = Y_B - Y_A = 2 \qquad z = Z_B - Z_A = 5$$

and if we introduce a scale factor *k* for each point (*E*, *F* or *G*), depending on the length of '*t*' in each case, we have

$$k \text{ for } E = 0.5 \quad k \text{ for } F = 2 \quad k \text{ for } G = -2$$

Then the scaled steps are *kx*, *ky*, and *kz* giving scaled steps respectively of

–2.5, 1, 2.5 for *E*, –10, 4, 10 for *F*, and 10, –4, –10 for *G*.

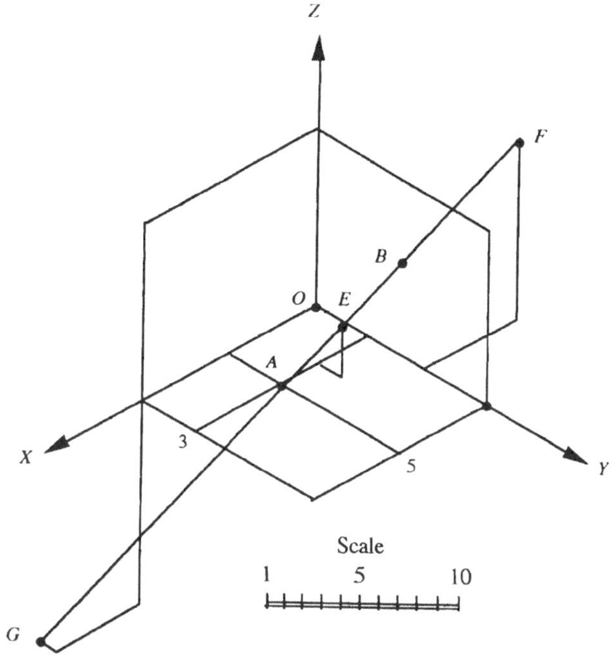

Figure 5.9

These are steps from the point $A(5, 3, 0)$ giving the coordinates of E

$$(5-2.5,\ 3+1,\ 0+2.5) = (2.5,\ 4,\ 2.5).$$

In the same way we obtain the coordinates of F and G from
$F(5-10, 3+4, 0+10) = (-5, 7, 10)$, and $G(5+10, 3-4, 0-10) = (15, -1, -10)$.

Note: Both methods of finding the coordinates of points on a line are used. Formulae (5.18) are necessary to solve problems such as the point of intersection of rays observed by theodolite.

5.10 The angle between two rays

One of the most important and useful formulae of *three-dimensional coordinate geometry* is that for the angle beween two rays in terms of their direction cosines. Consider the two rays AB and AC, of Figure 5.10, with direction cosines L, M, N, and L', M', N' respectively. Then

$$L = \frac{X_B - X_A}{AB} \qquad M = \frac{Y_B - Y_A}{AB} \qquad N = \frac{Z_B - Z_A}{AB}$$

and

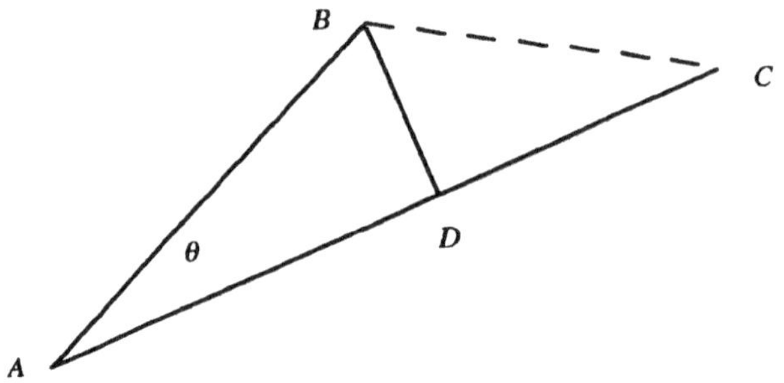

Figure 5.10

$$L' = \frac{X_C - X_A}{AC} \qquad M' = \frac{Y_C - Y_A}{AC} \qquad N' = \frac{Z_C - Z_A}{AC}$$

The angle $\theta = BAC$ between the rays is given by

$$\cos\theta = LL' + MM' + NN' \tag{5.19}$$

To prove this formula consider Figure 5.10 in which the line BD is perpendicular to AC. By the cosine formulae of plane trigonometry (see (3.4)) we have

$$BC^2 = AB^2 + AC^2 - 2\,AB.AC\cos\theta \tag{5.20}$$

To prove formula (5.19) we just have to recast this formula (5.20) in terms of the direction cosines. From (5.12) we have

$$\begin{aligned} BC^2 &= (X_C - X_B)^2 + (Y_C - Y_B)^2 + (Z_C - Z_B)^2 \\ AB^2 &= (X_B - X_A)^2 + (Y_B - Y_A)^2 + (Z_B - Z_A)^2 \\ AC^2 &= (X_C - X_A)^2 + (Y_C - Y_A)^2 + (Z_C - Z_A)^2 \end{aligned}$$

From (5.20)

$$2\,AB.AC\cos\theta = AB^2 + AC^2 - BC^2$$

The right hand side is

$$\begin{aligned} &+(X_B - X_A)^2 + (Y_B - Y_A)^2 + (Z_B - Z_A)^2 \\ &+(X_C - X_A)^2 + (Y_C - Y_A)^2 + (Z_C - Z_A)^2 \\ &-(X_C - X_B)^2 - (Y_C - Y_B)^2 - (Z_C - Z_B)^2 \end{aligned}$$

On multiplying out and collecting terms this expression simplifies to

$$2(X_B - X_A)(X_C - X_A) + 2(Y_B - Y_A)(Y_C - Y_A) + 2(Z_B - Z_A)(Z_C - Z_A)$$

equating to the left hand side and dividing throughout by $AB.AC$ we have

$$\cos\theta = \frac{(X_B - X_A)(X_C - X_A)}{AB.AC} + \frac{(Y_B - Y_A)(Y_C - Y_A)}{AB.AC} + \frac{(Z_B - Z_A)(Z_C - Z_A)}{AB.AC}$$

therefore

$$\cos\theta = LL' + MM' + NN' \tag{5.19}$$

The angle given by this formula follows a right handed rule. With the index finger pointing along *AB*, the second finger along *AC* and the thumb pointing at right angles to both *AB* and *AC*, the angle given by the formula is that obtained by a rotation about the thumb axis of *AB* into *AC*. It is positive clockwise and negative anticlockwise.

5.11 Parallel and perpendicular rays

If two rays are parallel, the angle between them is zero, therefore

$$LL' + MM' + NN' = 1$$

If the rays are perpendicular to each other (normal to each other), the angle between them is 90° therefore

$$LL' + MM' + NN' = 0$$

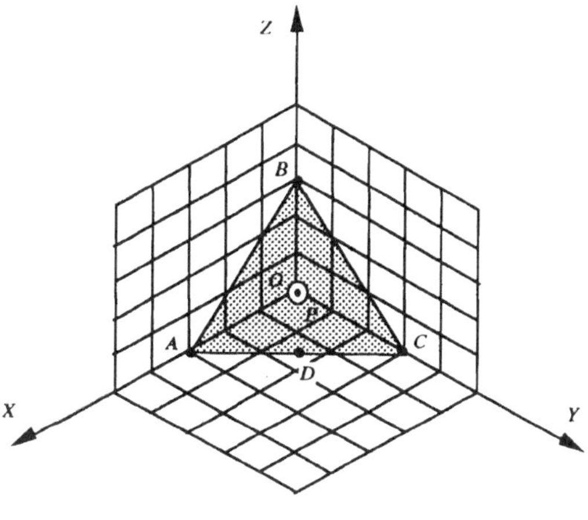

Figure 5.11

Exercise 1 In Figure 5.11 the points *A*, *B* and *C* have coordinates (3, 0, 0), (0, 0, 3) and (0, 3, 0) respectively. Verify, using the formula (5.19) that the angle at *A* between *AB* and *AC* is 60°.

It is clear that the triangle *ABC* is equilateral because

$$AB = \surd(9 + 0 + 9) = 4.2426$$

$$AC = \surd(9 + 9 + 0) = 4.2426$$

$$BC = \surd(0 + 9 + 9) = 4.2426$$

Therefore the three angles *A*, *B* and *C* are each 60°. The direction cosines of *AB* are

$$L = \frac{-3}{4.2426} = -0.7071 \qquad M = 0 \qquad N = \frac{3}{4.2426} = 0.7071$$

The direction cosines of *AC* are

$$L' = \frac{-3}{4.2426} = -0.7071 \qquad M' = \frac{3}{4.2426} = 0.7071 \qquad N' = 0$$

Therefore the angle A is given by

$$\cos A = LL' + MM' + NN'$$

$$= (-0.7071 \times -0.7071) + (0 \times 0.7071) + (0.7071 \times 0) = 0.5$$

$$A = 60^\circ$$

If two rays such as AB and CD of Figure 5.7 do not intersect, we can still give meaning to the angle between them. Consider rays which do intersect, one parallel to AB and the other parallel to CD. The angle between them lies in a plane containing both rays. This is the angle between the original rays.

Exercise 2 Prove that the angle between the rays AB and CD of Figure 5.7 is 31.427°.
In Section 5.8, the direction cosines of AB and CD have already been found to be

AB	−0.68027	0.27211	0.68027
CD	−0.62718	0.69686	0.34843

Therefore

$$\cos\theta = (-0.668027 \times -0.62718) + (0.27211 \times 0.69686) + (0.68027 \times 0.34843)$$

$$= 0.42665 + 0.18962 + 0.23703$$

$$= 0.8533$$

$$\cos\theta = 31.427^\circ$$

5.12 Projection of one ray on another

Consider Figure 5.10. Suppose we require the length of AD, which is called *the projection of AB on AC*. The direction cosines of AC are L', M', N' as before, then the length of AD is given by,

$$AD = AB\cos A = AB(LL' + MM' + NN')$$

$$= AB\left(\frac{(X_B - X_A)(X_C - X_A)}{AB.AC} + \frac{(Y_B - Y_A)(Y_C - Y_A)}{AB.AC} + \frac{(Z_B - Z_A)(Z_C - Z_A)}{AB.AC}\right)$$

$$= \frac{(X_B - X_A)(X_C - X_A)}{AC} + \frac{(Y_B - Y_A)(Y_C - Y_A)}{AC} + \frac{(Z_B - Z_A)(Z_C - Z_A)}{AC}$$

$$AD = (X_B - X_A)L' + (Y_B - Y_A)M' + (Z_B - Z_A)N' \qquad (5.21)$$

Exercise 1 If AD is the projection of AB on the line AC of Figure 5.11, show that the length of AD is 2.12. The result is readily known because triangle ABC is equilateral, and side $AC = 4.24$. Alternatively, applying the formula (5.21) we have

$$
\begin{aligned}
AD &= (X_B - X_A)L' + (Y_B - Y_A)M' + (Z_B - Z_A)N' \\
&= (0-3)\frac{-3}{4.24} + (0-0)\frac{3}{4.24} + (3-0)0 \\
&= \frac{9}{4.24} = 2.12
\end{aligned}
$$

The coordinates of point D can then be found from the equations of the line AD in this case

$$
\begin{aligned}
X_D &= X_A + AD.\,L' &&= (3 + 2.12 \times -3)/4.24 &&= 1.5 \\
Y_D &= Y_A + AD.\,M' &&= (0 + 2.12 \times 3)/4.24 &&= 1.5 \\
Z_D &= Z_A + AD.\,N' &&= 0 + 2.12 \times 0 &&= 0.0
\end{aligned}
$$

Exercise 2 From coordinates, calculate the length of BD of Figure 5.11. B and D are the points

$$(0, 0, 3) \text{ and } (1.5, 1.5, 0),$$

hence BD is given by

$$\sqrt{(1.5^2 + 1.5^2 + 3^2)} = \sqrt{13.5} = 3.67$$

5.13 Distance from a point to a ray

Alternatively the distance from a point to a ray can be found directly as follows. Suppose the length of BD is required. Then

$$BD = AB \sin\theta = AB\sqrt{(1 - \cos^2\theta)}$$

If we substitute in this formula for $\cos\theta$, BD can be written directly in terms of the direction cosines of AB and AC as follows

$$BD = AB\sqrt{((LM' - ML')^2 + (MN' - NM')^2 + (NL' - LN')^2)} \qquad (5.22)$$

Exercise 1 Verify the length of BD using this formula.
To assist with the arithmetic, it is convenient to list the direction cosines in order as follows.

L	M	N	L
–0.7071	0	0.7071	–0.7071
–0.7071	0.7071	0	–0.7071
L'	M'	N'	L'

The calculation is carried out by cross-multiplying pairs of numbers as indicated by the arrows. Down-arrow products take positive signs, and up-arrow products take negative signs. The terms in each bracket are

$$
\begin{aligned}
L.M' - M.L' &= (-0.7071 \times 0.7071) - (-0.7071 \times 0) &&= -0.5 \\
M.N' - N.M' &= (0 \times 0) - (0.7071 \times 0.7071) &&= -0.5 \\
N.L' - L.N' &= (0.7071 \times -0.7071) - (0 \times -0.7071) &&= -0.5
\end{aligned}
$$

$$\sin\theta = \sqrt{(1 - \cos^2\theta)} = \sqrt{(0.25 + 0.25 + 0.25)} = 0.8660$$

$$BD = 4.24 \times 0.8660 = 3.67.$$

Note: The terms in the brackets of (5.22) are *two by two determinants*. See Section 6.6 for more information

5.14 The common normal to two rays

Consider Figure 5.7 again. The rays AB and CD do not meet. It is often important to find the length of the shortest distance between them and the positions of each end of this shortest line. The existence of a common normal to each ray can be verified by inspecting the paper model representing Figure 5.7. Suppose the direction cosines of the common normal to each ray are U, V, and W. Then we have

$$LU + MV + NW = 0$$

$$L'U + M'V + N'W = 0$$

Dividing throughout by U gives the equations

$$L + M(V/U) + N(W/U) = 0$$

$$L' + M'(V/U) + N'(W/U) = 0 \qquad (5.23)$$

Which we can solve for V/U and W/U. Also

$$U^2 + V^2 + W^2 = 1$$

therefore

$$(V/U)^2 + (W/U)^2 + 1 = 1/U^2$$

so that U can be found, and therefore V and W. So the direction cosines of the common normal are now known.

Exercise 1 Find the direction cosines of the normal to the lines AB and CD. Let the ends of the normal be the points E and F on AB and CD respectively.
The direction cosines of AB and CD are

	L	M	N
AB	−0.680 27	0.272 11	0.680 27
CD	−0.627 18	0.696 86	0.348 43

The equations (5.23) in this case are therefore

$$-0.680\,27 + 0.272\,11\,(V/U) + 0.680\,27\,(W/U) = 0$$

$$-0.627\,18 + 0.696\,86\,(V/U) + 0.348\,43\,(W/U) = 0$$

Solution of these equations gives the result

$$(V/U) = 0.5000 \quad (W/U) = 0.799\,99$$

Then we obtain U from

$$(V/U)^2 + (W/U)^2 + 1 = 1/U^2 = 1 + (0.5000)^2 + (0.799\,99)^2$$

$$1/U^2 = 1.889\,984 \quad U^2 = 0.529\,105 \quad U = 0.727\,396$$

which gives

$$V = 0.5U = 0.363\,698 \quad \text{and } W = 0.79999U = 0.581\,910$$

Note: We have chosen the positive value for U. If the negative value is chosen, all direction cosines change sign and the final result is the normal facing in the opposite direction. If a particular direction is required, the two original lines and their normal will form either a right or left handed set of directions when chosen in order: line one, line two and their normal. The system is right handed if, when looking along the normal, a clockwise turn will bring line one into line two.

5.15 The length of the normal *EF*

The length of this normal *EF* is the projection of any line, such as *AC*, on the line with direction cosines U, V, and W just found. Note one point has to lie on *AB* and another on *CD*. Thus, in this case, the length of *EF* is given by

$$\begin{aligned} \mathrm{EF} &= (X_C - X_A)U + (Y_C - Y_A)V + (Z_C - Z_A)W \\ &= (4)\,0.727\,393 + (-3)\;0.363\,698 + (2)\,0.581\,910 \\ &= 2.982 \end{aligned}$$

Exercise 1 Show that the projections of *BD*, *CB* and *AD*, on *EF* are respectively all equal to 2.982.

From *BD* we have (0) 0.727 396 + (5) 0.363 698 + (2) 0.581 910 = 2.982
From *CB* we have (9) 0.727 396 + (–5) 0.363 698 + (–3) 0.581 910 = 2.982
From *AD* we have (–5) 0.727 396 + (7) 0.363 698 + (7) 0.581 910 = 2.982

5.16 Terminals of the common normal *EF*

To find the points E and F we write the equations of *AE* and *CF* respectively as

$$\begin{aligned} X_E &= X_A + AE.L \quad & X_F &= X_C + CF.L' \\ Y_E &= Y_A + AE.M \quad & Y_F &= Y_C + CF.M' \\ Z_E &= Z_A + AE.N \quad & Z_F &= Z_C + CF.N' \end{aligned} \qquad (5.24)$$

Two of the direction cosines of *EF*, U and V, may be expressed in terms of the coordinates of E and F and the length of *EF* as

$$U = (X_F - X_E)/EF \quad \text{and} \quad V = (Y_F - Y_E)/EF$$

or

$$U.EF = (X_F - X_E) \quad \text{and} \quad V.EF = (Y_F - Y_E) \qquad (5.25)$$

Note: We do not require to use the third direction cosine W. Substituting from (5.24) in (5.25) we have

$$\begin{aligned} U.EF &= X_C + CF.L' - X_A - AE.L \\ V.EF &= Y_C + CF.M' - Y_A - AE.M \end{aligned} \qquad (5.26)$$

These are two equations in the unknowns *AE* and *CF* which we can solve, and then from equations (5.24) obtain the coordinates of E and F.

Exercise 1 Calculate the coordinates of E and F for the given data. Substituting the numerical values in (5.26) we have

$$0.727\,396 \times 2.982 = 9 + CF \times -0.627\,18 - 5 + AE \times 0.680\,27$$

$$0.363\,698 \times 2.982 = 0 + CF \times 0.696\,86 - 3 + AE \times -0.272\,11 \quad (5.27)$$

and rearranging

$$-0.627\,18\,CF + 0.68027\,AE = -1.830\,715$$

$$0.696\,86\,CF - 0.272\,11\,AE = 4.084\,668$$

which solve to give $CF = 7.517$ and $AE = 4.239$.

Exercise 2 Show that the coordinates of E and F are

	X	Y	Z
E	2.116	4.153	2.884
F	4.286	5.238	4.619

The actual working is left as an exercise for the reader. These results can be checked by calculating the distance EF from these coordinates, and verifying that it is 2.982.

5.17 The equation of a plane

In Figure 5.12 a plane ABC is shown. The line OB is normal to the plane, or in other words the angle ABO is 90°. A and C are any points on the plane. If the direction cosines of OB are respectively L, M and N and the length of $OB = P$, then the equation of the plane is

$$LX + MY + NZ = P \quad (5.28)$$

OB is the projection of OA on OB. Therefore by equation (5.21) we have

$$P = L(X_A - 0) + M(Y_A - 0) + N(Z_A - 0) = LX_A + MY_A + NZ_A$$

Since A is any point on the plane, all points in the plane satisfy the equation

$$LX + MY + NZ = P \quad (5.28)$$

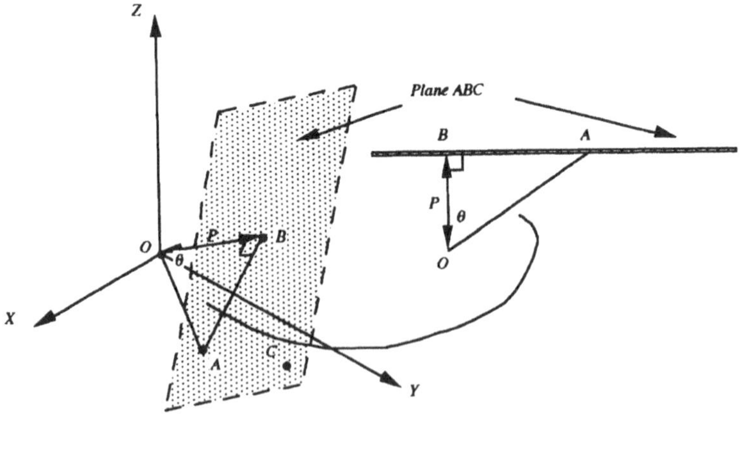

Figure 5.12

Usually a plane is defined by three points which do not lie on a straight line, and the equation is found from their coordinates. Assuming that the coordinates of A, B and C are given, and because the three points lie in this plane, their coordinates satisfy equation (5.28) and we have

$$LX_A + MY_A + NZ_A = P$$
$$LX_B + MY_B + NZ_B = P$$
$$LX_C + MY_C + NZ_C = P$$

which in matrix from is

$$\begin{bmatrix} X_A & Y_A & Z_A \\ X_B & Y_B & Z_B \\ X_C & Y_C & Z_C \end{bmatrix} \begin{bmatrix} L \\ M \\ N \end{bmatrix} = \begin{bmatrix} P \\ P \\ P \end{bmatrix}$$

It might be thought that there are four unknowns L, M, N and P and only three equations for solution. But it should be remembered that the direction cosines are not all independent because

$$L^2 + M^2 + N^2 = 1$$

Provided the plane does not pass through the origin of coordinates, we can divide each equation throughout by P to give

$$LX/P + MY/P + NZ/P = 1$$

or

$$(L/P)X + (M/P)Y + (N/P)Z = 1$$

The three equations in new unknowns (L/P) etc. become

$$\begin{bmatrix} X_A & Y_A & Z_A \\ X_B & Y_B & Z_B \\ X_C & Y_C & Z_C \end{bmatrix} \begin{bmatrix} L/P \\ M/P \\ N/P \end{bmatrix} = \begin{bmatrix} 1 \\ 1 \\ 1 \end{bmatrix}$$

and we solve for (L/P), (M/P) and (N/P) by any method.

Now we have

$$(L/P)^2 + (M/P)^2 + (N/P)^2 = 1/P^2$$

Therefore we can obtain $1/P^2$ and therefore P. Using this value we can finally calculate L, M and N from L/P, M/P and N/P.

Exercise 1 In Figure 5.11 the points A, B and C have coordinates (3, 0, 0), (0, 0, 3) and (0, 3, 0) respectively. Derive the equation of the plane ABC. We tabulate the data anticlockwise in order A, C, B.

	X	Y	Z
A	3	0	0
C	0	3	0
B	0	0	3

The equations to be solved are

$$3L/P + 0M/P + 0N/P = 1$$
$$0L/P + 3M/P + 0N/P = 1$$
$$0L/P + 0M/P + 3N/P = 1$$

which gives at once that

$$L/P = M/P = N/P = 1/3$$
$$1/P^2 = 3/9 = 1/3$$
$$P = \sqrt{3}$$

Then

$$L = M = N = \sqrt{3}/3 = 1/\sqrt{3}$$

Note: The solid formed by the axes planes and the plane *ABC* is the same as a glass corner cube used to reflect electronic signals in surveying. The geometrical path length of the signals in the cube is *P*. In the cube we have chosen, the side length *AB* equals 4.243 units. If we scale this to be one unit, then *P* becomes 0.408 units. This gives the rule that for a cube of side length *D* the geometrical path of signals is 0.408 *D*. In practice, the optical path taken is affected by *R* the refractive index of the glass to give the actual length of *RP*.

5.18 The equation of a sphere

The derivation of the equation of a sphere in three dimensions follows easily from that of a circle in two dimensions (see Section 4.10). A sphere is a surface defined by the fact that all points on it are a fixed distance from its centre. Let the centre have coordinates (a, b, c) and the radius be r. Any point (x, y, z) on the surface of the sphere is related to the radius and centre by

$$(x-a)^2 + (y-b)^2 + (z-c)^2 = r^2 \tag{5.29}$$

Expanding out we have

$$x^2 - 2ax + a^2 + y^2 - 2by + b^2 + z^2 - 2cz + c^2 - r^2 = 0$$
$$x^2 + y^2 + z^2 - 2ax - 2by - 2cz + a^2 + b^2 + c^2 - r^2 = 0$$

which can be written as

$$x^2 + y^2 + z^2 + 2ex + 2fy + 2gz + h = 0 \tag{5.30}$$

where the centre is the point $(-e, -f, -g)$ and the radius is $\sqrt{[(e^2 + f^2 + g^2) - h]}$. The equation (5.30) is the *equation of a sphere*. Clearly if the coordinates of the centre and the radius are known, the sphere is defined.

5.19 The equation of a sphere from four known points

A minimum of four points on the surface, not all of which lie in a plane, is needed to define a sphere. If the coordinates of these four points are measured by surveying or other techniques, the equation can be established.

Exercise 1 Find the equation of the sphere *ABCD* which fits the following data. A tabular layout is convenient.

Table 5.2

Point	x	y	z	x^2	y^2	z^2
A	9	9	4	81	81	16
B	6	9	7	36	81	49
C	6	5	9	36	25	81
D	10	2	4	100	4	16

Substituting the values of Table 5.2 in equation (5.30) we obtain four equations in e, f, g and h.

$$18e + 18f + 8g + h = -178$$
$$12e + 18f + 14g + h = -166$$
$$12e + 10f + 18g + h = -142$$
$$20e + 4f + 8g + h = -120$$

When these equations are solved by any method described in Chapter 12, which the reader may verify by substitution, we find that

$$e = -6 \quad f = -5 \quad g = -4 \quad h = 52$$

Thus the centre is at (6, 5, 4) and the radius r is given by

$$r = \sqrt{[(a^2 + b^2 + c^2) - h]}$$
$$r = \sqrt{[(36 + 25 + 16) - 52]} = 5$$

5.20 Fitting to a circle in three dimensions

In engineering and cartographical work, it often happens that a circle has to be fitted to points whose coordinates are established from surveying techniques or from digitising a map. The cartographic problem is the simpler because all the data lie in the plane of the two-dimensional coordinate system and the procedure is as in Section 4.10 The surveying problem, however, is often more complex because the circle does not always lie in a horizontal or vertical plane. For example in pipe work, a circular section through the pipe will very often be in a plane tilted with respect to the vertical and horizontal planes.

For example, pipes, such as those shown in Figure 5.13 can be treated as thin cylinders, whose cross-sections are circular. To know the attitude of the pipe axes and their point of intersection is often very important in assembly problems, such as in oil platforms, chemical works and the like. In practice, the process usually involves fitting to more than the minimum of data points by the method of least squares. However, it is always necessary to obtain good provisional values

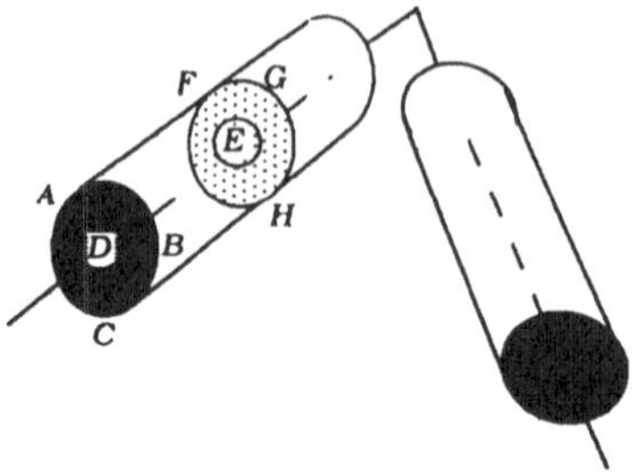

Figure 5.13

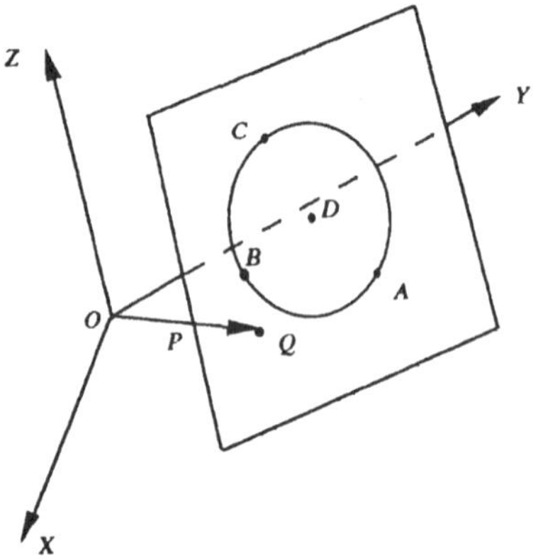

Figure 5.14

for the direction of a pipe axis and its radius, before a least squares fitting commences.

It is assumed that the coordinates of at least three points lying on a circular section of a pipe have been obtained by some survey method. Figure 5.14 shows a circle *ABC* with centre *D* lying in a plane tilted with respect to the right handed rectangular cartesian axes *OX*, *OY*, *OZ*. The point *Q* also lies in the plane *ABCD*. *Q* is the point where the line from the coordinate origin *O*, perpendicular to the plane, cuts the plane. The line *OQ* is of length *P*, and has direction cosines with the respect to the *OX*, *OY*, and *OZ* axes of *L*, *M*, and *N* respectively. The equation of the plane is given by

$$LX + MY + NZ = P \qquad (5.28)$$

By the same method as in Section 5.17 the first stage is to find the equation of the plane ABC from the following data

Point	X	Y	Z
A	10.964	4.640	12.816
B	8.678	1.214	11.799
C	5.412	2.730	14.045

This equation is

$$0.341\,77X - 0.469\,68Y + 0.814\,00Z = 12.000 \qquad (5.31)$$

(**Note:** These direction cosines are also those of the axis of the cylinder.)

To find the centre of the circle, we use the fact that the plane passing through the mid point of a chord of the circle, and normal to the chord, passes through the circle centre. Two such planes and the plane of the circle itself (5.31) will intersect at the centre.

From the chord *AB* we obtain its length and direction cosines which are

$$AB = 4.242\,35 \quad L = -0.538\,85 \quad M = -0.807\,57 \quad N = -0.239\,73$$

The length P of the normal from the origin to the required plane through the mid point of AB is obtained from

$$-0.538\,85\,(X_B + X_A) + -0.807\,57\,(Y_B + Y_A) + -0.239\,73\,(Z_B + Z_A) = 2P$$

in this case $P = -10.606\,24$. The equation of the required plane is therefore

$$0.538\,85X + 0.807\,57Y + 0.239\,73Z = 10.606\,24 \qquad (5.32)$$

In a similar manner the equation of the plane normal to the mid point of the chord BC is found to be

$$-0.769\,60X + 0.357\,23Y + 0.529\,25Z = 2.121\,56 \qquad (5.33)$$

Solving equations (5.31), (5.32) and (5.33) gives the coordinates of the centre of the circle ABC to be

$$(8.188, 3.683, 13.429)$$

The three values of the radius to A, B and C all give $r = 3$.

The equations of the axis of the cylinder are therefore compiled from the coordinates of D and the direction cosines of the plane ABC as follows:

$$\left.\begin{aligned} X &= X_D + tL = 8.188 + 0.34177t \\ Y &= Y_D + tM = 3.683 - 0.469\,68t \\ Z &= Z_D + tN = 13.429 + 0.814\,00t \end{aligned}\right\} \qquad (5.34)$$

A point E which is 6 units along the axis from D and nearer the origin will have coordinates, given by putting $t = -6$ in equations (5.34), as follows

$$(6.138, 6.502, 8.545)$$

This point E is the centre of the circle FGH between three points on another ring of the pipe (see Figure 5.13).

Exercise 1 Verify from the data below that FGH is a circle in a plane parallel to ABC and E is the centre of this circle.

Point	X	Y	Z
F	4.339	0.607	5.900
G	3.577	-0.535	5.561
H	2.488	-0.031	6.309

Assuming that the plane FGH is parallel to ABC and a distance 6 units away from it, its equation will be

$$0.341\,77X - 0.469\,68Y + 0.814\,00Z = 6.000 \qquad (5.35)$$

Substituting the values of F, G and H in this equation verifies that these points lie in this plane.

5.21 Creation of a stereoscopic image

It is quite common nowadays to employ a *stereoscopic optical model* instead of a solid model to illustrate some plan or construction proposal. Two slightly differing drawings are made which, when viewed by lens or other stereoscope, create the illusion of depth.

Consider Figure 5.15 which shows the plan and elevation of a simple tower consisting of points *A*, *B*, ..., *J*. The (x, y, z) coordinates of these key points on the tower would have been obtained by some surveying method. The objective is to create by computer the two views (Figure 5.16) which would be obtained if the tower had been photographed, and to view them subsequently to obtain a stereoscopic image, as with two photographs.

Points *L* and *R* of Figure 5.15 are the two projection centres and the projection plane is the line *dd'* orthogonal to the page. Let the equation of this plane be

$$Lx + My + Nz = P \tag{5.28}$$

We choose $P = 38$, and because the plane is parallel to the $y\ O\ z$ plane, $L = 1$, $M = 0$ and $N = 0$, therefore the equation reduces to

$$x = 38 \tag{5.36}$$

From the first viewpoint *L* we create the image to be viewed by the left eye to form the stereopair. This is shown in Figure 5.16. Consider a typical ray *LDd*, where *d* lies on the projection plane. If *t* is a scaling parameter, the first equation of this ray is

$$x_d = x_L + t\,L \tag{5.37}$$

Suppose the coordinates of the viewpoints *L* and *R* and of the tower points *A* to *J* are as listed in Table 5.3.

Table 5.3

	x	y	z
Viewpoints			
L	0	5	47
R	0	0	47
Tower points			
A	23	8	30
B	23	17	30
C	31	17	30
D	31	8	30
E	23	8	40
F	23	17	40
G	31	17	40
H	31	8	40
J	27	12.5	45

We select $x_L = 0$, and because all x values of the projected points are equal to 38, we have

$$x_d = 38$$

Also

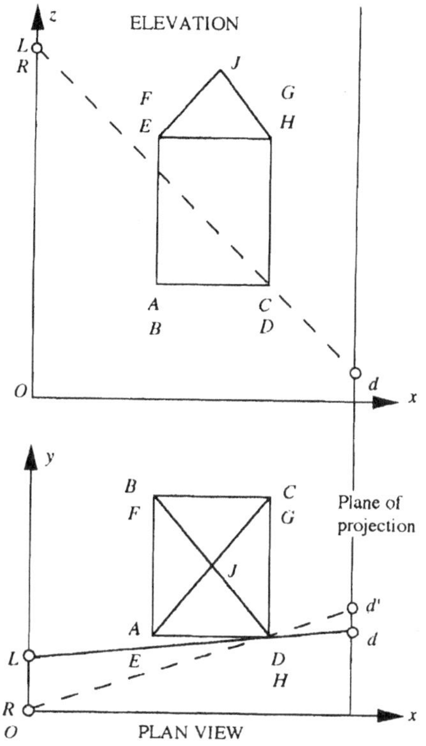

Figure 5.15

$$\frac{y_d - y_L}{y_D - y_L} = \frac{x_d}{x_D}$$

so

$$y_d - y_L = \frac{(y_D - y_L)x_d}{x_D} = \frac{38(y_D - y_L)}{x_D} = \frac{38 \times 3}{31} = 3.68$$

$$y_d = y_L + 3.68 = 5 + 3.68 = 8.68$$

All other values of y are calculated in the same way to give the results in Table 5.4.

The values of z coordinates are obtained in a similar way from typically

$$z_d - z_L = \frac{(z_D - z_L)x_d}{x_D} = \frac{38(z_D - z_L)}{x_D} = \frac{-38 \times 17}{31} = -20.84 \qquad (5.38)$$

$$z_d = z_L - 20.84 = 47 - 20.84 = 26.16$$

Table 5.4

Projected points (left)	y	z
a	9.96	18.91
b	24.83	18.91
c	19.71	26.16
d	8.68	26.16
e	9.96	35.43
f	24.83	35.43
g	19.71	38.42
h	8.68	38.42
j	15.56	44.19

Table 5.5

Projected points (right)	y	z
a'	13.22	18.91
b'	28.09	18.91
c'	20.84	26.16
d'	9.81	26.16
e'	13.22	35.43
f'	28.09	35.43
g'	20.84	38.42
h'	9.81	38.42
j'	17.59	44.19

All other points are treated similarly to give the results in column 3 of Table 5.4. From these coordinates (y, z) the drawing of the left eye view is plotted at any suitable scale.

The whole process is repeated for the right eye position R and values of Table 5.5 obtained, from which the right eye view is plotted at the same scale. If copies are made of the views of Figure 5.16, each on a separate piece of paper, and viewed with a simple stereoscope, a three-dimensional impression is obtained. Many persons can fuse the images without a stereoscope.

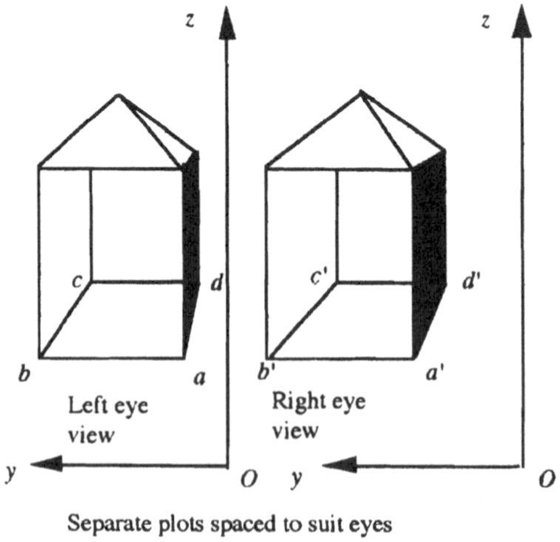

Figure 5.16

Calculating by hand or spreadsheet is tedious and impractical if many thousands of coordinated points are involved such as in a terrain model or complex road intersection. However, computers are ideally suited to such problems. Animated dynamic views can be obtained with amazing speed by modern processors, the results being depicted on the computer screen and viewed by a suitable stereoscope or synchronised flicker device.

SUMMARY OF KEY WORDS

isometric view, perspective views, right handed, directed step, vector, components, direction cosines, the projection of AB on AC, equation of a sphere, stereoscopic optical model

SUMMARY OF FORMULAE

COORDINATE TRANSFORMATIONS

$$x = r\cos V\cos U \quad y = r\cos V\sin U \quad z = r\sin V \tag{5.2}$$

$$r = \sqrt{(x^2 + y^2 + z^2)} \tag{5.3}$$

$$\tan U = \frac{y}{x} \tag{5.4}$$

$$\tan V = \frac{z}{\sqrt{\left(x^2 + y^2\right)}} \tag{5.5}$$

A useful check is

$$r = (x\cos U + y\sin U)\cos V + z\sin V \tag{5.6}$$

COORDINATE DIFFERENCES

$$AB^2 = (X_B - X_A)^2 + (Y_B - Y_A)^2 + (Z_B - Z_A)^2 \qquad (5.1)$$

$$\Delta x = x_B - x_A \text{ for } x \quad \Delta y = y_B - y_A \text{ for } y \quad \Delta z = z_B - z_A \text{ for } z \qquad (5.7)$$

$$\Delta x = r \cos V \cos U \quad \Delta y = r \cos V \sin U \quad \Delta z = r \sin V \qquad (5.8)$$

$$r = \sqrt{(\Delta x^2 + \Delta y^2 + \Delta z^2)} \qquad (5.9)$$

$$\tan U = \frac{\Delta y}{\Delta x} \qquad (5.10)$$

$$\tan V = \frac{\Delta z}{\sqrt{\left(\Delta x^2 + \Delta y^2\right)}} \qquad (5.11)$$

A useful check is

$$r = (\Delta x \cos U + \Delta y \sin U) \cos V + \Delta z \sin V \qquad (5.12)$$

DIRECTION COSINES AND LINES

$$L^2 + M^2 + N^2 = 1 \qquad (5.14)$$

$$X_P = X_A + t.L$$
$$Y_P = Y_A + t.M$$
$$Z_P = Z_A + t.N \qquad (5.18)$$

$$\cos \theta = LL' + MM' + NN' \qquad (5.19)$$

$$AD = (X_B - X_A)L' + (Y_B - Y_A)M' + (Z_B - Z_A)N' \qquad (5.21)$$

$$BD = AB \sqrt{((LM' - ML')^2 + (MN' - NM')^2 + (NL' - LN')^2)} \qquad (5.22)$$

EQUATION OF A PLANE

$$LX + MY + NZ = P \qquad (5.28)$$

EQUATION OF A SPHERE

$$x^2 + y^2 + z^2 + 2ex + 2fy + 2gz + h = 0 \qquad (5.30)$$

Chapter 6
Areas and Volumes

6.1 Areas of plane figures

The calculation of the area of a plane figure is important to surveying and cartography, especially for land tax evaluation and other financial matters. In practice, most areas are calculated from the plane coordinates of points spaced round the perimeter of the area to be quantified. We look first at the methods of determining the area of a plane triangle and later to more complex problems.

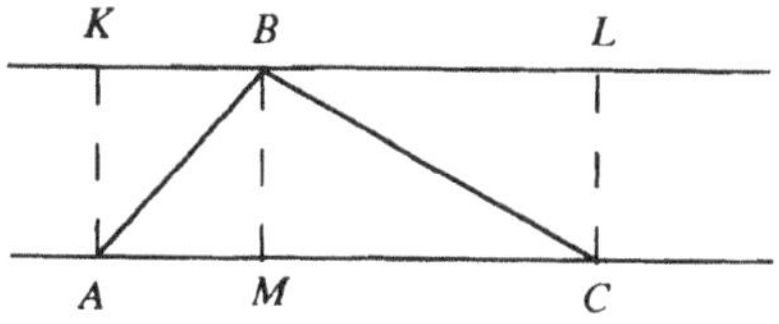

Figure 6.1

6.2 Area of a triangle

Consider the triangle *ABC* of Figure 6.1. *KL* and *AC* are parallel lines and *AKLC* is a rectangle. By inspection it can be seen that the area of the triangle *ABC* is half that of the rectangle *AKLC*. Let the area of the triangle *ABC* be Δ. Thus we see that

$$\Delta = \tfrac{1}{2}\, AC.MB$$

$$\Delta = \tfrac{1}{2}\, bc \sin A \qquad (6.1)$$

Also from (3.9) we have

$$\Delta = \tfrac{1}{2}\, ba \sin C \text{ and } \Delta = \tfrac{1}{2}\, ac \sin B$$

Another useful expression for the area of triangle especially in a three dimensional coordinate system is *Hero's formula* given in (3.31)

$$\Delta = \sqrt{[s(s-a)(s-b)(s-c)]} \qquad (6.2)$$

$$\text{where } 2s = a + b + c$$

Exercise 1 Calculate the area of the triangle whose sides are 6, 7 and 8 units long.

$$2s = 21 \quad s = 10.5 \quad s - a = 4.5 \quad s - b = 3.5 \quad s - c = 2.5$$

$$\Delta = \sqrt{[10.5 \times 4.5 \times 3.5 \times 2.5]} = 20.33 \text{ sq units.}$$

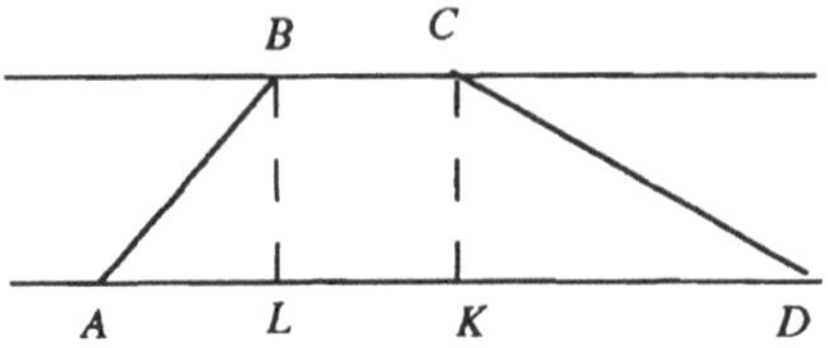

Figure 6.2

6.3 Area of a trapezium

A *trapezium* is a four-sided figure (quadrilateral) with two opposite sides parallel. One is shown in Figure 6.2 in which BC is parallel to AD. The area Δ of the trapezium $ABCD$ is given by

$$\Delta = \tfrac{1}{2}BL(BC + AD)$$

Consider the sum of the areas of the triangles ABL and KCD, and the rectangle $BCKL$, then

$$\begin{aligned} \Delta &= \tfrac{1}{2}AL.BL + \tfrac{1}{2}KD.BL + KL.BL \\ &= \tfrac{1}{2}BL(AL + KD + 2KL) \\ &= \tfrac{1}{2}BL(AL + KD + KL + BC) \\ &= \tfrac{1}{2}BL(AD + BC) \end{aligned}$$

Exercise 1 Calculate the area of the trapezium $ABCD$ in which

$$BL = 4 \text{ cm} \quad AD = 23 \text{ cm} \quad BC = 6 \text{ cm}$$

$$\Delta = 2(23 + 6) = 2 \times 29 = 58 \text{ sq cm}$$

6.4 Area of a triangle in terms of plane coordinates

Now consider the triangle ABC in Figure 6.3. Let its area be Δ. Then if we consider all areas to be positive for the moment

$$\begin{aligned} \Delta &= \text{area } ABDF - \text{ area } BDEC - \text{ area } CEFA \\ 2\Delta &= FD(FA + DB) - ED(EC + DB) - FE(FA + EC) \\ &= FA(FD - FE) + DB(FD - ED) - EC(ED + FE) \\ &= FA.ED + DB.FE - EC.FD \end{aligned}$$

But $\quad FA = y_A \quad DB = y_B \quad EC = y_C$

$$ED = x_B - x_C \quad FE = x_C - x_A \quad FD = x_B - x_A$$

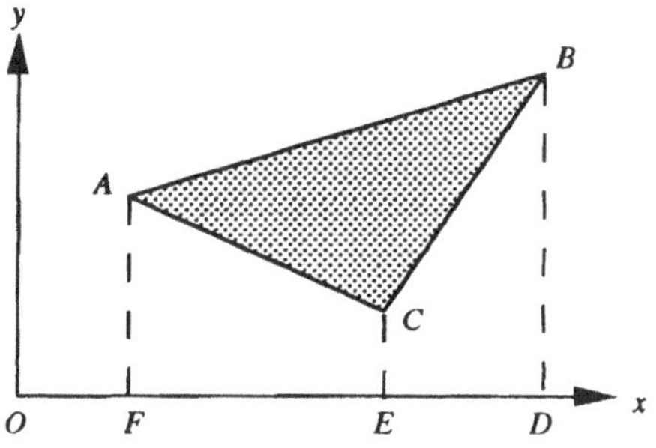

Figure 6.3

therefore

$$2\Delta = y_A(x_B - x_C) + y_B(x_C - x_A) - y_C(x_B - x_A)$$

and finally

$$2\Delta = y_A x_B - y_A x_C + y_B x_C - y_B x_A - y_C x_B + y_C x_A$$

$$2\Delta = y_A x_B + y_B x_C + y_C x_A - x_A y_B - x_B y_C - x_C y_A \qquad (6.3)$$

At first sight there may not be a pattern to this result, but if we write the coordinates down as an array in two lines repeating the first pair as follows, the pattern will appear

$$\begin{array}{cccc} y_A & y_B & y_C & y_A \\ x_A & x_B & x_C & x_A \end{array}$$

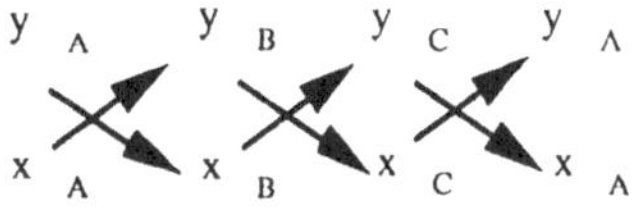

The rule is to follow a cross multiplication rule indicated by the arrows: down products are positive and up products are negative. Note that the coordinates of the first point A are listed twice.

Exercise 1 Calculate the area of the triangle ABC formed by the points whose coordinates are $A(1, 3)$ $B(5, 4)$ $C(3, 2)$. Writing the coordinates in two rows gives

$$\begin{array}{cccc} 1 & 5 & 3 & 1 \\ 3 & 4 & 2 & 3 \end{array}$$

and the area of the triangle as

$$\begin{aligned} 2\Delta &= (1 \times 4) + (5 \times 2) + (3 \times 3) - (3 \times 5) - (4 \times 3) - (2 \times 1) \\ &= 4 + 10 + 9 - 15 - 12 - 2 \\ &= -6 \end{aligned}$$

The negative sign arises because the points ABC have been taken clockwise.

Because an anticlockwise angle is positive in the coordinate system, an anticlockwise order for the points would yield a positive result.

Exercise 2 Calculate the area of the triangle in the order *ACB*. Here we have

$$\begin{matrix} 1 & 3 & 5 & 1 \\ 3 & 2 & 4 & 3 \end{matrix}$$

$$\begin{aligned} 2\Delta &= (1 \times 2) + (3 \times 4) + (5 \times 3) - (3 \times 3) - (2 \times 5) - (4 \times 1) \\ &= 2 + 12 + 15 - 9 - 10 - 4 \\ &= +6 \end{aligned}$$

Note: In this book simple numbers have been chosen to illustrate theory. In practice, the coordinates of points can be quite large. In such cases it is usual to modify these coordinates before performing a calculation. The method is to move the origin to a point whose coordinates are the average of all values, say i points. If this new origin, called the *centroid*, has the coordinates

$$\bar{x} = \text{average of all } x_i \text{ values,} \quad \bar{y} = \text{average of all } y_i \text{ values}$$

all coordinates are reduced to new values X, Y using

$$X_i = x_i - \bar{x} \quad Y_i = y_i - \bar{y}$$

and the area calculated using the new values. This device is simply to reduce the size of the numbers in calculation.

Exercise 3 Calculate the area of the triangle *ABC*, where the coordinates are *A*(51,73), *B*(55,74), *C*(53,72). Here we have the centroid

$$\bar{x} = (51 + 55 + 53)/3 = 53 \text{ and } \bar{y} = (73 + 74 + 72)/3 = 73$$

and the new coordinates reduced to the centroid are

$$\begin{matrix} -2 & +2 & 0 & -2 \\ 0 & +1 & -1 & 0 \end{matrix}$$

$$\begin{aligned} 2\Delta &= -2 - 2 - 2 \\ &= -6 \end{aligned}$$

which is the same as if the original values had been used without modification.

6.5 Extension to polygon

The above rule applies also to any closed polygon no matter how large. It is important to list the coordinates in the correct order in which the points are joined to form the perimeter of the figure so that correct signs are obtained for areas that turn backwards. A clockwise or anticlockwise convention of numbering points will produce the correct numerical value.

We will derive the result for the quadrilateral *ABCD* of Figure 6.4. The area of the polygon is the sum of the areas of the triangles *ABC* and *ACD*. Writing down their coordinates we have

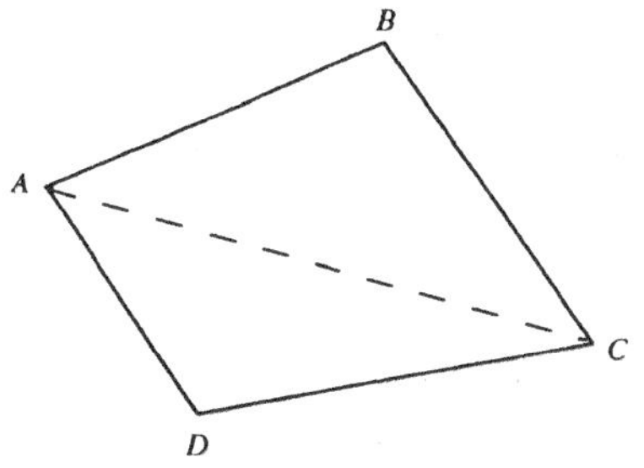

Figure 6.4

$$\begin{matrix} y_A & y_B & y_C & y_A \\ x_A & x_B & x_C & x_A \end{matrix}$$

and

$$\begin{matrix} y_A & y_C & y_D & y_A \\ x_A & x_C & x_D & x_A \end{matrix}$$

giving 2Δ twice the sum of the two triangular areas as

$$2\Delta = y_A x_B + y_B x_C + y_C x_A - x_A y_B - x_B y_C - x_C y_A$$
$$+ y_A x_C + y_C x_D + y_D x_A - x_A y_C - x_C y_D - x_D y_A$$

$$2\Delta = y_A x_B + y_B x_C - x_A y_B - x_B y_C$$
$$+ y_C x_D + y_D x_A - x_C y_D - x_D y_A$$

which after rearranging gives the same result as from the coordinates of the four points listed as

$$\begin{matrix} y_A & y_B & y_C & y_D & y_A \\ x_A & x_B & x_C & x_D & x_A \end{matrix}$$

Note: When calculating an area with a map digitiser, the operator traces round the perimeter of an area whose coordinates are sampled at some preset rate. A modified version of the above formula is then used to calculate the enclosed area from the digitised coordinates. The same formula is used to calculate cross-sectional areas in engineering works.

6.6 Determinants

Although the cross multiplication rule for calculating the area of a triangle is convenient for hand calculation, a mathematical notation for the process is required, so that problems can be treated in an orderly manner. To deal with such and similar problems the algebra of *determinants* was devised. The notation used is similar to that employed in matrices (see Chapter 7) but is not to be confused with it. At first sight the process may seem a little perplexing. We simply have a set of useful rules to carry out arithmetic. These rules can be programmed in a computer and are available in spreadsheets for actual calculations. Some understanding of the structure is needed therefore to use the notation properly. Unlike matrices, determinants deal only with square *arrays of numbers*.

A square array of numbers, D, is written within parallel vertical lines as follows

$$D = \begin{vmatrix} x_A & y_A \\ x_B & y_B \end{vmatrix}$$

D is called a *determinant of the second order* and is given the meaning

$$D = x_A y_B - x_B y_A \tag{6.4}$$

Exercise 1 Evaluate the determinant D where

$$D = \begin{vmatrix} 1 & 2 \\ 3 & 4 \end{vmatrix}$$

$$D = (1 \times 4) - (3 \times 2) = 4 - 6 = -2$$

Exercise 2 Evaluate the determinant D where

$$D = \begin{vmatrix} y_B & 1 \\ y_C & 1 \end{vmatrix}$$

Then

$$D = (y_B \times 1) - (y_C \times 1) = y_B - y_C$$

The process can be extended to deal with any size of square array. For the moment we need only to consider a *determinant of the third order*, for example D where

$$D = \begin{vmatrix} x_A & y_A & 1 \\ x_B & y_B & 1 \\ x_C & y_C & 1 \end{vmatrix}$$

This determinant D has the meaning

$$D = x_A y_B + x_B y_C + x_C y_A - y_A x_B - y_B x_C - y_C x_A \tag{6.5}$$

which is twice the area of the triangle ABC obtained by the cross multiplication rule explained in Section 6.4. To explain how the determinant is written out in full, or *expanded*, we need to define a few terms. The third-order determinant D can be written in terms of second-order determinants, D_1 D_2 and D_3 as follows

$$\mathrm{D} = x_A D_1 - y_A D_2 + 1.D_3$$

where

$$D_1 = \begin{vmatrix} y_B & 1 \\ y_C & 1 \end{vmatrix} \qquad D_2 = \begin{vmatrix} x_B & 1 \\ x_C & 1 \end{vmatrix} \qquad D_3 = \begin{vmatrix} x_B & y_B \\ x_C & y_C \end{vmatrix}$$

Examining these determinants and D we see that

D_1 is the determinant formed from D omitting the first row and the first column of D. It is called a *minor* determinant of D. Because it is clearly related to x_A, it is called "the minor of the element x_A".

D_2 is the determinant formed from D omitting its first row and second column, and it is called "the minor of the element y_A".

D_3 is the determinant formed from D omitting the first row and the third column of D. It is called "the minor of the element 1".

The rule then for expansion of a determinant is:

Take the elements of the first row, multiply each of these elements by its minor, attaching + and – signs alternately.

Exercise 3 Evaluate the determinant

$$D = \begin{vmatrix} 1 & 4 & 2 \\ 2 & 3 & 4 \\ 5 & 1 & 3 \end{vmatrix}$$

$$D = 1\begin{vmatrix} 3 & 4 \\ 1 & 3 \end{vmatrix} - 4\begin{vmatrix} 2 & 4 \\ 5 & 3 \end{vmatrix} + 2\begin{vmatrix} 2 & 3 \\ 5 & 1 \end{vmatrix}$$

$$D = 1(9-4) - 4\,(6-20) + 2\,(2-15)$$
$$= 5 + 56 - 26 = 35$$

Exercise 4 Evaluate the determinant again by expanding it down the first column instead of along the first row. We have

$$D = 1\begin{vmatrix} 3 & 4 \\ 1 & 3 \end{vmatrix} - 2\begin{vmatrix} 4 & 2 \\ 1 & 3 \end{vmatrix} + 5\begin{vmatrix} 4 & 2 \\ 3 & 4 \end{vmatrix}$$

$$D = 1(9-4) - 2(12-2) + 5(16-6)$$
$$= 5 - 20 + 50 = 35$$

Exercise 5 Evaluate the determinant again by expanding it down the third column instead of along the first row. We have

$$D = 2\begin{vmatrix} 2 & 3 \\ 5 & 1 \end{vmatrix} - 4\begin{vmatrix} 1 & 4 \\ 5 & 1 \end{vmatrix} + 3\begin{vmatrix} 1 & 4 \\ 2 & 3 \end{vmatrix}$$

$$D = 2(2-15) - 4(1-20) + 3(3-8)$$
$$= -26 + 76 - 15 = 35$$

Note: We can expand a determinant using *any row or column* and its minors.

Exercise 6 Evaluate the determinant

$$D = \begin{vmatrix} 1 & 3 & 1 \\ 5 & 4 & 1 \\ 3 & 2 & 1 \end{vmatrix}$$

This is best expanded down column 3 to give

$$D = (5 \times 2) - (3 \times 4) - ((1 \times 2) - (3 \times 3)) + (1 \times 4) - (5 \times 3)$$
$$= 10 - 12 - 2 + 9 + 4 - 15$$
$$= -6$$

This same result was obtained by the cross multiplying rule in Section 6.4 Exercise 1.

Note: Many operations on determinants are possible which do not concern us here. The reader should refer to a more advanced text to learn about these if required.

6.7 The tetrahedron

A *tetrahedron* is a four-sided figure with plane sides. The best-known example is a pyramid which usually has three sides equal. Some cartons of milk have all four sides equal. In surveying, the volumes of solid (such as part of a hill which is to be excavated to make way for a road) have to be calculated. Generally, the solid can be divided into many tetrahedra whose separate volumes are calculated. The coordinates of the corners of the tetrahedra can be obtained from surveying, photogrammetry or map measurements. Therefore the calculation of the volume of a tetrahedron is fundamental to the wider calculation of solids in general.

Consider the tetrahedron *ABCD* shown in Figure 6.5. It is worthwhile making a paper model of this solid to understand the basis of the mathematics that follows. The solid is constructed from the shape indicated in the figure which should be drawn to scale when making the model. Shape One is cut out and folded into Shape Two. Note that the points are lettered in an anticlockwise manner according to their positions in the *xy* plane.

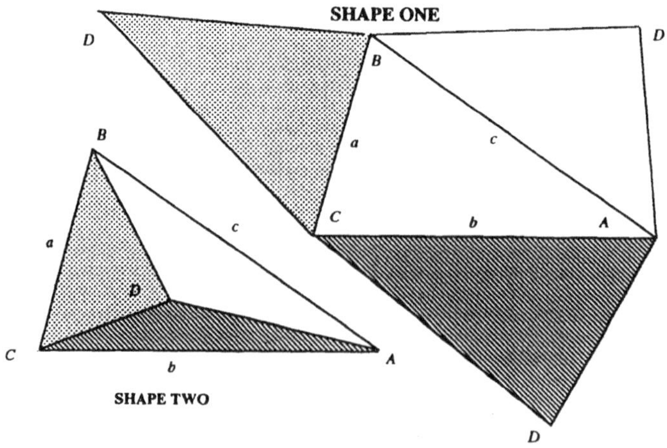

Figure 6.5

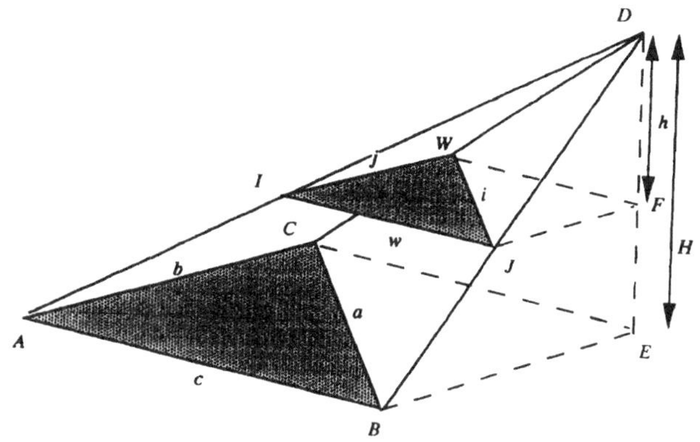

Figure 6.6

In Figure 6.6, the line $DE = H$, perpendicular to the plane of the triangle ABC, is the height of the tetrahedron. Plane IJW, a plane parallel to ABC between ABC and point D, cuts DE at F. Let $DF = h$. Let the areas of triangles ABC and IJW be Δ and Δ_s respectively. Then

$$\frac{\Delta_s}{\Delta} = \frac{h^2}{H^2} \qquad (6.6)$$

This follows because the respective sides of the triangles ABC and IJW are proportional to H and h, for

angle I = angle A

and from similar triangles

$$\frac{j}{b} = \frac{DW}{DC} = \frac{h}{H} \quad \text{and} \quad \frac{w}{c} = \frac{DJ}{DB} = \frac{h}{H}$$

and

$$\Delta = \tfrac{1}{2}bc \sin A \qquad \Delta_S = \tfrac{1}{2}jw \sin I$$

therefore since $I = A$

$$\frac{\Delta_s}{\Delta} = \frac{jw}{bc} = \frac{h^2}{H^2}$$

$$\Delta_s = \Delta\frac{h^2}{H^2} \qquad (6.6)$$

Exercise 1 Show that the area of the triangle IJW is $\tfrac{1}{4}\Delta$ if $h = \tfrac{1}{2}H$.

$$\Delta_s = \Delta\frac{h^2}{H^2} = \Delta\frac{H^2}{4H^2} = \tfrac{1}{4}\Delta$$

6.8 Volume of the tetrahedron

The volume of the tetrahedron can be thought of as the sum of the areas of all triangles like IJW of thickness dh. We can think of dh as very small. The volume V is therefore given by the integral (see Chapter 9)

$$V = \int_0^H \Delta_s \delta h$$

$$= \int_0^H \Delta \frac{h^2}{H^2} \delta h$$

$$= \Delta \frac{1}{H^2} \int_0^H h^2 \delta h$$

$$= \Delta \frac{1}{H^2} \left[\frac{h^3}{3} \right]_0^H$$

$$= \Delta \frac{1}{H^2} \frac{H^3}{3}$$

$$V = \tfrac{1}{3} \Delta H \qquad (6.7)$$

Thus we have the very simple result that the volume of a tetrahedron is equal to one third of the area of its base times the height. *The result is true also of a cone because the areas of successive slices parallel to the base are proportional to the square of the height.*

Note: In the derivation we selected the triangle ABC to be the base and DE as height. The triangle on any of the four faces could have been chosen together with the appropriate perpendicular to that face as height.

6.9 Volume of the tetrahedron from coordinates

The volume V of a tetrahedron $ABCD$ is given very neatly, in terms of the coordinates of its four corners, by the *fourth-order determinant* (see also Section 8.18).

We will now derive this important result. Again we treat the plane ABC as base and DE as the height. The plane ABC is not generally parallel to any coordinate plane.

$$6V = \begin{vmatrix} x_A & y_A & z_A & 1 \\ x_B & y_B & z_B & 1 \\ x_C & y_C & z_C & 1 \\ x_D & y_D & z_D & 1 \end{vmatrix}$$

Area of a triangle in three dimensions

First we need to find the formula for the area of triangle ABC in this general case. To follow the logic of the next explanation, hold a flat object, such as a piece of card, in front of your eyes against the background of a room corner imagined to form a right handed set of coordinate axes. Also see Figure 6.7 where we have

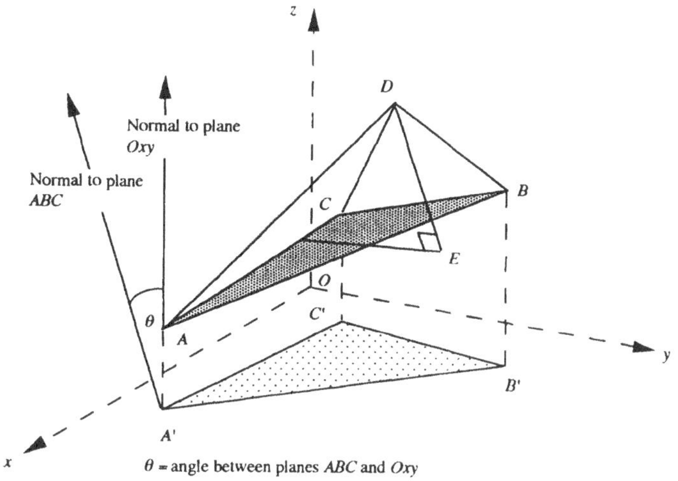

Figure 6.7

chosen the bottom corner of a room for origin.

Let $N = \cos\theta$ be the direction cosine of the normal to the plane *ABC* with respect to the axis *Oz* (see Figure 6.7) Let the orthogonal projections of the points *A*, *B* and *C* on the *Oxy* plane be *A′ B′* and *C′* respectively, then Δ', the area of this triangle, *A′B′C′* is given by

$$2\Delta' = \begin{vmatrix} x_{A'} & y_{A'} & 1 \\ x_{B'} & y_{B'} & 1 \\ x_{C'} & y_{C'} & 1 \end{vmatrix}$$

But *A′*, *B′* and *C′* have the same *x* and *y* values as *A*, *B* and *C* so

$$2\Delta' = \begin{vmatrix} x_A & y_A & 1 \\ x_B & y_B & 1 \\ x_C & y_C & 1 \end{vmatrix}$$

Since *N* is the cosine of the angle between the normal to the plane *ABC* and the *Oz* axis, it is also the cosine of the angle between the plane *ABC* and the *Oxy* plane. Thus the area of triangle *ABC* is given by

$$\Delta' = N\Delta$$

therefore

$$2\Delta = \frac{2\Delta'}{N} \tag{6.8}$$

Now from (5.28) the equation of the plane *ABC* is

$$Lx + My + Nz = P$$

From three points in the plane we obtain L, M, N and P (see Section 5.17) and therefore we can calculate Δ.

The length of the perpendicular H from the fourth point D to the plane ABC is given by

$$H = Lx_D + My_D + Nz_D - P \tag{6.9}$$

Thus we can calculate the volume of the tetrahedron from the formula (6.7)

$$V = \tfrac{1}{3}\Delta H \tag{6.7}$$

All that remains is to recast these expressions into the neat determinant formula for volume. Consider the equation of the plane

$$Lx + My + Nz = P$$

or

$$Lx + My + Nz - P = 0 \tag{6.10}$$

This is the general form of a linear equation

$$Tx + Qy + Rz + S = 0 \tag{6.11}$$

We can put

$$T = kL \quad Q = kM \quad R = kN \quad S = -kP$$

where k is a constant. That equations (6.10) and (6.11) are the same can be seen by substituting in (6.11) thus

$$kLx + kMy + kNz + (-kP) = 0 \tag{6.12}$$

and dividing by k gives equation (6.10).

Because (see (5.14))

$$L^2 + M^2 + N^2 = 1$$
$$T^2 + Q^2 + R^2 = k^2$$
$$k = \sqrt{(T^2 + Q^2 + R^2)}$$

we have

$$L = \frac{T}{k} \qquad M = \frac{Q}{k} \qquad N = \frac{R}{k} \qquad P = \frac{-S}{k}$$

If we know the coordinates of three points A, B and C then equation (6.11) is satisfied by their coordinates. Any other point (x, y, z), also satisfies (6.11) giving the equations in the four unknowns P, Q, R and S.

$$Tx + Qy + Rz + S = 0$$
$$Tx_A + Qy_A + Rz_A + S = 0$$
$$Tx_B + Qy_B + Rz_B + S = 0$$
$$Tx_C + Qy_C + Rz_C + S = 0$$

For these equations to be *consistent* (see Section 12.6) the determinant of their coefficients must be zero therefore

$$\begin{vmatrix} x & y & z & 1 \\ x_A & y_A & z_A & 1 \\ x_B & y_B & z_B & 1 \\ x_C & y_C & z_C & 1 \end{vmatrix} = 0 \tag{6.13}$$

Expanding the determinant along the first row we have

$$Tx + Qy + Rz + S = 0$$

where

$$T = \begin{vmatrix} y_A & z_A & 1 \\ y_B & z_B & 1 \\ y_C & z_C & 1 \end{vmatrix}$$

$$Q = -\begin{vmatrix} x_A & z_A & 1 \\ x_B & z_B & 1 \\ x_C & z_C & 1 \end{vmatrix}$$

$$R = \begin{vmatrix} x_A & y_A & 1 \\ x_B & y_B & 1 \\ x_C & y_C & 1 \end{vmatrix}$$

$$S = -\begin{vmatrix} x_A & y_A & z_A \\ x_B & y_B & z_B \\ x_C & y_C & z_C \end{vmatrix} \tag{6.14}$$

and from

$$k = \sqrt{(T^2 + Q^2 + R^2)}$$

and

$$L = \frac{T}{k} \qquad M = \frac{Q}{k} \qquad N = \frac{R}{k} \qquad P = \frac{-S}{k}$$

we can express L etc. in terms of third-order determinants. The perpendicular from D to the plane ABC is given by (see equation (6.9))

$$H = Lx_D + My_D + Nz_D - P$$

so we also have

$$kH = Tx_D + Qy_D + Rz_D + S$$

This is the determinant formed by putting the coordinates of D in (6.11) and

$$2\Delta N = \begin{vmatrix} x_A & y_A & 1 \\ x_B & y_B & 1 \\ x_C & y_C & 1 \end{vmatrix} = R$$

$$D = \frac{R}{2N} = \frac{kN}{2N} = \frac{k}{2}$$

Finally the volume V is given by

$$V = \tfrac{1}{3}\Delta H = \tfrac{1}{6}kH = \tfrac{1}{6}(Tx_D + Qy_D + Rz_D + S)$$

or

$$6V = \begin{vmatrix} x_D & y_D & z_D & 1 \\ x_A & y_A & z_A & 1 \\ x_B & y_B & z_B & 1 \\ x_C & y_C & z_C & 1 \end{vmatrix}$$

If this determinant is re-ordered in the neater form as

$$6V = -\begin{vmatrix} x_A & y_A & z_A & 1 \\ x_B & y_B & z_B & 1 \\ x_C & y_C & z_C & 1 \\ x_D & y_D & z_D & 1 \end{vmatrix} \tag{6.15}$$

its sign changes. Usually its numerical value is all that is required. We will now work some examples based on the following data.

Point	x	y	z
A	11	6	2
B	1	9	0
C	3	2	0.5
D	6	10	7

The following exercises involving determinants are quite arduous if worked by hand using a small calculator. If a spreadsheet is available, such as Excel, they are quite simple, because the function MDETERM() is available to evaluate determinants.

Exercise 1 Find the equation of the plane ABC by the method of determinants. (It is easier to use the method of Section 5.17 if only a hand calculator is available.)
The determinants T, Q, R and S are evaluated as follows.

$$T = \begin{vmatrix} 6 & 2 & 1 \\ 9 & 0 & 1 \\ 2 & 0.5 & 1 \end{vmatrix}$$

$$Q = -\begin{vmatrix} 11 & 2 & 1 \\ 1 & 0 & 1 \\ 3 & 0.5 & 1 \end{vmatrix}$$

$$R = \begin{vmatrix} 11 & 6 & 1 \\ 1 & 9 & 1 \\ 3 & 2 & 1 \end{vmatrix}$$

$$S = -\begin{vmatrix} 11 & 6 & 2 \\ 1 & 9 & 0 \\ 3 & 2 & 0.5 \end{vmatrix}$$

Therefore

$$T = -12.5 \quad Q = +1 \quad R = 64 \quad S = +3.5$$

and

$$k = \sqrt{(T^2 + Q^2 + R^2)} = 65.2169$$

giving

$$L = -0.191\,668 \quad M = +0.015333 \quad N = 0.981340 \quad P = -0.053\,667$$

Therefore the equation of the plane in the form

$$Tx + Qy + Rz + S = 0$$

is

$$-12.5x + y + 64z + 3.5 = 0$$

and in the form

$$Lx + My + Nz - P = 0$$

is

$$-0.191\,668x + 0.015\,333\,4y + 0.981\,340\,04z + 0.053\,667 = 0$$

Exercise 2 Find the area of the triangle *ABC*. The area of triangle *ABC* is given by

$$2\Delta = k = 65.2169$$

$$\Delta = 32.61$$

Check $\Delta = \dfrac{R}{2N}$

Exercise 3 Find the area of the triangle *ABC* using Hero's formula for Δ. Hero's formula (3.31) for the area of a triangle *ABC* in terms of its sides a, b, and c is

$$\Delta = \sqrt{s(s-a)(s-b)(s-c)}$$

where

$$2s = a + b + c$$

The sides obtained from coordinates are

$$AB = c = 10.63$$
$$BC = a = 7.30$$
$$CA = b = 9.07$$

$$AB = c = 10.63 \quad s - c = 2.87$$
$$BC = a = 7.30 \quad s - a = 6.20$$
$$CA = b = 9.07 \quad s - b = 4.43$$
$$2s = 27.00 \quad s = 13.50$$
$$\text{Area} = 32.62$$

Within the expected precision, this agrees with the result obtained above by the coordinates method.

Exercise 4 Find the height H of D above the plane ABC. The height is given by substituting the coordinates of D in the equation of the plane, therefore

$$H = -(0.191\,668 \times 6) + (0.015\,333\,4 \times 10) + (0.981\,340\,04 \times 7) + 0.053\,667$$
$$= 5.926\,37$$

Exercise 5 Calculate the volume of the tetrahedron $ABCD$: (a) from first principles and (b) by the determinant formula.

(a) From first principles

$$V = \tfrac{1}{3}\Delta H = \frac{32.608 \times 5.926\,37}{3} = 64.416$$

(b) By determinant we have

$$6V = -\begin{vmatrix} 11 & 6 & 2 & 1 \\ 1 & 9 & 0 & 1 \\ 3 & 2 & 0.5 & 1 \\ 6 & 10 & 7 & 1 \end{vmatrix}$$

$$= +\ 386.5$$
$$V = 64.416$$

Note: Because the determinant (6.15) gives the volume of the tetrahedron $ABCD$, if these points all lie in the same plane, the volume is zero, hence the condition for coplanarity of the four points is

$$\begin{vmatrix} x_A & y_A & z_A & 1 \\ x_B & y_B & z_B & 1 \\ x_C & y_C & z_C & 1 \\ x_D & y_D & z_D & 1 \end{vmatrix} = 0$$

This relationship is important in photogrammetry.

Exercise 6 Verify that the following points are coplanar

	X	Y	Z
A	1	2	3.000
B	2	4	0.770
C	4	2	1.456
D	2	1	3.343

All that is required is to evaluate the determinant

$$\begin{vmatrix} 1 & 2 & 3.000 & 1 \\ 2 & 4 & 0.770 & 1 \\ 4 & 2 & 1.456 & 1 \\ 2 & 1 & 3.343 & 1 \end{vmatrix}$$

It is left as an exercise for the reader to show that it is zero.

SUMMARY OF KEY WORDS

Hero's formula, trapezium, centroid, determinants,
determinant of the second order,
determinant of the third order, minor determinant, tetrahedron,
fourth order determinant

SUMMARY OF FORMULAE

$$\Delta = \tfrac{1}{2} bc \sin A \tag{6.1}$$

$$D = \begin{vmatrix} x_A & y_A \\ x_B & y_B \end{vmatrix}$$

$$D = x_A y_B - x_B y_A \tag{6.4}$$

$$D = \begin{vmatrix} x_A & y_A & 1 \\ x_B & y_B & 1 \\ x_C & y_C & 1 \end{vmatrix}$$

$$D = x_A y_B + x_B y_C + x_C y_A - y_A x_B - y_B x_C - y_C x_A \tag{6.5}$$

$$D = x_A D_1 - y_A D_2 + 1.D_3$$

where

$$D_1 = \begin{vmatrix} y_B & 1 \\ y_C & 1 \end{vmatrix} \qquad D_2 = \begin{vmatrix} x_B & 1 \\ x_C & 1 \end{vmatrix} \qquad D_3 = \begin{vmatrix} x_B & y_B \\ x_C & y_C \end{vmatrix}$$

$$V = \tfrac{1}{3} \Delta H \tag{6.7}$$

$$6V = -\begin{vmatrix} x_A & y_A & z_A & 1 \\ x_B & y_B & z_B & 1 \\ x_C & y_C & z_C & 1 \\ x_D & y_D & z_D & 1 \end{vmatrix} \tag{6.15}$$

Chapter 7 Matrices

7.1 Matrices

Matrix algebra was invented to deal with the arithmetic procedures needed to solve large numbers of linear equations and related problems. Special rules were devised for the four arithmetic processes of addition, subtraction, multiplication and division, in handling *arrays* of numbers instead of individual values. Of course, the procedures ultimately involve many repetitive calculations, but these are largely unseen by the user. These rules are expressed in a neat algebraic form for which some new and some old forms of notation are employed. It is usual to distinguish matrix algebra from ordinary algebra by always printing letters in bold type, and in hand-written work the letters are underlined.

We first give some idea of the power and usefulness of matrix algebra by working out a simple problem. Later, a more formal approach is given in the hope that the reader will see that it is very worthwhile making some effort to learn the few rules that apply, especially when it is realised that many computer spreadsheet systems require a user to know the structure of matrix algebra, and thus obtain freedom from the tedium of doing the arithmetic.

7.2 An introduction to matrix algebra

Consider the simultaneous linear equations representing two lines in a plane

$$x + 2y = 5 \tag{7.1}$$

$$x + 6y = 13 \tag{7.2}$$

Provided these equations do not represent parallel lines they will have a solution. That is there are values of x and y which satisfy both equations at once. They are called *simultaneous linear equations* because they do not involve powers of the unknowns x and y other than the first, and they have to be solved together. By contrast an equation involving squares of x and y is called a *second-order equation*. An example of a second-order equation is

$$x^2 + 2y^2 = 5$$

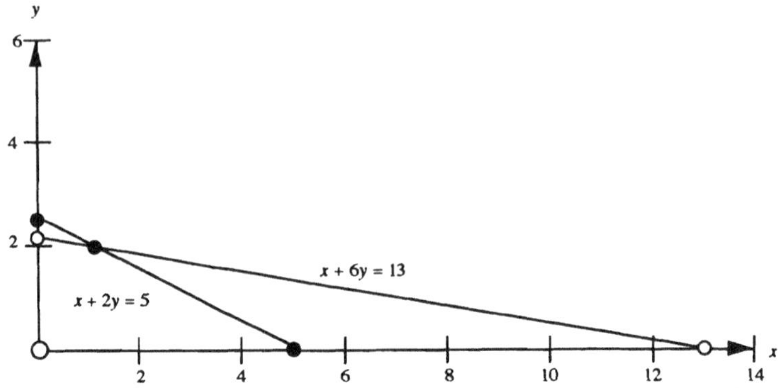

Figure 7.1

In this case there may be more than one solution for x and y. In the case of linear equations there is *only one* solution or no solution at all.

Exercise 1 Plot the equations (7.1) and (7.2) on graph paper and read off the coordinates of the point where they cut each other.

To plot the equations select easy values for x and y, such as zero, and calculate the corresponding values from the equations.For example

Equation (7.1)

$$\text{If } x = 0 \text{ then } y = 5/2 = 2.5 \quad \text{If } y = 0 \text{ then } x = 5$$

Equation (7.2)

$$\text{If } x = 0 \text{ then } y = 13/6 = 2.17 \quad \text{If } y = 0 \text{ then } x = 13$$

Figure 7.1 shows the plotted lines representing (7.1) and (7.2). They intersect at the point whose coordinates are (1, 2). Thus the solution to the linear equations is $x = 1$ and $y = 2$.These results are verified by substitution back into the equations: for

$$x + 2y = (1) + 2(2) = 1 + 4 = 5$$

and

$$x + 6y = (1) + 6(2) = 1 + 12 = 13$$

7.3 Solution by elimination

Another way to solve the equations without drawing a graph is by eliminating one variable at time. Consider the equations again

$$x + 2y = 5 \qquad (7.1)$$

$$x + 6y = 13 \qquad (7.2)$$

A simple method of solution is first to express x in terms of y from equation (7.2) and then substitute this into equation (7.1) to give the result for y. Thus from (7.2)

$$x = 13 - 6y$$

and substituting in (7.1) we have

$$(13 - 6y) + 2y = 5$$
$$13 - 6y + 2y = 5$$
$$8 = 4y$$
$$y = 2$$

Then from (7.2)

$$x + 6(2) = 13$$
$$x = 1$$

The mathematical language called 'matrix algebra' was devised to put this elimination procedure on a formal basis, and to simplify the whole process of handling linear equations often numbering many hundreds. Consider the expression

$$4y = 8$$

The solution for y is obtained from

$$y = \frac{8}{4} = 2$$

It is neater to write the reciprocal of 4 as 4^{-1}, and the calculation as

$$y = 4^{-1}.8 = 2$$

In general, a linear equation can be written

$$ax = b$$

and the solution

$$x = a^{-1}.b$$

Exercise 1 Write the solution to the equation $3z = 9$ in the general form. We have

$$z = 3^{-1}.9 = 3$$

The originator of matrix algebra thought that it would be convenient to treat large sets of linear equations in this neat manner. Consider the two linear equations again

$$x + 2y = 5 \qquad (7.1)$$
$$x + 6y = 13 \qquad (7.2)$$

Suppose we write these as

$$\boldsymbol{Ax} = \boldsymbol{b} \qquad (7.3)$$

We use heavy type to say they have a different meaning to the simpler

$$ax = b$$

Before explaining the meaning of the notation used in equations (7.3) we state that their solution is

$$\boldsymbol{x} = \boldsymbol{A}^{-1}.\boldsymbol{b} \qquad (7.4)$$

This very simple algebraic language is valid no matter how many linear equations there are. Often, in practical problems in geodesy, there can be many thousands of equations to solve. Now we explain the notation used in equation (7.3).

7.4 Matrix notation

First of all, we separate the coefficients from the unknowns in equations (7.1) and (7.2) and write them as a separate *square array* of numbers within square brackets, which we call $\boldsymbol{A}$. Thus

$$\boldsymbol{A} = \begin{bmatrix} 1 & 2 \\ 1 & 6 \end{bmatrix}$$

This array of numbers within the square brackets, is called a square *matrix* because it is to be subjected to the rules of matrix algebra which we will explain. The *place* of the number in the matrix is very important. Like a coordinate system, the rows and columns are numbered, in this case from the top left corner along and down. For example the number 2 in $\boldsymbol{A}$ above is the element $\boldsymbol{a}_{12}$, being in the *first* row and the *second* column, and the number 6 is element $\boldsymbol{a}_{22}$. A whole row, such as row 1, is written $\boldsymbol{a}_{1*}$ and a whole column, such as column 2, as $\boldsymbol{a}_{*2}$. The asterisk is used as a counter for all the numbers in that row or column. Note that the numbers of the matrix are not connected to each other in any way. All that matters is their *position* in the array.

Next, we write the unknowns x and y in a *column matrix* called $\boldsymbol{x}$

$$\boldsymbol{x} = \begin{bmatrix} x \\ y \end{bmatrix}$$

and finally the numerical part as another column matrix called $\boldsymbol{b}$.

$$\boldsymbol{b} = \begin{bmatrix} 5 \\ 13 \end{bmatrix}$$

Assembling all these three matrices together we have

$$\boldsymbol{Ax} = \boldsymbol{b}$$

or in full

$$\begin{bmatrix} 1 & 2 \\ 1 & 6 \end{bmatrix}\begin{bmatrix} x \\ y \end{bmatrix} = \begin{bmatrix} 5 \\ 13 \end{bmatrix}$$

All that remains is to explain how the square matrix $\boldsymbol{A}$ is *multiplied* by the column matrix $\boldsymbol{x}$ to give the left hand sides of the equations (7.1) and (7.2).

Note: Care must be taken not to mistake the column matrix $\boldsymbol{x}$, in heavy type, for the unknown 'x' in its first row. Perhaps a better notation is to write the elements of $\boldsymbol{x}$ as x_1 and x_2. Both forms of notation are common.

7.5 Matrix multiplication

The rule is to adopt a *row times column multiplication*. That is, we take each term of the first *row* of matrix $\boldsymbol{A}$ and multiply it by the corresponding term in the *column* $\boldsymbol{x}$ and add the results. This is best explained by the example. The left side of equation (7.1)

$$1x + 2y$$

is the result of multiplying each element of row one of $\boldsymbol{A}$ by each element of column $\boldsymbol{x}$, and adding the results, and the left side of equation (2)

$$x + 6y$$

is the result of multiplying row 2 of $\boldsymbol{A}$ by column $\boldsymbol{x}$. The box layout below should explain the (row × column) multiplication

$$\begin{bmatrix} 1 & 2 \\ 1 & 6 \end{bmatrix}\begin{bmatrix} x \\ y \end{bmatrix}$$

Begin the multiplication with row 1 ($\boldsymbol{a}_{1*}$) highlighted in the box below. The asterisk means we use the *positions* 1 and 2 *along* the row in succession, multiplying them by successive *positions down* the column $\boldsymbol{x}_{*1}$.

$$\boxed{1 \quad 2} \; \boxed{\begin{matrix} x \\ y \end{matrix}}$$

Thus we obtain the row-column product

$$[x + 2y]$$

which is placed in position $\boldsymbol{b}_{11}$ of the resulting column.

The multiplication now treats row 2 ($\boldsymbol{a}_{2*}$) highlighted in the box below. The asterisk means we use the positions 1 and 2 along the row in succession, multiplying them by successive positions down the column $\boldsymbol{x}_{*1}$. Thus we have

$$\boxed{1 \quad 6} \; \boxed{\begin{matrix} x \\ y \end{matrix}}$$

and again we obtain the row-column product

$$[x + 6y]$$

which is placed in position $\boldsymbol{b}_{21}$ of the resulting column. The complete result is

$$\begin{bmatrix} (x+2y) \\ (x+6y) \end{bmatrix}$$

But we have been given that

$$\boldsymbol{b} = \begin{bmatrix} 5 \\ 13 \end{bmatrix}$$

So, equating each element from equivalent positions, we have the equations in traditional form

$$x + 2y = 5 \qquad (7.1)$$

$$x + 6y = 13 \qquad (7.2)$$

Note: To be able to do this row-column multiplication, there must be the *same number of columns* in $\boldsymbol{A}$ *as there are rows* in $\boldsymbol{x}$. Then we can say that the matrices *are conformable for multiplication.*

It may seem to the beginner that this is a complicated process and not the simple procedure suggested earlier. However, because it can be programmed into a computer very easily it is an exceedingly powerful arithmetic tool. To use the tool requires some understanding of the procedures and is well worth the effort of learning it. The next exercise should drive home the method.

Exercise 1 Write down in longhand the result of the following matrix multiplication.

$$\begin{bmatrix} 4 & -2 & 4 \\ -2 & 5 & -4 \\ 4 & -4 & 14 \end{bmatrix} \begin{bmatrix} x \\ y \\ z \end{bmatrix}$$

Answer

$$\begin{bmatrix} 4x - 2y + 4z \\ -2x + 5y - 4z \\ 4x - 4y + 14z \end{bmatrix}$$

Exercise 2 Multiply the following matrices together

$$\begin{bmatrix} 1 & 2 \\ 3 & 4 \end{bmatrix} \begin{bmatrix} 5 & 6 \\ 7 & 8 \end{bmatrix}$$

Call them $\boldsymbol{A}$ and $\boldsymbol{B}$ and the result $\boldsymbol{C}$. Thus we require

$$\boldsymbol{AB} = \boldsymbol{C}$$

The box layout is

$$\boxed{1 \quad 2}\ \boxed{\begin{matrix} 5 \\ 7 \end{matrix}} \qquad \boxed{1 \quad 2}\ \boxed{\begin{matrix} 6 \\ 8 \end{matrix}}$$

$$\boxed{3 \quad 4}\ \boxed{\begin{matrix} 5 \\ 7 \end{matrix}} \qquad \boxed{3 \quad 4}\ \boxed{\begin{matrix} 6 \\ 8 \end{matrix}}$$

The row-column multiplications can be written

$$\begin{matrix} \boldsymbol{a}_{1*}\boldsymbol{b}_{*1} & \boldsymbol{a}_{1*}\boldsymbol{b}_{*2} \\ \boldsymbol{a}_{2*}\boldsymbol{b}_{*1} & \boldsymbol{a}_{2*}\boldsymbol{b}_{*2} \end{matrix}$$

meaning

$$\begin{bmatrix} (1\times 5)+(2\times 7) & (1\times 6)+(2\times 8) \\ (3\times 5)+(4\times 7) & (3\times 6)+(4\times 8) \end{bmatrix}$$

$$= \begin{bmatrix} (5+14) & (6+16) \\ (15+28) & (18+32) \end{bmatrix}$$

$$= \begin{bmatrix} 19 & 22 \\ 43 & 50 \end{bmatrix}$$

$$= \begin{bmatrix} \boldsymbol{c}_{11} & \boldsymbol{c}_{12} \\ \boldsymbol{c}_{21} & \boldsymbol{c}_{22} \end{bmatrix}$$

In practical use, we need only state that we want the result of the matrix multiplication

$$\boldsymbol{AB} = \boldsymbol{C}$$

And, in using a spreadsheet such as Excel, simply call

$$\boldsymbol{C} = \text{MMULT}(\boldsymbol{A},\boldsymbol{B})$$

having selected an array location of correct size, here (2 × 2) (two by two), for the resulting matrix $\boldsymbol{C}$.

7.6 The inverse of a matrix

To continue with the solution of equations (7.1) and (7.2), consider equation (7.4) again. The solution is given by

$$\boldsymbol{x} = \boldsymbol{A}^{-1}.\boldsymbol{b} \qquad (7.4)$$

An example of a (2 × 2) matrix $\boldsymbol{A}^{-1}$, in full, is

$$\begin{bmatrix} x \\ y \end{bmatrix} = \begin{bmatrix} 1.5 & -0.5 \\ -0.25 & 0.25 \end{bmatrix} \begin{bmatrix} 5 \\ 13 \end{bmatrix}$$

Before we explain how to find the matrix $\boldsymbol{A}^{-1}$ of equation (7.4) we continue to work out the result by matrix multiplication. Thus we obtain

$$x = 1.5 \times 5 - 0.5 \times 13 = 7.5 - 6.5 = 1$$

and

$$y = -0.25 \times 5 + 0.25 \times 13 = -1.25 + 3.25 = 2$$

These results are expressed as the column matrix $\boldsymbol{x}$

$$\begin{bmatrix} 1 \\ 2 \end{bmatrix}$$

Before explaining the key factor in the process, how to find the *inverse* of **A**, consider the product of the matrix *A* and its inverse. That is

$$\boldsymbol{A}.\boldsymbol{A}^{-1}$$

In full we have the matrix product

$$\begin{bmatrix} 1 & 2 \\ 1 & 6 \end{bmatrix} \begin{bmatrix} 1.5 & -0.5 \\ -0.25 & 0.25 \end{bmatrix}$$

giving the results

$1 \times 1.5 + 2 \times (-0.25) = 1.5 - 0.5 = 1$ $\quad 1 \times (-0.5) + 2 \times 0.25 = -0.5 + 0.5 = 0$

$1 \times 1.5 + 6 \times (-0.25) = 1.5 - 1.5 = 0$ $\quad 1 \times (-0.5) + 6 \times 0.25 = -0.5 + 1.5 = 1$

These results are expressed as a matrix

$$\begin{bmatrix} 1 & 0 \\ 0 & 1 \end{bmatrix}$$

This matrix is called the *unit matrix **I***. It plays the same role in matrix algebra as the *number one* (unity) does in ordinary algebra. Compare this with ordinary algebra.

Exercise 1 Carry out the following multiplication

$$\begin{bmatrix} 1 & 0 \\ 0 & 1 \end{bmatrix} \begin{bmatrix} a & b \\ d & c \end{bmatrix}$$

We obtain the matrix

$$\begin{bmatrix} a & b \\ d & c \end{bmatrix}$$

This shows that the unit matrix ***I*** acts like the number 1 (unity) in ordinary arithmetic or algebra. Therefore

$$\boldsymbol{IA} = \boldsymbol{A}$$

Again, compare this with ordinary algebra, if

$$ab = c$$

then

$$a = \frac{c}{b} = b^{-1}.c$$

and, with ordinary arithmetic, if

$$3 \times 4 = 12$$

then

$$3 = \frac{12}{4} = 4^{-1}.12$$

Exercise 2 Using the same figures as before show that

$$\boldsymbol{A}^{-1}.\boldsymbol{A} = \boldsymbol{I}$$

We have

$$\begin{bmatrix} 1.5 & -0.5 \\ -0.25 & 0.25 \end{bmatrix}\begin{bmatrix} 1 & 2 \\ 1 & 6 \end{bmatrix}$$

giving the result

$$\begin{bmatrix} 1 & 0 \\ 0 & 1 \end{bmatrix} = \boldsymbol{A}^{-1}.\boldsymbol{A}$$

This result tells us the *order* of multiplying a square matrix by its inverse does not matter because

$$\boldsymbol{A}.\boldsymbol{A}^{-1} = \boldsymbol{A}^{-1}.\boldsymbol{A} = \boldsymbol{I} \qquad (7.5)$$

The size of $\boldsymbol{I}$ is the same as $\boldsymbol{A}$.

Exercise 3 Multiply the following matrices together, in either order, and show that the result in both cases is approximately the unit matrix $\boldsymbol{I}$. (Small rounding errors in the arithmetic give only an *approximation* to $\boldsymbol{I}$.)

$$\begin{bmatrix} 4 & -2 & 4 \\ -2 & 5 & -4 \\ 4 & -4 & 14 \end{bmatrix} = \boldsymbol{A}$$

$$\begin{bmatrix} 0.375 & 0.083 & -0.083 \\ 0.083 & 0.278 & 0.056 \\ -0.083 & 0.056 & 0.111 \end{bmatrix} = \boldsymbol{A}^{-1}$$

To obtain $\boldsymbol{A}.\boldsymbol{A}^{-1}$ work down the rows of $\boldsymbol{A}$ in succession as follows.

Row 1 of $\boldsymbol{A}$ times column 1 of $\boldsymbol{A}^{-1}$ gives

$$4 \times 0.375 - 2 \times 0.083 + 4 \times (-0.083) = 1.5 - 0.166 - 0.332 = 1.002$$

Row 1 of $\boldsymbol{A}$ times column 2 of $\boldsymbol{A}^{-1}$ gives

$$4 \times 0.083 - 2 \times 0.278 + 4 \times 0.056 = 0.332 - 0.556 + 0.224 = 0$$

Row 1 of $\boldsymbol{A}$ times column 3 of $\boldsymbol{A}^{-1}$ gives

$$4 \times (-0.083) - 2 \times 0.056 + 4 \times 0.111 = -0.332 - 0.112 + 0.444 = 0$$

Thus the first row of the inverse of $\boldsymbol{A}^{-1}$ is

$$1.002 \quad 0 \quad 0$$

Row 2 of $\boldsymbol{A}$ times column 1 of $\boldsymbol{A}^{-1}$ gives

$$-2 \times 0.375 + 5 \times 0.083 - 4 \times (-0.083) = -0.75 + 0.415 + 0.332 = -0.003$$

Continuing in this way we obtain the approximate unit matrix

$$\begin{bmatrix} 1.002 & 0 & 0 \\ -0.003 & 1 & 0.002 \\ 0.006 & 0.004 & 0.998 \end{bmatrix}$$

The approximation is due to rounding effects on the elements of the inverse.

7.7 The inverse of a square matrix

In classical matrix algebra only a square matrix may have an inverse. To find the inverse of the (2 × 2) matrix **A** we proceed as follows. Let the elements of the inverse **B** be a, b, c and d as shown.

$$\begin{bmatrix} 1 & 2 \\ 1 & 6 \end{bmatrix} = \boldsymbol{A} \qquad \begin{bmatrix} a & b \\ c & d \end{bmatrix} = \boldsymbol{B}$$

The product of these two matrices has to be the (2 × 2) unit matrix **I** where **I** is

$$\begin{bmatrix} 1 & 0 \\ 0 & 1 \end{bmatrix}$$

Then we have

$$\boldsymbol{AB} = \boldsymbol{I}$$

or

$$\begin{bmatrix} 1 & 2 \\ 1 & 6 \end{bmatrix} \begin{bmatrix} a & b \\ c & d \end{bmatrix} = \begin{bmatrix} 1 & 0 \\ 0 & 1 \end{bmatrix}$$

Matrices are equal when their elements are equal, thus the product of the first row of **A** and the first column of **B** has to be 1. Therefore we obtain the equation

$$a + 2c = 1 \qquad (7.6)$$

Next, the product of the first row of **A** and the *second* column of **B** has to be 0. Thus

$$b + 2d = 0 \qquad (7.7)$$

Next, the product of the *second* row of **A** and the first column of **B** has to be 0. Thus

$$a + 6c = 0 \qquad (7.8)$$

Finally, the product of the *second* row of **A** and the second column of **B** has to be 1. Therefore

$$b + 6d = 1 \qquad (7.9)$$

From (7.8)

$$a = -6c$$

Substituting in (7.6) we have

$$-6c + 2c = 1$$

$$c = -1/4 = -0.25$$

and from (7.8)

$$a = 6/4 = 3/2 = 1.5$$

Also from (7.7)

$$b = -2d$$

and substituting in (7.9) we have

$$-2d + 6d = 1$$

$$d = 1/4 = 0.25$$

Finally from (7.7)

$$b = -2d = -0.5$$

So all four elements of the inverse have been found.

$$\begin{bmatrix} 1.5 & -0.5 \\ -0.25 & 0.25 \end{bmatrix} = \boldsymbol{B} = \boldsymbol{A}^{-1}$$

Note: In practice, special arithmetic routines are used by computers to find the inverse by methods developed by the mathematicians Gauss and Cholesky in the nineteenth century. (See Chapter 12 for more information.) Most spreadsheet and computer languages have such routines mounted. For example in Excel we need only to call

MINVERSE(**A**)

to obtain the inverse $\boldsymbol{A}^{-1}$.

7.8 Inverse of a (2 × 2) matrix by Cramer's rule

The inversion of a (2 × 2) matrix can be achieved by a simple rule which is worth learning. The matrix $\boldsymbol{B}$ is to be inverted, where

$$\boldsymbol{B} = \begin{bmatrix} e & f \\ g & h \end{bmatrix}$$

Then the inverse $\boldsymbol{B}^{-1}$ is given by

$$\frac{1}{K}\begin{bmatrix} h & -f \\ -g & e \end{bmatrix}$$

where

$$K = eh - fg \tag{7.10}$$

The number K is called the *determinant* of the matrix $\boldsymbol{B}$ and is obtained from the products obtained by cross mutiplying the diagonals.

Exercise 1 Obtain the inverse of the matrix

$$\begin{bmatrix} 2 & 4 \\ 1 & 6 \end{bmatrix}$$

by Cramer's rule.

$$K = 2 \times 6 - 4 \times 1 = 8$$

and the inverse is

$$1/8\begin{bmatrix} 6 & -4 \\ -1 & 2 \end{bmatrix}$$

$$= \begin{bmatrix} 6/8 & -4/8 \\ -1/8 & 2/8 \end{bmatrix}$$

$$= \begin{bmatrix} 0.75 & -0.5 \\ -0.125 & 0.25 \end{bmatrix}$$

Note: Cramer's rule can be extended to give a theoretical way to invert any square matrix. It is however a grossly inefficient, and is never used in practice other than for advanced *theoretical studies* of the structure of matrices. (See Chapter 12.)

7.9 Singular matrices

Not all square matrices have an inverse.

Exercise 1 Find the inverse of the matrix $\boldsymbol{M}$ by Cramer's rule where

$$\boldsymbol{M} = \begin{bmatrix} e & f \\ 3e & 3f \end{bmatrix}$$

The determinant K is

$$K = 3ef - 3ef = 0 \text{ then } \frac{1}{K} = \frac{1}{0}$$

which is not defined in arithmetic. Therefore the matrix $\boldsymbol{M}$ has no inverse. It is called a *singular* matrix.The problem can arise from the same equation being written as a linear combination of another by mistake (here equation 2 is three times the first).Generally speaking, if a problem is properly posed, singular matrices should not arise. For further discussion of this matter see a specialist book such as *Theory and Problems of Matrices*, F Ayres, Schaum Outline Series.

7.10 More matrix multiplication

The (row × column) multiplication rule applies to matrices generally. For matrix multiplication to be possible, there must be the same number of columns in the first matrix as there are rows in the second. The numbers in each row and column are called the *elements* of the matrix. They are usually assigned subscripts: the row number is written first followed by the column number. Thus the general (3 × 3), pronounced 'three by three', matrix $\boldsymbol{A}$ has elements $\boldsymbol{a}_{ij}$, and a (3 × 2) matrix $\boldsymbol{B}$ has elements $\boldsymbol{b}_{jk}$ as follows

$$\begin{bmatrix} a_{11} & a_{12} & a_{13} \\ a_{21} & a_{22} & a_{23} \\ a_{31} & a_{32} & a_{33} \end{bmatrix}\begin{bmatrix} b_{11} & b_{12} \\ b_{21} & b_{22} \\ b_{31} & b_{32} \end{bmatrix}$$

The result of the multiplication is a (3 × 2) matrix. This has the dimensions of $\boldsymbol{i}\ \boldsymbol{k}$, the outside indices in the product of

$$\boldsymbol{a_{ij}b_{jk}}$$

It is important to examine the dimensions of matrices to see that they are conformable for multiplication. This can be told at once from the individual dimensions. In this example the multiplication is

$$(3 \times 3)\ (3 \times 2)$$

The result is a matrix given by the outer numbers, in this case (3 × 2).

Note: It would be impossible to *change the order of the above matrices* in the product because there are only two columns in ***B*** but three rows in ***A***. The dimensions are

$$(3 \times \mathbf{2})(\mathbf{3} \times 3)$$

Because the two inner dimensions are not the same, the operation of matrix multiplication is impossible. Thus although the product ***AB*** is defined, ***BA*** is not, or we say it is *not conformable* for multipication. If on the other hand both ***A*** and ***B*** are square, then both products ***AB*** and ***BA*** are conformable for multiplication. If these products *are equal*, we say that the two matrices *commute*. By contrast ordinary numbers *always* commute, because

$$3 \times 4 = 4 \times 3$$

The same is true of ordinary algebra, for

$$ab = ba$$

The expressions 'premultiply' and postmultiply' are used of matrices. For example if it is said that 'matrix ***A*** postmultiplies matrix ***B***', this means the product ***BA***.

Exercise 1 Show that the matrices ***M*** and ***N*** conform for multiplication but do not commute, that is ***MN*** ≠ ***NM***.

$$\boldsymbol{MN} = \begin{bmatrix} 1 & 2 \\ 1 & 1 \end{bmatrix}\begin{bmatrix} 3 & 1 \\ 1 & 2 \end{bmatrix}$$

$$= \begin{bmatrix} 5 & 5 \\ 4 & 3 \end{bmatrix}$$

And

$$\boldsymbol{NM} = \begin{bmatrix} 3 & 1 \\ 1 & 2 \end{bmatrix}\begin{bmatrix} 1 & 2 \\ 1 & 1 \end{bmatrix}$$

$$= \begin{bmatrix} 4 & 7 \\ 3 & 4 \end{bmatrix}$$

$$\boldsymbol{MN} \neq \boldsymbol{NM}$$

Exercise 2 What are the dimensions of the matrix formed by premultiplying a (10 × 3) matrix by a (7 × 10) matrix? The result is given by

$$(7 \times 10)\,(10 \times 3) \;= (7 \times 3)$$

7.11 Other matrix operations: addition and subtraction

So far we have dealt only with the two important matrix operations of multiplication and inversion (the equivalent of division). The operations of addition and subtraction are much simpler. Provided the matrices are of the same dimensions they can be added or subtracted. Thus for example, if $\boldsymbol{A}$ and $\boldsymbol{B}$ are two matrices, we can obtain a third matrix $\boldsymbol{C}$ by addition from

$$\boldsymbol{C} = \boldsymbol{A} + \boldsymbol{B} \tag{7.11}$$

The process merely involves the addition of respective elements. A typical element of $\boldsymbol{C}$ such as $\boldsymbol{c}_{ij}$ is given by

$$\boldsymbol{c}_{ij} = \boldsymbol{a}_{ij} + \boldsymbol{b}_{ij}$$

Exercise 1 Add the matrices $\boldsymbol{A}$ and $\boldsymbol{B}$ to give matrix $\boldsymbol{C}$ where

$$\boldsymbol{A} = \begin{bmatrix} 1 & 2 \\ 3 & 4 \end{bmatrix} \quad \boldsymbol{B} = \begin{bmatrix} 5 & 6 \\ 7 & 8 \end{bmatrix}$$

then

$$\boldsymbol{C} = \begin{bmatrix} (1+5) & (2+6) \\ (3+7) & (4+8) \end{bmatrix}$$

$$= \begin{bmatrix} 6 & 8 \\ 10 & 12 \end{bmatrix}$$

Exercise 2 Subtract matrix $\boldsymbol{B}$ from matrix $\boldsymbol{A}$ to give $\boldsymbol{D}$.

$$\boldsymbol{D} = \boldsymbol{A} - \boldsymbol{B} \tag{7.12}$$

$$\boldsymbol{D} = \begin{bmatrix} (1-5) & (2-6) \\ (3-7) & (4-8) \end{bmatrix}$$

$$= \begin{bmatrix} -4 & -4 \\ -4 & -4 \end{bmatrix}$$

$$= -4\begin{bmatrix} 1 & 1 \\ 1 & 1 \end{bmatrix}$$

The last form of the result shows that a common factor (–4) can be taken from a matrix and placed outside. Such a common factor is called a *scalar* because *it alters the size of each element of the matrix.*

Exercise 3 Simplify the following matrix

$$\begin{bmatrix} -4 & 0 \\ 0 & -4 \end{bmatrix}$$

Clearly this can be written as

$$-4\begin{bmatrix} 1 & 0 \\ 0 & 1 \end{bmatrix} = -4\boldsymbol{I}$$

7.12 The transpose of a matrix

If the rows and columns of a matrix are exchanged the matrix is said to be *transposed.*

For example if

$$\boldsymbol{C} = \begin{bmatrix} 6 & 8 \\ 10 & 12 \end{bmatrix}$$

the transpose of $\boldsymbol{C}$, written as $\boldsymbol{C}^T$, is

$$\boldsymbol{C}^T = \begin{bmatrix} 6 & 10 \\ 8 & 12 \end{bmatrix}$$

Occasionally the transpose is written as

$$\boldsymbol{C}^{!}$$

The transpose is commonly used in printed books to save space. For example if the column matrix $\boldsymbol{x}$ is given by

$$\boldsymbol{x} = \begin{bmatrix} x \\ y \\ z \end{bmatrix}$$

its transpose uses only one line, because

$$\boldsymbol{x}^T = \begin{bmatrix} x & y & z \end{bmatrix}$$

7.13 Further operations on matrices

The use of brackets in matrix statements is the same as in ordinary algebra. For example

$$\boldsymbol{A} + (\boldsymbol{B} - \boldsymbol{C}) = \boldsymbol{A} + \boldsymbol{B} - \boldsymbol{C} = (\boldsymbol{A} + \boldsymbol{B}) - \boldsymbol{C}$$

$$\boldsymbol{A}(\boldsymbol{B} + \boldsymbol{C}) = \boldsymbol{AB} + \boldsymbol{AC} \qquad (7.13)$$

In these examples the matrices are said to be *associative.*

Exercise 1 Verify that the above associative rules apply to the matrices ***A***, ***B*** and ***C***.

$$\boldsymbol{A}=\begin{bmatrix}1 & 2\\ 3 & 4\end{bmatrix} \quad \boldsymbol{B}=\begin{bmatrix}1 & 0\\ 0 & 1\end{bmatrix} \quad \boldsymbol{C}=\begin{bmatrix}5 & 6\\ 7 & 8\end{bmatrix}$$

Note: For these matrices to be conformable for all three operations of addition, subtraction and multiplication they must all be square.

7.14 Operations on matrix products: reversal rule

We now consider operations on matrices which are entirely different from ordinary arithmetic and algebra: the transposition and inversion of matrix products.

Transposition of a product

Consider the product of two matrices

$$\boldsymbol{C}=\boldsymbol{AB}$$

If we now transpose *C* we have

$$\boldsymbol{C}^T=(\boldsymbol{AB})^T$$

Clearly we can multiply ***AB*** then transpose the result, but it is also true that

$$\boldsymbol{C}^{\mathrm{T}}=\boldsymbol{B}^{\mathrm{T}}\boldsymbol{A}^{\mathrm{T}} \qquad (7.14)$$

The rule is that the order of multiplication is *reversed* in transposition.

Exercise 1 Verify the *reverse order rule* for the transposition of the following matrices

$$\boldsymbol{A}=\begin{bmatrix}1 & 2\\ 3 & 4\end{bmatrix} \quad \boldsymbol{C}=\begin{bmatrix}5 & 6\\ 7 & 8\end{bmatrix}$$

Inversion of a product

The same rule also applies to the inverses of matrices. Consider the product of two matrices

$$\boldsymbol{C}=\boldsymbol{AB}$$

If we now invert ***C*** we have

$$\boldsymbol{C}^{-1}=(\boldsymbol{AB})^{-1}$$

Clearly we can multiply ***AB*** then invert the result, but it is also true that

$$\boldsymbol{C}^{-1}=\boldsymbol{B}^{-1}\boldsymbol{A}^{-1} \qquad (7.15)$$

The rule is that the order of multiplication is reversed in inversion.

Exercise 2 Verify the *reverse order rule* for the inversion of the following matrices

$$\boldsymbol{A}=\begin{bmatrix}1 & 2\\ 3 & 4\end{bmatrix} \quad \boldsymbol{C}=\begin{bmatrix}5 & 6\\ 7 & 8\end{bmatrix}$$

Note: The reversal rule applies generally for transposition and inversion. If $\boldsymbol{D} = \boldsymbol{ABC}$

$$\boldsymbol{D}^{\mathrm{T}} = \boldsymbol{C}^{\mathrm{T}}\boldsymbol{B}^{\mathrm{T}}\boldsymbol{A}^{\mathrm{T}} \tag{7.16}$$

$$\boldsymbol{D}^{-1} = \boldsymbol{C}^{-1}\boldsymbol{B}^{-1}\boldsymbol{A}^{-1} \tag{7.17}$$

7.15 Some applications of matrix algebra

For a number of reasons, matrix notation and methods are very useful to deal with problems other than the solution of linear equations. The very existence of computer programs and commands in spreadsheets to work matrix operations has added to their use. We now give a few important examples.

The length of a vector

In three dimensions, the distance S from the origin O to a point $P(x, y, z)$ is given by

$$S^2 = x^2 + y^2 + z^2$$

If we write (x, y, z) as a column matrix $\boldsymbol{x}$ then S is given by

$$S^2 = \boldsymbol{x}^{\mathrm{T}}\boldsymbol{x} \tag{7.18}$$

The matrix dimensions of $\boldsymbol{x}^{\mathrm{T}}$ and $\boldsymbol{x}$ are (1×3) and (3×1) so their product has dimension

$$(1 \times 3)(3 \times 1) = (1 \times 1), \text{ a single number}$$

$$\boldsymbol{x}^{\mathrm{T}}\boldsymbol{x} = \begin{bmatrix} x & y & z \end{bmatrix} \begin{bmatrix} x \\ y \\ z \end{bmatrix}$$

$$\begin{aligned} &= \begin{bmatrix} xx + yy + zz \end{bmatrix} \\ &= x^2 + y^2 + z^2 \\ &= S^2 \end{aligned}$$

Exercise 1 Write down in full the matrix which results from the product

$$\boldsymbol{x}\boldsymbol{x}^{\mathrm{T}}$$

In full this product is

$$\begin{bmatrix} x \\ y \\ z \end{bmatrix} \begin{bmatrix} x & y & z \end{bmatrix}$$

The dimensions of this column and row are $(3 \times 1)(1 \times 3)$, therefore the result is a (3×3) symmetric matrix

$$\begin{bmatrix} xx & xy & xz \\ yx & yy & yz \\ zx & zy & zz \end{bmatrix}$$

A matrix of this form is very important in statistical work. Clearly

$$\boldsymbol{x}\boldsymbol{x}^{\mathrm{T}} \neq \boldsymbol{x}^{\mathrm{T}}\boldsymbol{x}$$

7.16 Rotations in three dimensions

One of the most important applications of matrix algebra to cartography and surveying is to deal with the rotation of rigid bodies. In photogrammetry, a three-dimensional model is often rotated and, in cartography, a map may need a changed orientation. Much of the problem of datum definition in geodesy is based on rotations of axes of coordinate systems. In our discussions here, we deal only with rotations related to the known coordinate axes. Consider a solid object PQR located within three, right handed, cartesian axes Ox, Oy and Oz, as in Figure 7.2. Figure 7.2 shows the plan view in the Oxy plane. The Oz axis is pointing outwards from the page as you read it. Consider that the solid is rotated about the Oz axis in an anticlockwise manner from position PQ to position P_1Q_1. The only thing that has happened is that the body has been rotated positively within the system. Therefore

$$OP = OP_1 \qquad OQ = OQ_1$$

Also P_1Q_1 makes an angle B with the direction of PQ, and the length of PQ = length of P_1Q_1.

The purpose of the rotation matrix is to enable us to transform the original coordinates of the solid, such as for the point P, to the new coordinates in its new position P_1. Let $OP = OP_1 = r$ and let the respective coordinates be

$$P(x_P, y_P, z_P) \quad \text{and} \quad P_1(x_{P1}, y_{P1}, z_{P1})$$

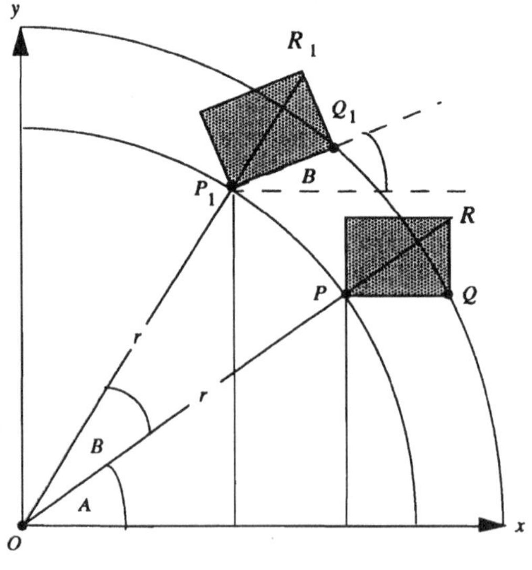

Figure 7.2

Because the rotation is about the Oz axis there will be no change to the z coordinates of P as it moves to P_1. Therefore

$$z_{P1} = z_P$$

Now

$$x_P = r\cos A$$
$$y_P = r\sin A$$
$$x_{P_1} = r\cos(A+B)$$
$$y_{P_1} = r\sin(A+B)$$
$$x_{P_1} = r\cos A\cos B - r\sin A\sin B$$
$$y_{P_1} = r\sin A\cos B + r\cos A\sin B$$
$$x_{P_1} = x_P\cos B - y_P\sin B$$
$$y_{P_1} = y_P\cos B + x_P\sin B$$

Rearranging the order we have

$$x_{P_1} = \cos B.x_P - \sin B.y_P$$
$$y_{P_1} = \sin B.x_P + \cos B.y_P$$

Adding in the z coordinate we have

$$\begin{aligned} x_{P_1} &= \cos B.x_P - \sin B.y_P + 0.z_P \\ y_{P_1} &= \sin B.x_P + \cos B.y_P + 0.z_P \\ z_{P_1} &= \quad 0.x_P + \quad 0.y_P + 1.z_P \end{aligned} \tag{7.19}$$

The zero coefficients have been added so that we can recast these equations in matrix form as follows

$$\begin{bmatrix} x_{P_1} \\ y_{P_1} \\ z_{P_1} \end{bmatrix} = \begin{bmatrix} \cos B & -\sin B & 0 \\ \sin B & \cos B & 0 \\ 0 & 0 & 1 \end{bmatrix} \begin{bmatrix} x_P \\ y_P \\ z_P \end{bmatrix} \tag{7.20}$$

This is of the form

$$\boldsymbol{X} = \boldsymbol{R}_z\boldsymbol{x}$$

where $\boldsymbol{X}$ is the vector of the new coordinates, and $\boldsymbol{x}$ of the old coordinates.

Exercise 1 The point P has coordinates (10, 8, 3). Find its new coordinates if $B = 30°$. Substituting in equations (7.19) for $\cos B = \sqrt{3}/2 = 0.8660$ and $\sin B = 0.5$ gives

$$x_{P1} = 0.8660 \times 10 - 0.5 \times 8 = 4.66$$
$$y_{P1} = 0.5 \times 10 + 0.8660 \times 8 = 11.928$$
$$z_{P1} = 3$$

7.17 Structure of the rotation matrix

To ensure that a correct rotation matrix is formed, its structure is worth noting. In the above case, a rotation of the body was made about the *Oz* axis yielding a matrix of the form

$$\boldsymbol{R}_z = \begin{bmatrix} C & -S & 0 \\ S & C & 0 \\ 0 & 0 & 1 \end{bmatrix}$$

In this matrix C = cosine and S = sine of the angle of rotation. The signs of C and S depend on the quadrant of the angle.

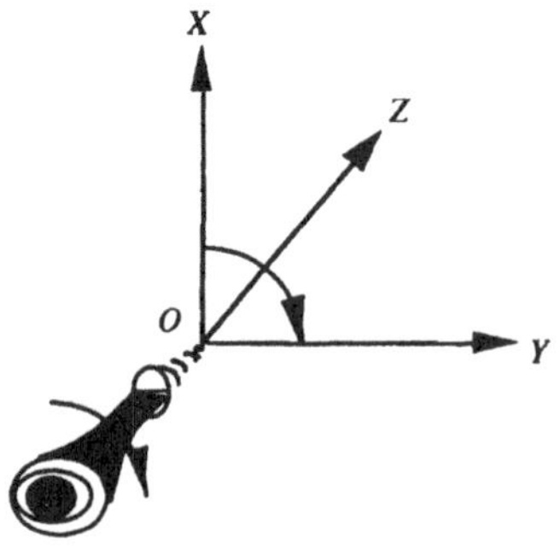

Figure 7.3

Note on the sign of rotation

To grasp the idea of the sign of a rotation consider Figure 7.3, which shows a screw about to be driven into a piece of wood by a screwdriver. The three axes (*Ox*, *Oy*, *Oz*) are also shown. If the screwdriver is rotated in a clockwise sense as we look along *Oz* the screw will enter the wood. This is a *positive* rotation. Conversely if the screw is to be extracted from the wood an anticlockwise or *negative* rotation is needed. Now compare Figure 7.2 with 7.3. Because in Figure 7.2 we are looking *down* the *Oz* axis we see the rotation from the other aspect to that of Figure 7.3, and the positive rotation *appears* anticlockwise in Figure 7.2.

Exercise 1 Rotate the point (4.66, 11.928, 3) of Exercise 1 in Section 7.16 back to its original position at *P*. Because the rotation would remove a screw pointing along the *Oz* axis, the rotation is negative, therefore

$C = \cos(-30°) = \cos 30° = 0.8660$ and $S = \sin(-30°) = -\sin 30° = -0.5$

and

$$\boldsymbol{RR}_z = \begin{bmatrix} 0.8660 & 0.5 & 0 \\ -0.5 & 0.8660 & 0 \\ 0 & 0 & 1 \end{bmatrix}$$

Thus

$$x_P = 0.8660 \times 4.66 + 0.5 \times 11.928 = 10$$
$$y_P = -0.5 \times 4.66 + 0.8660 \times 11.928 = 8$$
$$z_P = 3$$

Note: We have denoted the matrix for the reversal as $\boldsymbol{RR}_z$ because it is different from the matrix for the forward rotation $\boldsymbol{R}_z$. Inspection shows that

$$\boldsymbol{RR}_z = \boldsymbol{R}_z^T$$

the transpose of matrix for the forward case.

7.18 Rotation of solid body about the *Oy* and *Oz* axes

To derive the equations for the effect of a rotation of a solid about the *Oy* axis, we consider Figure 7.4.

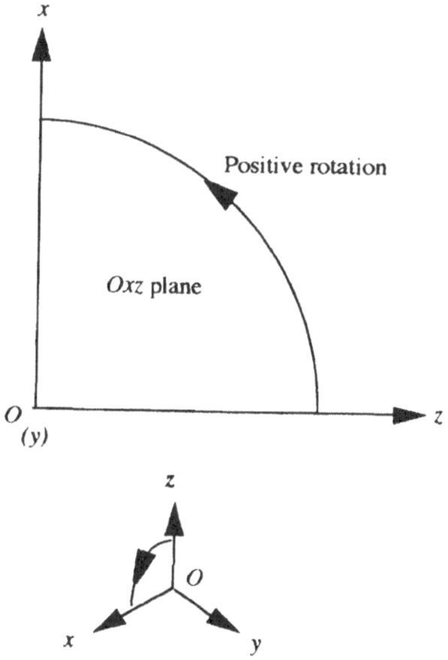

Figure 7.4

Here the *Oy* axis is pointing out of the page towards the reader, as was the case for the *Oz* axis of the previous case in Figure 7.2. A positive rotation of the solid is anticlockwise again. The order of coordinates has to be changed to accord with the figure. That is we quote the coordinates in the new order (z, x, y). Thus we can write down the structure of the rotation as before to be

$$\begin{bmatrix} z_{P_1} \\ x_{P_1} \\ y_{P_1} \end{bmatrix} = \begin{bmatrix} C & -S & 0 \\ S & C & 0 \\ 0 & 0 & 1 \end{bmatrix} \begin{bmatrix} z_P \\ x_P \\ y_P \end{bmatrix} \qquad (7.21)$$

Writing these out in full we have

$$z_{P_1} = \cos B.z_P - \sin B.x_P + 0.y_P$$
$$x_{P_1} = \sin B.z_P + \cos B.x_P + 0.y_P$$
$$y_{P_1} = 0.z_P + 0.x_P + 1.y_P$$

Rearranging rows and columns to the usual order gives

$$x_{P_1} = \cos B.x_P + 0.y_P + \sin B.z_P$$
$$y_{P_1} = 0.x_P + 1.y_P + 0.z_P$$
$$z_{P_1} = -\sin B.x_P + 0.y_P + \cos B.z_P$$

and in matrix form

$$\begin{bmatrix} x_{P_1} \\ y_{P_1} \\ z_{P_1} \end{bmatrix} = \begin{bmatrix} C & 0 & S \\ 0 & 1 & 0 \\ -S & 0 & C \end{bmatrix} \begin{bmatrix} x_P \\ y_P \\ z_P \end{bmatrix} \tag{7.22}$$

By a similar argument we find that the equations for rotation about the *Ox* axis are

$$\begin{bmatrix} x_{P_1} \\ y_{P_1} \\ z_{P_1} \end{bmatrix} = \begin{bmatrix} 1 & 0 & 0 \\ 0 & C & -S \\ 0 & S & C \end{bmatrix} \begin{bmatrix} x_P \\ y_P \\ z_P \end{bmatrix} \tag{7.23}$$

Summarising the three structures we have

$$\boldsymbol{R}_x = \begin{bmatrix} 1 & 0 & 0 \\ 0 & C & -S \\ 0 & S & C \end{bmatrix}$$
$$\boldsymbol{R}_y = \begin{bmatrix} C & 0 & S \\ 0 & 1 & 0 \\ -S & 0 & C \end{bmatrix}$$
$$\boldsymbol{R}_z = \begin{bmatrix} C & -S & 0 \\ S & C & 0 \\ 0 & 0 & 1 \end{bmatrix}$$

Clearly the only matter requiring attention is the sign of S.

Note about rotation of axes

If we had rotated the *axes* about one axis, instead of the *solid body* within a set of fixed axes, the sign of the rotation angle changes. The effect is only to change the signs of S in all three matrices. Both types of problem arise.

7.19 Reverse case of rotation

A common problem often arises in which two sets of known coordinates are given, and the rotation between them has to be found. One way to treat the problem is to calculate the respective bearings of the lines joining these points to the respective origins. The difference of these bearings is therefore the rotational angle required. However, a more convenient and direct solution in terms of the coordinates is usually adopted. Consider the basic rotation equations for a rotation about the *Oz* axis. These are

$$x_{P_1} = \cos B . x_P - \sin B . y_P + 0 . z_P$$
$$y_{P_1} = \sin B . x_P + \cos B . y_P + 0 . z_P$$
$$z_{P_1} = 0 . x_P + 0 . y_P + 1 . z_P$$

Rearranging the order gives

$$x_{P_1} = x_P . \cos B - y_P . \sin B + 0 . z_P$$
$$y_{P_1} = x_P . \sin B + y_P . \cos B + 0 . z_P$$
$$z_{P_1} = 0 . x_P + 0 . y_P + 1 . z_P$$

The *z* coordinates can be omitted because they are unaffected by the rotation so we solve the simpler problem

$$x_{P_1} = x_P . \cos B - y_P . \sin B$$
$$y_{P_1} = x_P . \sin B + y_P . \cos B$$

writing $C = \cos B$ and $S = \sin B$ we have

$$\begin{aligned} x_P . C - y_P . S &= x_{P_1} \\ x_P . S + y_P . C &= y_{P_1} \end{aligned} \qquad (7.24)$$

In matrix form this is

$$\begin{bmatrix} x_P & -y_P \\ y_P & x_P \end{bmatrix} \begin{bmatrix} C \\ S \end{bmatrix} = \begin{bmatrix} x_{P_1} \\ y_{P_1} \end{bmatrix} \qquad (7.25)$$

which is of the form

$$\boldsymbol{Ax} = \boldsymbol{b}$$

whose solution is

$$\boldsymbol{x} = \boldsymbol{A}^{-1}\boldsymbol{b}$$

Exercise 1 The coordinates of *P* on two systems are respectively

(10, 8, 3) and (4.66, 11.928, 3)

Assuming that the different values are due only to a rotation *B* about the *Oz* axis, show that this angle of rotation is 30°.

From (7.24) we have

$$\begin{bmatrix} 10 & -8 \\ 8 & 10 \end{bmatrix}\begin{bmatrix} C \\ S \end{bmatrix} = \begin{bmatrix} 4.660 \\ 11.928 \end{bmatrix}$$

which gives the solution $C = 0.8660, S = 0.5000$. Since both C and S are positive, the angle B lies in the first quadrant, therefore $B = 30°$.
Note: Without the two separate values for C and S we cannot tell in which quadrant B lies.

7.20 Orthogonal matrices

Consider the effect of a rotation on the coordinates, expressed as

$$\boldsymbol{X} = \boldsymbol{R}\,\boldsymbol{x}$$

Multiplying both sides of the equation by the inverse of $\boldsymbol{R}$ we have

$$\boldsymbol{R}^{-1}\boldsymbol{X} = \boldsymbol{R}^{-1}\boldsymbol{R}\,\boldsymbol{x}$$

but

$$\boldsymbol{R}^{-1}\boldsymbol{R} = \boldsymbol{I}$$

and

$$\boldsymbol{I}\,\boldsymbol{x} = \boldsymbol{x}$$

so

$$\boldsymbol{R}^{-1}\boldsymbol{X} = \boldsymbol{x}$$

or

$$\boldsymbol{x} = \boldsymbol{R}^{-1}\boldsymbol{X}$$

The length of the line from the origin to the original position of point P is given by

$$\boldsymbol{x}^{\mathrm{T}}\boldsymbol{x}$$

and to the new position by

$$\boldsymbol{X}^{\mathrm{T}}\boldsymbol{X}$$

This length does not change due to a rotation so

$$\boldsymbol{x}^{\mathrm{T}}\boldsymbol{x} = \boldsymbol{X}^{\mathrm{T}}\boldsymbol{X}$$

but

$$\boldsymbol{X} = \boldsymbol{R}\,\boldsymbol{x}$$

so

$$\boldsymbol{x}^{\mathrm{T}}\boldsymbol{x} = (\boldsymbol{R}\,\boldsymbol{x})^{\mathrm{T}}\boldsymbol{R}\,\boldsymbol{x}$$
$$= \boldsymbol{x}^{\mathrm{T}}\boldsymbol{R}^{\mathrm{T}}\boldsymbol{R}\,\boldsymbol{x}$$

For this to be possible

$$\boldsymbol{R}^{\mathrm{T}}\boldsymbol{R} = \boldsymbol{I}$$

but

$$\boldsymbol{R}^{-1}\boldsymbol{R} = \boldsymbol{I}$$

therefore

$$\boldsymbol{R}^{\mathrm{T}} = \boldsymbol{R}^{-1} \qquad (7.26)$$

Thus the *inverse* of a matrix which causes only a rotation is equal to its *transpose*. This is a very important result, because many matrices used in cartography and surveying are of this type. The practical significance of the result is that the inversion is the very simple operation of transposition. Matrices of this type are called *orthogonal* matrices.

Exercise 1 Test to see if the following matrix is orthogonal.

$$\boldsymbol{R} = \begin{bmatrix} \cos A & -\sin A \\ \sin A & \cos A \end{bmatrix}$$

Now

$$\boldsymbol{R}^{\mathrm{T}} = \begin{bmatrix} \cos A & \sin A \\ -\sin A & \cos A \end{bmatrix}$$

then

$$\boldsymbol{R}^{\mathrm{T}}\boldsymbol{R} = \begin{bmatrix} \cos A & \sin A \\ -\sin A & \cos A \end{bmatrix} \begin{bmatrix} \cos A & -\sin A \\ \sin A & \cos A \end{bmatrix}$$

$$\boldsymbol{R}^{\mathrm{T}}\boldsymbol{R} = \begin{bmatrix} (\cos A \cos A + \sin A \sin A) & (-\cos A \sin A + \sin A \cos A) \\ (-\sin A \cos A + \sin A \cos A) & (\cos A \cos A + \sin A \sin A) \end{bmatrix}$$

but

$$(\cos A \cos A + \sin A \sin A) = \cos^2 A + \sin^2 A = 1$$

therefore

$$\boldsymbol{R}^{\mathrm{T}}\boldsymbol{R} = \begin{bmatrix} 1 & 0 \\ 0 & 1 \end{bmatrix}$$

$$= \boldsymbol{I}$$

But

$$\boldsymbol{R}^{-1}\boldsymbol{R} = \boldsymbol{I}$$

therefore

$$\boldsymbol{R}^{-1} = \boldsymbol{R}^{\mathrm{T}}$$

The matrix is orthogonal.

Note: An orthogonal matrix has the properties that the matrix product of any row or column by itself is always unity, and the matrix product of a row (column) by another row (column) is always zero. For

$$\cos A \cos A + \sin A \sin A = \cos^2 A + \sin^2 A = 1$$

and

$$-\sin A \cos A + \sin A \cos A = 0$$

7.21 Coordinate transformations in three dimensions

Many problems in surveying and cartography, but especially in photogrammetry, involve coordinates in three dimensions. Systems have to be related to each other by datum shifts, by scaling, and by rotation about any of three axes. It is not possible to consider all aspects of this major topic in this book. However the basic and simplest operations merely involve the standard operations of matrix addition, subtraction, multiplication and inversion. An example will suffice to illustrate methods.

Exercise 1 The points A, B, C, D on system (x, y, z) have to be transformed to a new system (X, Y, Z). The change parameters are, *in order*,
(a) a datum shift to point (–1, –2, –3) as origin,
(b) a rotation of the axes about the y axis of 10°,
(c) a scale change of 1.5.

Point	x	y	z
A	11	6	0
B	1	9	0
C	3	2	0
D	7	10	6

Consider one point only for the moment. To save space in printing, we write the column vector of coordinates in the form

$$\boldsymbol{x}^{\mathrm{T}} = (x, y, z)$$

and the datum shift vector as

$$\boldsymbol{x}_0^{\mathrm{T}} = (1,2,3)$$

The translation to the new origin is written

$$\boldsymbol{x}_1^{\mathrm{T}} = \boldsymbol{x}^{\mathrm{T}} + \boldsymbol{x}_0^{\mathrm{T}} \qquad (7.27)$$

$$= (x+1, y+2, z+3) \text{ in this case}$$

Exercise 2 Change all points $ABCD$ to the new origin (–1, –2, –3). We have

Point	x_1	y_1	z_1
A	11+1	6+2	0+3
B	1+1	9+2	0+3
C	3+1	2+2	0+3
D	7+1	10+2	6+3

Point	x_1	y_1	z_1
A	12	8	3
B	2	11	3
C	4	4	3
D	8	12	9

Since the rotation of the axes is to be about the Oy axis the y coordinates will be unchanged. The x and z coordinates will be affected by the rotation through angle A. The matrix to achieve such an effect is

$$\boldsymbol{R}_y = \begin{bmatrix} \cos A & 0 & -\sin A \\ 0 & 1 & 0 \\ \sin A & 0 & \cos A \end{bmatrix}$$

The suffix 'y' indicates that rotation is about the y axis.

Note: This matrix is similar to (7.22). Only the sign of sin A changes because we are treating *axes* rotation.

A similar rotation about the x axis would be

$$\boldsymbol{R}_x = \begin{bmatrix} 1 & 0 & 0 \\ 0 & \cos A & \sin A \\ 0 & -\sin A & \cos A \end{bmatrix}$$

and about the z axis

$$\boldsymbol{R}_z = \begin{bmatrix} \cos A & \sin A & 0 \\ -\sin A & \cos A & 0 \\ 0 & 0 & 1 \end{bmatrix}$$

Note: These matrices apply to the *rotation of the axes*. If the solid is rotated, the matrices are similar except that the signs of 'sin A' are changed throughout.

The effect of the rotation on the coordinates of a point is given by

$$\boldsymbol{x}_2 = \boldsymbol{R}_y \boldsymbol{x}_1$$

Exercise 3 Transform the new coordinates of point $A(12, 8, 3)$ if $A = 10°$ about the y axis. The rotation matrix is

$$\boldsymbol{R}_y = \begin{bmatrix} \cos 10° & 0 & -\sin 10° \\ 0 & 1 & 0 \\ \sin 10° & 0 & \cos 10° \end{bmatrix}$$

$$\boldsymbol{R}_y = \begin{bmatrix} 0.9848 & 0 & -0.1736 \\ 0 & 1 & 0 \\ 0.1736 & 0 & 0.9848 \end{bmatrix}$$

Using the direct multiplication we have

$$\boldsymbol{x}_2 = \boldsymbol{R}_y \boldsymbol{x}_1$$

$$\begin{bmatrix} x_2 \\ y_2 \\ z_2 \end{bmatrix} = \begin{bmatrix} 0.9848 & 0 & -0.1736 \\ 0 & 1 & 0 \\ 0.1736 & 0 & 0.9848 \end{bmatrix} \begin{bmatrix} 12 \\ 8 \\ 3 \end{bmatrix}$$

$$\begin{bmatrix} x_2 \\ y_2 \\ z_2 \end{bmatrix} = \begin{bmatrix} 11.3 \\ 8 \\ 5.04 \end{bmatrix}$$

As an illustration and check, using the transposed multiplication we have

$$\boldsymbol{x}_2^{\mathrm{T}} = \boldsymbol{x}_1^{\mathrm{T}} \boldsymbol{R}_y^{\mathrm{T}}$$

$$\boldsymbol{R}_y^{\mathrm{T}} = \begin{bmatrix} 0.9848 & 0 & 0.1736 \\ 0 & 1 & 0 \\ -0.1736 & 0 & 0.9848 \end{bmatrix}$$

$$\begin{bmatrix} x_2 & y_2 & z_2 \end{bmatrix} = \begin{bmatrix} 11.3 & 8 & 5.04 \end{bmatrix} \begin{bmatrix} 0.9848 & 0 & 0.1736 \\ 0 & 1 & 0 \\ -0.1736 & 0 & 0.9848 \end{bmatrix}$$

$$\begin{bmatrix} x_2 & y_2 & z_2 \end{bmatrix} = \begin{bmatrix} 11.3 & 8 & 5.04 \end{bmatrix}$$

The final scale change is brought about by multiplying the results by the scale factor 1.5 to give

$$\begin{bmatrix} X & Y & Z \end{bmatrix} = 1.5 \begin{bmatrix} 11.3 & 8 & 5.04 \end{bmatrix}$$

$$\begin{bmatrix} X & Y & Z \end{bmatrix} = \begin{bmatrix} 16.95 & 12 & 7.56 \end{bmatrix}$$

Note: When a large number of points, say 100, have to be transformed, their values are assembled in a matrix, say $\boldsymbol{A}$ of dimensions (100 × 3). The resulting matrix, say $\boldsymbol{B}$, is also (100 × 3). The rotation matrix $\boldsymbol{R}$ is of dimensions (3 × 3). The transformation then is

$$\boldsymbol{B} = (\boldsymbol{R}\boldsymbol{A}^{\mathrm{T}})^{\mathrm{T}}$$

The dimensions are

$$\begin{aligned} (100 \times 3) &= ((3 \times 3)(3 \times 100))^{\mathrm{T}} \\ &= (3 \times 100)^{\mathrm{T}} \\ &= (100 \times 3) \end{aligned}$$

The results of the above transformations on the four points A, B, C and D are

Point	X	Y	Z
A	16.95	12	7.56
B	2.17	16.5	4.95
C	5.13	6	5.47
D	9.47	18	15.38

A useful partial check on results is to take out the lengths of some lines before and after transformations to see if they are 1.5 times longer after the transformation.

Lengths	After	Before	Scale
AB	15.66	10.44	1.5
BC	10.92	7.28	1.5
CA	13.42	8.94	1.5

7.22 The reverse transformations

Because the rotation matrices used in these transformations are orthogonal, their inverses are obtained by simple transposition. If there is a scale factor present the matrix can be factorised into a product of two matrices easily inverted (see below).

Diagonal matrices

A matrix which has at least one value on its diagonal and zeros elsewhere is called a *diagonal matrix*. For example **I** is a diagonal matrix. Another example is the matrix

$$\boldsymbol{D} = \begin{bmatrix} a & 0 \\ 0 & b \end{bmatrix}$$

A diagonal matrix is easy to invert for

$$\boldsymbol{D}^{-1} = \begin{bmatrix} a^{-1} & 0 \\ 0 & b^{-1} \end{bmatrix}$$

where

$$a^{-1} = 1/a \quad \text{and} \quad b^{-1} = 1/b$$

That this is true can be seen by multiplication

$$\boldsymbol{D}\boldsymbol{D}^{-1} = \boldsymbol{D}^{-1}\boldsymbol{D} = \boldsymbol{I}$$

Matrix factorisation

A lookout should be kept for any matrix which is the product of a diagonal and an orthogonal matrix, because its factors are easily inverted and the reversal rule applied. Many scaled transformations in surveying and geodesy are of this type.

Example A non-orthogonal matrix $\boldsymbol{N}$ is the product of a diagonal matrix $\boldsymbol{D}$ and an orthogonal matrix $\boldsymbol{M}$. Its effect on a transformed vector $\boldsymbol{x}$ is to rotate it through

60° and to rescale the ordinate y by a factor of 2. (This can be verified also by drawing.) We have

$$DM = N$$

with respective values

$$DM = \begin{bmatrix} 1 & 0 \\ 0 & 2 \end{bmatrix} \begin{bmatrix} \frac{1}{2} & \frac{\sqrt{3}}{2} \\ -\frac{\sqrt{3}}{2} & \frac{1}{2} \end{bmatrix} = \begin{bmatrix} \frac{1}{2} & \frac{\sqrt{3}}{2} \\ -\sqrt{3} & 1 \end{bmatrix} = N$$

The vector $\boldsymbol{x}$, given by $\boldsymbol{x}^T = [4 \quad 3]$, is transformed by the orthogonal matrix $\boldsymbol{M}$ without changing its length (5); i.e. it rotates it through 60°. The diagonal matrix $\boldsymbol{D}$ enlarges the ordinate from 4 to 8, and the length of the vector to 8.54.

Exercise 1 Verify, directly or via the product $\boldsymbol{DM}$, that the inverse of $\boldsymbol{N}$ is

$$\begin{bmatrix} \frac{1}{2} & -\frac{\sqrt{3}}{4} \\ +\frac{\sqrt{3}}{2} & \frac{1}{4} \end{bmatrix}$$

(1) The inverse of $\boldsymbol{N}$ directly by Cramers rule is

$$\frac{1}{D} \begin{bmatrix} 1 & -\frac{\sqrt{3}}{2} \\ +\sqrt{3} & \frac{1}{2} \end{bmatrix}$$

where

$$D = \frac{1}{2} + \frac{3}{2} = 2$$

therefore

$$N^{-1} = \begin{bmatrix} \frac{1}{2} & -\frac{\sqrt{3}}{4} \\ +\frac{\sqrt{3}}{2} & \frac{1}{4} \end{bmatrix}$$

(2) The inverse of **N** from the product rule is

$$N^{-1} = M^{-1} D^{-1}$$

$$\begin{bmatrix} \frac{1}{2} & -\frac{\sqrt{3}}{2} \\ +\frac{\sqrt{3}}{2} & \frac{1}{2} \end{bmatrix} \begin{bmatrix} 1 & 0 \\ 0 & \frac{1}{2} \end{bmatrix} = N^{-1}$$

therefore

$$N^{-1} = \begin{bmatrix} \frac{1}{2} & -\frac{\sqrt{3}}{4} \\ +\frac{\sqrt{3}}{2} & \frac{1}{4} \end{bmatrix}$$

SUMMARY OF KEY WORDS

Matrix algebra, arrays, simultaneous linear equations, square array, column matrix, row times column multiplication, conformable for multiplication, inverse of a matrix, unit matrix, determinant, singular matrix, commute, transposed matrix, scalar, reverse order rule, associative, orthogonal matrix, diagonal matrix

SUMMARY OF FORMULAE

Multiplication	$\boldsymbol{Ax} = \boldsymbol{b}$	(7.3)
Division	$\boldsymbol{x} = \boldsymbol{A}^{-1}.\boldsymbol{b}$	(7.4)
Inversion	$\boldsymbol{A}.\boldsymbol{A}^{-1} = \boldsymbol{A}^{-1}.\boldsymbol{A} = \boldsymbol{I}$	(7.5)
Determinant	$K = eh - fg$	(7.10)
	singular if $K = 0$	
Addition	$\boldsymbol{C} = \boldsymbol{A} + \boldsymbol{B}$	(7.11)
Subtraction	$\boldsymbol{D} = \boldsymbol{A} - \boldsymbol{B}$	(7.12)
Associative law	$\boldsymbol{A}(\boldsymbol{B} + \boldsymbol{C}) = \boldsymbol{AB} + \boldsymbol{AC}$	(7.13)
Transpose If $\boldsymbol{C} = \boldsymbol{AB}$	$\boldsymbol{C}^{\mathrm{T}} = \boldsymbol{B}^{\mathrm{T}}\boldsymbol{A}^{\mathrm{T}}$	(7.14)
If $\boldsymbol{C} = \boldsymbol{AB}$ then	$\boldsymbol{C}^{-1} = \boldsymbol{B}^{-1}\boldsymbol{A}^{-1}$	(7.15)
If $\boldsymbol{D} = \boldsymbol{ABC}$ then	$\boldsymbol{D}^{\mathrm{T}} = \boldsymbol{C}^{\mathrm{T}}\boldsymbol{B}^{\mathrm{T}}\boldsymbol{A}^{\mathrm{T}}$	(7.16)
If $\boldsymbol{D} = \boldsymbol{ABC}$ then	$\boldsymbol{D}^{-1} = \boldsymbol{C}^{-1}\boldsymbol{B}^{-1}\boldsymbol{A}^{-1}$	(7.17)
Length of vector	$S^2 = \boldsymbol{x}^{\mathrm{T}}\boldsymbol{x}$	(7.18)
Orthogonal matrix	$\boldsymbol{R}^{\mathrm{T}} = \boldsymbol{R}^{-1}$	(7.26)
	$\boldsymbol{x}_1^{\mathrm{T}} = \boldsymbol{x}^{\mathrm{T}} - \boldsymbol{x}_0^{\mathrm{T}}$	(7.27)

Chapter 8
Vectors

8.1 Vectors

Unlike a *scalar* which only has a *size*, a *vector* has both *size* and *direction*. For example if we say that 'Glasgow is 600 kilometres from London', the distance '600' is a scalar quantity. There are many places the same distance from London all lying on a circle of radius 600 km. However if we also say that 'Glasgow is 600 km north west of London', by giving information which has both *magnitude* and *direction,* we have given vector information. In this case only one place, Glasgow, can satisfy the definition.

The *algebra of vectors* evolved to deal with problems in electromagnetism, in which a magnetic force, an electrical force and a mechanical force act as an orthogonal triplet in three-dimensional space. Vector algebra was also found to be useful in many other fields such as photogrammetry and geodesy. Its notation is very neat and lends itself well to theoretical analysis. However, when calculations are required, matrix methods are employed. It is debatable whether vector methods are really needed to handle simple three-dimensional problems in cartography and surveying. However because the literature includes many articles using vector algebra, for completeness we give an introduction to the topic here.

8.2 Vector algebra

The notation used in vector algebra is very like matrix algebra, with which it overlaps to some extent. *It is assumed that the reader has studied Chapters 5 and 7 before reading on.* In fact most of the results of Chapter 5 can be recast in vector form. Initially the algebra of vectors may seem unnecessarily complicated to the beginner. The various ideas are developed to give meaning to the *multiplication of vectors* which is the really effective operation of the algebra. As with all mathematics, it is vital that the notation is properly understood.

In Figure 8.1(a) we illustrate a vector PQ. It is denoted in bold type by the letter $\mathbf{a}$. The *size* of $\mathbf{a}$ is the length of PQ denoted by the *modulus sign* $|\mathbf{a}|$. To show the directional property we define $\bar{\mathbf{a}}$ to be a unit vector in the direction of PQ.Then

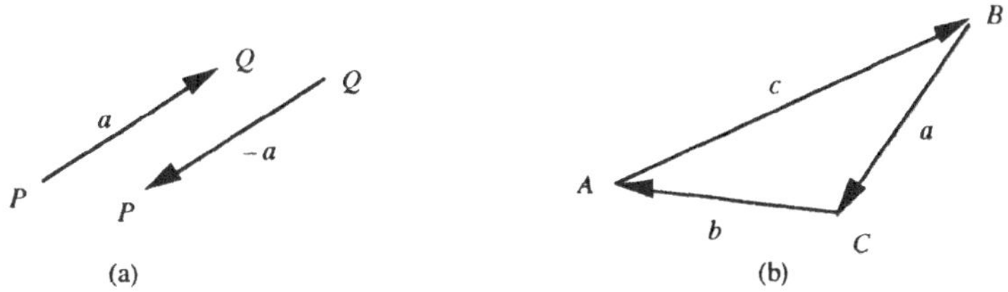

Figure 8.1

$$\boldsymbol{a} = |\boldsymbol{a}|\overline{\boldsymbol{a}} \tag{8.1}$$

Exercise 1 Write the following information in vector form.

'Glasgow is 600 km north west of London'

We define

$$\overline{\boldsymbol{a}} = 1 \text{ km in a direction north west}$$

The length of the vector $\boldsymbol{a}$ is 600 × 1 km therefore

$$|\boldsymbol{a}| = 600$$

Therefore

$$\boldsymbol{a} = |\boldsymbol{a}|\,\overline{\boldsymbol{a}} = 600\,\overline{\boldsymbol{a}}$$

The return journey *looked at from the London end* is

$$-\boldsymbol{a} = -600\,\overline{\boldsymbol{a}}$$

and in words

'600 km south east'

Note: '600' is written in ordinary type because it is a scalar quantity having size alone.

Exercise 2 Write in vector form the statement that 'the ceiling of the room is eight feet above the floor'. Let

$$\overline{\boldsymbol{a}} = 1 \text{ foot vertically}$$

then the length of the vector is

$$|\boldsymbol{a}| = 8$$

and the height of the room is

$$\boldsymbol{h} = |\boldsymbol{a}|\,\overline{\boldsymbol{a}} = 8\overline{\boldsymbol{a}}$$

8.3 Triangle *ABC*

Now consider the sides of the triangle *ABC* of Figure 8.1(b) as vectors $\boldsymbol{a}$, $\boldsymbol{b}$, and $\boldsymbol{c}$. We see that

$$\boldsymbol{a} + \boldsymbol{b} + \boldsymbol{c} = \mathbf{0}$$

where **0** is the *null vector*. The order in which we carry out the addition does not matter because a route such as

$$\boldsymbol{a} + \boldsymbol{c} + \boldsymbol{b}$$

would bring us back to B. See also Section 5.6.

8.4 Free vectors and position vectors

If two vectors are *parallel* and the same *size* they are said to be *equal* no matter where they are located in space. Such vectors are *free vectors*. If on the other hand we wish to specify their location within a three-dimensional coordinate system, *position vectors* have to be specified from some point. For example we could specify the position of P (see Figure 8.1) relative to the origin O as the *position vector* $\boldsymbol{p}$. Thus a point Q would have its position given by

$$\boldsymbol{q} = \boldsymbol{p} + \boldsymbol{a}$$

where $\boldsymbol{a}$ is the vector PQ.

Exercise 1 Express the position vector of Q in terms of the separate unit vectors.

$$\boldsymbol{q} = |\boldsymbol{q}|\,\bar{\boldsymbol{q}} = |\boldsymbol{p}|\,\bar{\boldsymbol{p}} + |\boldsymbol{a}|\,\bar{\boldsymbol{a}}$$

8.5 Orthogonal unit vectors

For greatest practical convenience it is best to express a vector in terms of unit vectors lying in the three orthogonal axes Ox, Oy and Oz. These axes form a right handed set. That is, if we look along the Oz axis, a clockwise rotation about Oz will turn the Ox axis into the Oy axis. We can express $\boldsymbol{a}$ by its components as follows

$$\begin{aligned} \boldsymbol{a} &= \boldsymbol{a}_x + \boldsymbol{a}_y + \boldsymbol{a}_z \\ &= |\boldsymbol{a}_x|\bar{\boldsymbol{a}}_x + |\boldsymbol{a}_y|\bar{\boldsymbol{a}}_y + |\boldsymbol{a}_z|\bar{\boldsymbol{a}}_z \end{aligned} \tag{8.2}$$

where $\bar{\boldsymbol{a}}_x$ is one unit in the direction of the Ox axis,
$\bar{\boldsymbol{a}}_y$ is one unit in the direction of the Oy axis,
$\bar{\boldsymbol{a}}_z$ is one unit in the direction of the Oz axis.

The unit vectors along the axes are also often written as **i**, **j**, **k**. Thus we also have the notation

$$\bar{\boldsymbol{a}}_x = \mathbf{i} \quad \bar{\boldsymbol{a}}_y = \mathbf{j} \quad \bar{\boldsymbol{a}}_z = \mathbf{k} \tag{8.3}$$

Exercise 1 Express the vector OP in terms of its orthogonal unit vectors, where the coordinates of P are (25, 55, 20). The vector $OP = \boldsymbol{p}$ given by

$$\boldsymbol{p} = 25\bar{\boldsymbol{a}}_x + 55\bar{\boldsymbol{a}}_y + 20\bar{\boldsymbol{a}}_z$$

or

$$\boldsymbol{p} = 25\mathbf{i} + 55\mathbf{j} + 20\mathbf{k}$$

Returning to the triangle ABC, in a similar way the vectors $\boldsymbol{b}$ and $\boldsymbol{c}$ are written

$$\boldsymbol{b} = \boldsymbol{b}_x + \boldsymbol{b}_y + \boldsymbol{b}_z \quad \text{and} \quad \boldsymbol{c} = \boldsymbol{c}_x + \boldsymbol{c}_y + \boldsymbol{c}_z$$

Thus we can also write for the sides of the triangle ABC

$$\boldsymbol{a} + \boldsymbol{b} + \boldsymbol{c} = \boldsymbol{a}_x + \boldsymbol{a}_y + \boldsymbol{a}_z + \boldsymbol{b}_x + \boldsymbol{b}_y + \boldsymbol{b}_z + \boldsymbol{c}_x + \boldsymbol{c}_y + \boldsymbol{c}_z$$
$$= \boldsymbol{a}_x + \boldsymbol{b}_x + \boldsymbol{c}_x + \boldsymbol{a}_y + \boldsymbol{b}_y + \boldsymbol{c}_y + \boldsymbol{a}_z + \boldsymbol{b}_z + \boldsymbol{c}_z$$

The net effect of moving round the triangle is the same as moving along each of the three axes by the respective components. In this case

$$\boldsymbol{a} + \boldsymbol{b} + \boldsymbol{c} = \mathbf{0}$$

Exercise 2 The triangle *ABC* has the vertices, *A* (1, 2, 3), *B* (3, 5, 6) and C (4, 6, 8). Show that the vectors representing the sides are linked by the expression

$$\boldsymbol{a} + \boldsymbol{b} + \boldsymbol{c} = \mathbf{0}$$

From the coordinates we have

$$\boldsymbol{c} = \text{AB} = (3 - 1)\,\mathbf{i} + (5 - 2)\,\mathbf{j} + (6 - 3)\,\mathbf{k} = 2\,\mathbf{i} + 3\,\mathbf{j} + 3\mathbf{k}$$
$$\boldsymbol{a} = \text{BC} = (4 - 3)\,\mathbf{i} + (6 - 5)\,\mathbf{j} + (8 - 6)\,\mathbf{k} = \mathbf{i} + \mathbf{j} + 2\mathbf{k}$$
$$\boldsymbol{b} = \text{CA} = (1 - 4)\,\mathbf{i} + (2 - 6)\,\mathbf{j} + (3 - 8)\,\mathbf{k} = -3\,\mathbf{i} - 4\,\mathbf{j} - 5\mathbf{k}$$

Hence we see that

$$\boldsymbol{a} + \boldsymbol{b} + \boldsymbol{c} = \mathbf{0}$$

8.6 Components and direction cosines

If the vector $\boldsymbol{a}$ makes angles with the respective axes *Ox*, *Oy* and *Oz*, of

$$\alpha,\ \beta \text{ and } \gamma$$

the lengths of the respective components of $\boldsymbol{a}$ are

$$a_x = |\boldsymbol{a}|\cos\alpha \quad a_y = |\boldsymbol{a}|\cos\beta \quad a_z = |\boldsymbol{a}|\cos\gamma \tag{8.4}$$

Let the unit vectors in the three axes be respectively **i**, **j** and **k**. Thus we write the components of $\boldsymbol{a}$ in the three axes as

$$\boldsymbol{a}_x = |\boldsymbol{a}|\,\mathbf{i}\cos\alpha \quad \boldsymbol{a}_y = |\boldsymbol{a}|\,\mathbf{j}\cos\beta \quad \boldsymbol{a}_z = |\boldsymbol{a}|\,\mathbf{k}\cos\gamma$$

It will be noticed that in traditional notation

$$\boldsymbol{a}_x = |\boldsymbol{a}|L\mathbf{i} \quad \boldsymbol{a}_y = |\boldsymbol{a}|M\mathbf{j} \quad \boldsymbol{a}_z = |\boldsymbol{a}|N\mathbf{k} \tag{8.5}$$

where *L*, *M* and *N* are the direction cosines of the line. See equations (5.13).

8.7 Length of a vector

From equations (8.5) and consideration of Figure 8.2 we see that the length of the vector $|\boldsymbol{a}|$ is given by Pythagoras's theorem (2.1) as

$$|\boldsymbol{a}| = \sqrt{(a_x a_x + a_y a_y + a_z a_z)} \tag{8.6}$$

because a_x, a_y and a_z are the lengths of the orthogonal vectors $\boldsymbol{a}_x$, $\boldsymbol{a}_y$ and $\boldsymbol{a}_z$ respectively. (Remember: *Scalars* in light type, *vectors* in bold type.)

Exercise 1 Calculate the direction cosines of the vector which has components

$$\boldsymbol{a}_x = 3\,\mathbf{i} \quad \boldsymbol{a}_y = 4\,\mathbf{j} \quad \boldsymbol{a}_z = 5\,\mathbf{k}$$

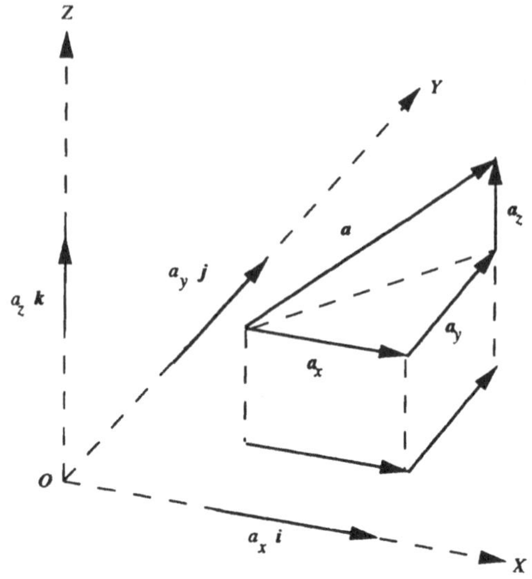

Figure 8.2

Then

$$a_x = 3 \quad a_y = 4 \quad a_z = 5$$

therefore from equation (8.6)

$$|\boldsymbol{a}| = \sqrt{(9 + 16 + 25)} = \sqrt{50} = 7.071$$

Now

$$L = \frac{a_x}{|\boldsymbol{a}|} \qquad M = \frac{a_y}{|\boldsymbol{a}|} \qquad N = \frac{a_z}{|\boldsymbol{a}|}$$

therefore

$$L = \frac{3}{7.071} = 0.4243 \qquad M = \frac{4}{7.071} = 0.5657 \qquad N = \frac{5}{7.071} = 0.7071$$

Check

$$L^2 + M^2 + N^2 = 1.0000$$

Exercise 2 Express the points A, B, C, and D as position vectors $\boldsymbol{a}$, $\boldsymbol{b}$, $\boldsymbol{c}$, $\boldsymbol{d}$, in terms of unit axis vectors **i**, **j** and **k** in an orthogonal system.

Point	x	y	z
A	11	6	2
B	1	9	0
C	3	2	0.5
D	6	10	7

The position vectors are

$$\boldsymbol{a} = 11\,\mathbf{i} + 6\,\mathbf{j} + 2\,\mathbf{k}$$
$$\boldsymbol{b} = \mathbf{i} + 9\,\mathbf{j} + 0\,\mathbf{k}$$
$$\boldsymbol{c} = 3\,\mathbf{i} + 2\,\mathbf{j} + 0.5\,\mathbf{k}$$
$$\boldsymbol{d} = 6\,\mathbf{i} + 10\,\mathbf{j} + 7\,\mathbf{k}$$

Exercise 3 Determine the lengths of the position vectors of Exercise 2.

$$|\boldsymbol{a}| = \sqrt{(121 + 36 + 4)} = 12.69$$
$$|\boldsymbol{b}| = \sqrt{(1 + 81)} = 9.06$$
$$|\boldsymbol{c}| = \sqrt{(9 + 4 + 0.25)} = 3.64$$
$$|\boldsymbol{d}| = \sqrt{(36 + 100 + 49)} = 13.60$$

Exercise 4 Determine the space vectors *AB*, *AC* and *AD*.

From the position vectors we have

$$AB = \boldsymbol{p} = \boldsymbol{b} - \boldsymbol{a} = -10\,\mathbf{i} + 3\,\mathbf{j} - 2\,\mathbf{k}$$
$$AC = \boldsymbol{q} = \boldsymbol{c} - \boldsymbol{a} = -8\,\mathbf{i} - 4\,\mathbf{j} - 1.5\,\mathbf{k}$$
$$AD = \boldsymbol{r} = \boldsymbol{d} - \boldsymbol{a} = -5\,\mathbf{i} + 4\,\mathbf{j} + 5\,\mathbf{k}$$

Exercise 5 Show that the vectors $\boldsymbol{p}$, $\boldsymbol{q}$, and $\boldsymbol{s}$ form a closed triangle.

$$AB = \boldsymbol{p} = -10\,\mathbf{i} + 3\,\mathbf{j} - 2\,\mathbf{k}$$
$$CA = \boldsymbol{q} = 8\,\mathbf{i} + 4\,\mathbf{j} + 1.5\,\mathbf{k}$$
$$BC = \boldsymbol{s} = 2\,\mathbf{i} - 7\,\mathbf{j} + 0.5\,\mathbf{k}$$

The vectors form a closed set because

$$\boldsymbol{p} + \boldsymbol{q} + \boldsymbol{s} = \mathbf{0}$$

8.8 Multiplication of a vector by a scalar

The multiplication of a vector by a scalar simply changes its *size*. It has no effect on its direction. For example

$$3\mathbf{a} = 3\,|\boldsymbol{a}|\,\bar{\boldsymbol{a}}$$

8.9 Multiplication of vectors

The rules for the multiplication of vectors were developed to solve problems in electromagnetism and other similar fields. There are two types of multiplication designed to produce particular results experienced in science. In *scalar multiplication*, the *dot product* of two vectors, the result is a *scalar, not a vector.* In *vector multiplication*, the result is a *vector*. As soon as possible, we will demonstrate to the reader their effectiveness.

8.10 Scalar multiplication or 'dot product'

The dot product, or scalar multiplication, of two vectors $\boldsymbol{a}$ and $\boldsymbol{b}$ inclined to each other at an angle U is defined to be

$$\boldsymbol{a} \cdot \boldsymbol{b} = |\boldsymbol{a}|\ |\boldsymbol{b}| \cos U \tag{8.7}$$

Exercise 1 If $U = 0°$ what is the dot product of two vectors of length 3 and 5 units? Does the order of multiplication matter?

Answers: 15 square units because $\cos 0° = 1$, and 'no', therefore

$$\boldsymbol{a} \cdot \boldsymbol{b} = \boldsymbol{b} \cdot \boldsymbol{a}$$

Exercise 2 If $U = 90°$ what is the dot product of two vectors of length 3 and 5 units?
Answer: zero, because $\cos 90° = 0$.

8.11 Dot products of orthogonal unit vectors i, j, and k

The scalar products of pairs of like and unlike orthogonal unit vectors are of great importance in what follows below. Firstly, from the results of Exercises 1 and 2 in section 8.10, there is the result that the *dot product of **like** unit vectors is unity*, because they are parallel to each other. i.e.

$$\mathbf{i} \cdot \mathbf{i} = 1 \quad \mathbf{j} \cdot \mathbf{j} = 1 \quad \mathbf{k} \cdot \mathbf{k} = 1 \tag{8.8}$$

and the *dot product of **unlike** unit vectors is zero* whatever their order, because they are orthogonal to each other, i.e.

$$\mathbf{i} \cdot \mathbf{j} = 0 \quad \mathbf{i} \cdot \mathbf{k} = 0 \quad \mathbf{j} \cdot \mathbf{k} = 0 \tag{8.9}$$

$$\mathbf{j} \cdot \mathbf{i} = 0 \quad \mathbf{k} \cdot \mathbf{i} = 0 \quad \mathbf{k} \cdot \mathbf{j} = 0 \tag{8.10}$$

8.12 Dot product in terms of components

Consider the dot product

$$\mathbf{a} \cdot \mathbf{b} = |\boldsymbol{a}|\ |\boldsymbol{b}| \cos U$$

$$\mathbf{a} \cdot \mathbf{b} = (a_x \mathbf{i} + a_y \mathbf{j} + a_z \mathbf{k}) \cdot (b_x \mathbf{i} + b_y \mathbf{j} + b_z \mathbf{k}) \tag{8.11}$$

Here a_x etc. are scalars which may be positive or negative according to the direction of the vectors. Multiplying out the right hand side, by the rules of ordinary arithmetic, gives three terms in which $\mathbf{i} \cdot \mathbf{i}$, $\mathbf{j} \cdot \mathbf{j}$ and $\mathbf{k} \cdot \mathbf{k}$ occur, each of which is unity. There are six other terms in which unlike produts occur. These are all zero. Therefore

$$\mathbf{a} \cdot \mathbf{b} = a_x b_x + a_y b_y + a_z b_z \tag{8.12}$$

Note: If we write $\boldsymbol{a}$ and $\boldsymbol{b}$ as column matrices (see Section 7.15) the dot product is given by

$$\boldsymbol{a} \cdot \boldsymbol{b} \text{ (in } \textit{vector} \text{ notation)} = \boldsymbol{a}^T \boldsymbol{b} \text{ (in } \textit{matrix} \text{ notation)} \qquad (8.13)$$

and this is how the dot product is calculated.

Exercise 1 Using the dot product, find the length of the vector (**2i, 3j, 4k**). The length of the vector $\boldsymbol{a}$ is given by

$$|\boldsymbol{a}.\boldsymbol{a}| = |\boldsymbol{a}^T\boldsymbol{a}| \text{ (in matrix notation)}$$
$$= \sqrt{(4 + 9 + 16)} = \sqrt{29} = 5.38$$

Exercise 2 Find the angle between the vectors $\boldsymbol{p}$, $\boldsymbol{q}$, and $\boldsymbol{s}$ forming a triangle. From Exercise 5 of Section 8.7, we know they close on themselves.

$$AB = \boldsymbol{p} = -10\,\mathbf{i} + 3\,\mathbf{j} - 2\,\mathbf{k}$$
$$CA = \boldsymbol{q} = 8\,\mathbf{i} + 4\,\mathbf{j} + 1.5\,\mathbf{k}$$
$$BC = \boldsymbol{s} = 2\,\mathbf{i} - 7\,\mathbf{j} + 0.5\,\mathbf{k}$$

The lengths of the vectors are

$$|\boldsymbol{p}| = 10.63 \quad |\boldsymbol{q}| = 9.07 \quad |\boldsymbol{s}| = 7.30$$

The vector $AC = -CA = -\boldsymbol{q} = -8\,\mathbf{i} - 4\,\mathbf{j} - 1.5\,\mathbf{k}$
The angle A between AB and AC (see Section 5.10) is given by

$$\cos A = LL' + NN' + MM'$$

Now from (8.5)

$$L = \frac{p_x}{|\boldsymbol{p}|} \qquad M = \frac{p_y}{|\boldsymbol{p}|} \qquad N = \frac{p_z}{|\boldsymbol{p}|}$$

$$L' = \frac{q_x}{|\boldsymbol{q}|} \qquad M' = \frac{q_y}{|\boldsymbol{q}|} \qquad N' = \frac{q_z}{|\boldsymbol{q}|}$$

therefore

$$\cos A = LL' + NN' + MM'$$

$$= \frac{p_x}{|\boldsymbol{p}|}\frac{q_x}{|\boldsymbol{q}|} + \frac{p_y}{|\boldsymbol{p}|}\frac{q_y}{|\boldsymbol{q}|} + \frac{p_z}{|\boldsymbol{p}|}\frac{q_z}{|\boldsymbol{q}|}$$

therefore

$$\cos A = \frac{\mathbf{p} \cdot \mathbf{q}}{|\boldsymbol{p}||\boldsymbol{q}|} \qquad (8.14)$$

This is the very important result (5.19) expressed in vector notation and is consistent with the definitions of equation (8.7).
In the example

$$\cos A = \frac{(10 \times 8 - 3 \times 4 + 2 \times 1.5)}{10.63 \times 9.07} = 0.7364$$

$$A = 42.57°$$

The angle B between BA and BC is given by

$$\cos B = \frac{(-\mathbf{p}).\boldsymbol{s}}{|\boldsymbol{p}||\boldsymbol{s}|} = \frac{(10\times2+3\times7+2\times0.5)}{10.63\times7.30} = 0.5412$$

$$B = 57.23°$$

The angle C between CB and CA is given by

$$\cos C = \frac{(-\mathbf{s}).\boldsymbol{q}}{|\boldsymbol{s}||\boldsymbol{q}|} = \frac{(-2\times8+7\times4-0.5\times1.5)}{7.30\times9.07} = 0.1699$$

$$C = 80.22°$$

Exercise 3 Check that the angles of the triangle add up to 180°.

$$A + B + C = 42.57° + 57.23° + 80.22° = 180.02°$$

The discrepancy is due to rounding errors in the arithmetic.

Exercise 4 Use the sine rule (see (3.9)) to check the three values of $2R$, the diameter of the circumcircle of the triangle ABC. We have

$$\frac{BC}{\sin A} = \frac{7.30}{\sin 42.57°} = 10.79 = 2R$$
$$\frac{AC}{\sin B} = \frac{9.07}{\sin 57.23°} = 10.79 = 2R$$
$$\frac{AB}{\sin C} = \frac{10.63}{\sin 80.22°} = 10.79 = 2R$$

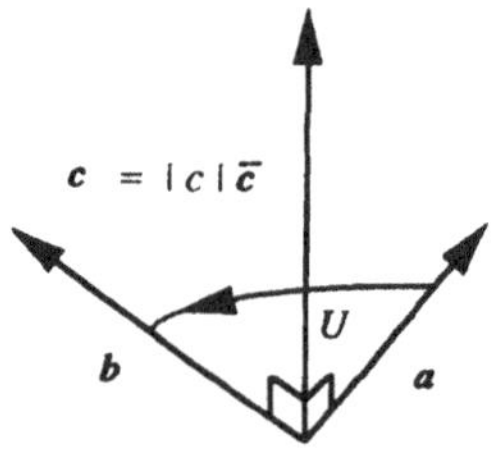

Figure 8.3

8.13 Vector multiplication or 'cross product'

The mathematical rules governing *vector multiplication* were devised to deal with problems in electromagnetism.The outcome of vector multiplication is *another vector,* lying in a direction which is orthogonal to the plane of the two original vectors. The three vectors form a right handed set as shown in Figure 8.3. Vector $\boldsymbol{a}$ makes an angle U with vector $\boldsymbol{b}$. Vector $\boldsymbol{c}$ is orthogonal to the plane formed by $\boldsymbol{a}$ and $\boldsymbol{b}$. For example if we multiply the vectors $\boldsymbol{a}$ and $\boldsymbol{b}$ in that order, vector $\boldsymbol{c}$ is obtained. To distingish vector from scalar multiplication we write

$$\boldsymbol{a} \times \boldsymbol{b} = \boldsymbol{c}$$

The vector multiplication, or *cross product*, of two vectors $\boldsymbol{a}$ and $\boldsymbol{b}$ inclined to each other at an angle U is defined to be

$$\boldsymbol{a} \times \boldsymbol{b} = |\boldsymbol{a}|\ |\boldsymbol{b}| \sin U\ \bar{\boldsymbol{c}} \qquad (8.15)$$

where $\bar{\boldsymbol{c}}$ is a unit vector in a direction orthogonal to the plane of $\boldsymbol{a}$ and $\boldsymbol{b}$. Looking along the direction of $\bar{\boldsymbol{c}}$ a clockwise rotation moves $\boldsymbol{a}$ into $\boldsymbol{b}$. The cross product is sometimes written as

$$\boldsymbol{a} \wedge \boldsymbol{b}$$

Exercise 1 If $U = 0°$ what is the cross product of two vectors of length 3 and 5 units long? Answer: zero because sin 0° = 0.

Exercise 2 If $U = 90°$ what is the cross product of two vectors of length 3 and 5 units long? Does the order of multiplication matter?
Answers: $15\bar{\boldsymbol{c}}$, because cos 90° = 1. Yes the order matters because to bring $\boldsymbol{b}$ into $\boldsymbol{a}$ requires a negative rotation

$$U = -90°$$

and since

$$\sin(-90°) = -1$$

$$\boldsymbol{b} \times \boldsymbol{a} = -15\bar{\boldsymbol{c}}$$

The general rule is

$$\boldsymbol{b} \times \boldsymbol{a} = -\boldsymbol{a} \times \boldsymbol{b} \qquad (8.16)$$

8.14 Cross products of orthogonal unit vectors i, j, and k

The cross products of pairs of like and unlike orthogonal unit vectors are of great importance in what follows below. Because like vectors are parallel, the angle between them is zero, and sin 0° = 0 so we have the rule that the *cross product of **like** unit vectors is a **zero** vector*. Therefore

$$\mathbf{i} \times \mathbf{i} = 0 \quad \mathbf{j} \times \mathbf{j} = 0 \quad \mathbf{k} \times \mathbf{k} = 0 \qquad (8.17)$$

Conversely the *cross product of* ***unlike*** *unit vectors is ± unit vector* depending on their order because sin 90° = 1

$$\mathbf{i} \times \mathbf{j} = \mathbf{k} \quad \mathbf{j} \times \mathbf{k} = \mathbf{i} \quad \mathbf{k} \times \mathbf{i} = \mathbf{j} \qquad (8.18)$$

but

$$\mathbf{j} \times \mathbf{i} = -\mathbf{k} \quad \mathbf{k} \times \mathbf{j} = -\mathbf{i} \quad \mathbf{i} \times \mathbf{k} = -\mathbf{j} \qquad (8.19)$$

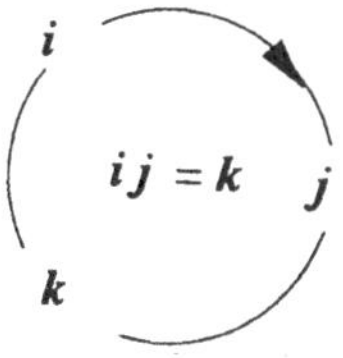

Note: It helps to remember these signs if we think of **i**, **j** and **k** written clockwise in a circle: adjacent *clockwise* combinations are positive, adjacent *anticlockwise* combinations are negative.

8.15 Cross product in terms of components

Consider the cross product

$$\boldsymbol{a} \times \boldsymbol{b} = |\boldsymbol{a}|\ |\boldsymbol{b}| \sin U\, \bar{\boldsymbol{c}} \tag{8.20}$$

$$\boldsymbol{a} \times \boldsymbol{b} = (a_x\mathbf{i} + a_y\mathbf{j} + a_z\mathbf{k}) \times (b_x\mathbf{i} + b_y\mathbf{j} + b_z\mathbf{k})$$

Multiplying the right hand side gives three terms in which $\mathbf{i} \times \mathbf{i}$, $\mathbf{j} \times \mathbf{j}$ and $\mathbf{k} \times \mathbf{k}$ occur, each of which is a zero vector. (**Note:** We use the notation $\times$ to remind us to use (8.17), (8.18) and (8.19)). There are six other terms in which unlike cross products occur. These are all plus or minus unit vectors, depending on the order. Omitting the zero terms, it is worthwhile multiplying out in full as follows

$$\begin{aligned}\boldsymbol{a} \times \boldsymbol{b} &= (a_x\mathbf{i} + a_y\mathbf{j} + a_z\mathbf{k}) \times (b_x\mathbf{i} + b_y\mathbf{j} + b_z\mathbf{k})\\ &= a_xb_y\, \mathbf{i}\times\mathbf{j} + a_xb_z\, \mathbf{i}\times\mathbf{k} + a_yb_x\, \mathbf{j}\times\mathbf{i}\\ &\quad + a_yb_z\, \mathbf{j}\times\mathbf{k} + a_zb_x\, \mathbf{k}\times\mathbf{i} + a_zb_y\, \mathbf{k}\times\mathbf{j}\end{aligned}$$

Applying the results of (8.17), (8.18) and (8.19) we have

$$\begin{aligned}\boldsymbol{a} \times \boldsymbol{b} &= a_xb_y\mathbf{k} - a_xb_z\mathbf{j} - a_yb_x\mathbf{k} + a_yb_z\mathbf{i} + a_zb_x\mathbf{j} - a_zb_y\,\mathbf{i}\\ &= +\,(a_yb_z - a_zb_y)\,\mathbf{i} + (a_zb_x - a_xb_z)\,\mathbf{j} + (a_xb_y - a_yb_x)\,\mathbf{k}\end{aligned}$$

which can be expressed as a determinant (see Section 6.6) by

$$\boldsymbol{a} \times \boldsymbol{b} = \begin{vmatrix} \mathbf{i} & \mathbf{j} & \mathbf{k} \\ a_x & a_y & a_z \\ b_x & b_y & b_z \end{vmatrix}$$

$$\boldsymbol{a} \times \boldsymbol{b} = \begin{vmatrix} a_y & a_z \\ b_y & b_z \end{vmatrix}\mathbf{i} - \begin{vmatrix} a_x & a_z \\ b_x & b_z \end{vmatrix}\mathbf{j} + \begin{vmatrix} a_x & a_y \\ b_x & b_y \end{vmatrix}\mathbf{k}$$

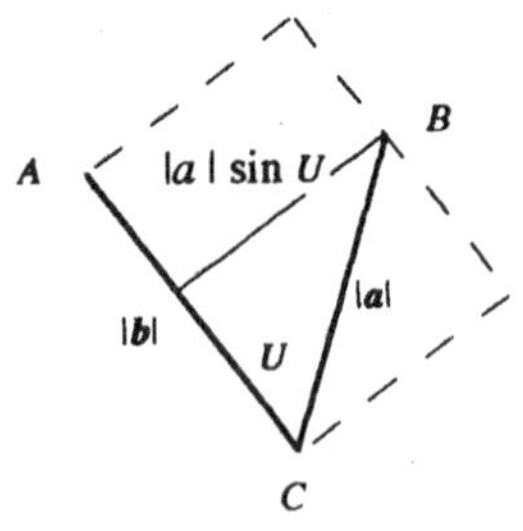

Figure 8.4

By inspection of Figure 8.4 the length of the vector $\boldsymbol{a} \times \boldsymbol{b}$ is given by

$$|\boldsymbol{a} \times \boldsymbol{b}| = |\boldsymbol{a}|\ |\boldsymbol{b}| \sin U \tag{8.21}$$

which is the area of the parallelogram formed by $\boldsymbol{a}$ and $\boldsymbol{b}$, and twice the area of the triangle *ABC*.

Exercise 1 Find the area of the triangle *ABC*, whose vertices are

Point	x	y	z
A	11	6	2
B	1	9	0
C	3	2	0.5

$$\boldsymbol{a} = CB \qquad \boldsymbol{b} = CA$$

$$a_x = x_B - x_C = -2 \quad a_y = y_B - y_C = 7 \quad a_z = z_B - z_C = -0.5$$
$$b_x = x_A - x_C = 8 \quad b_y = y_A - y_C = 4 \quad b_z = z_A - z_C = 1.5$$

$$\boldsymbol{a} \times \boldsymbol{b} = \begin{vmatrix} a_y & a_z \\ b_y & b_z \end{vmatrix} \mathbf{i} - \begin{vmatrix} a_x & a_z \\ b_x & b_z \end{vmatrix} \mathbf{j} + \begin{vmatrix} a_x & a_y \\ b_x & b_y \end{vmatrix} \mathbf{k}$$

$$\boldsymbol{a} \times \boldsymbol{b} = \begin{vmatrix} 7 & -0.5 \\ 4 & 1.5 \end{vmatrix} \mathbf{i} - \begin{vmatrix} -2 & -0.5 \\ 8 & 1.5 \end{vmatrix} \mathbf{j} + \begin{vmatrix} -2 & 7 \\ 8 & 4 \end{vmatrix} \mathbf{k}$$

$$= 12.5\,\mathbf{i} \ - 1.0\,\mathbf{j} \ - 64\,\mathbf{k}$$

$$|\boldsymbol{a} \times \boldsymbol{b}| = \sqrt{}(156.25 + 1 + 4096) = 65.2169$$

Area of triangle *ABC* = 32.608 sq units.

This result is derived in an entirely different way in Section 6.9, exercise 2.

8.16 Parallel and orthogonal vectors

If the length of the cross product vector is zero the vectors are parallel. (Contrast this result with the dot product where the reverse is true for the scalar outcome.)

8.17 The scalar triple product

The *scalar triple product* of three vectors $\boldsymbol{a}$, $\boldsymbol{b}$ and $\boldsymbol{c}$ is usefully defined as

$$\mathbf{a} \,.\, (\mathbf{b} \times \mathbf{c})$$

The outcome is a scalar, because the combined dot product and vector product is

$$\mathbf{a} \,.\, (\mathbf{b} \times \mathbf{c}) = \textbf{vector . vector = scalar}$$

Now let the vector product within the bracket be the vector $\boldsymbol{d}$; then

$$\boldsymbol{d} = \boldsymbol{b} \times \mathbf{c} = \begin{vmatrix} b_y & b_z \\ c_y & c_z \end{vmatrix} \mathbf{i} - \begin{vmatrix} b_x & b_z \\ c_x & c_z \end{vmatrix} \mathbf{j} + \begin{vmatrix} b_x & b_y \\ c_x & c_y \end{vmatrix} \mathbf{k}$$

$$= d_x\mathbf{i} + d_y\mathbf{j} + d_z\mathbf{k}$$

The dot product of $\boldsymbol{a}$ with $\boldsymbol{d}$ is

$$\mathbf{a}\,.\,\mathbf{d} = (a_x\mathbf{i} + a_y\mathbf{j} + a_z\mathbf{k})\,.\,(d_x\mathbf{i} + d_y\mathbf{j} + d_z\mathbf{k})$$

$$= a_xd_x + a_yd_y + a_zd_z$$

$$= a_x\begin{vmatrix} b_y & b_z \\ c_y & c_z \end{vmatrix} - a_y\begin{vmatrix} b_x & b_z \\ c_x & c_z \end{vmatrix} + a_z\begin{vmatrix} b_x & b_y \\ c_x & c_y \end{vmatrix}$$

therefore

$$\boldsymbol{a}\,.\,(\boldsymbol{b}\times\boldsymbol{c}) = \begin{vmatrix} a_x & a_y & a_z \\ b_x & b_y & b_z \\ c_x & c_y & c_z \end{vmatrix}$$

This is the *volume* of the solid whose edges are represented by the vectors $\boldsymbol{a}$, $\boldsymbol{b}$ and $\boldsymbol{c}$. This solid, which is called a *parallelepiped,* is illustrated in Figure 8.5. We now derive this important result which was derived in a different way in Section 6.9.

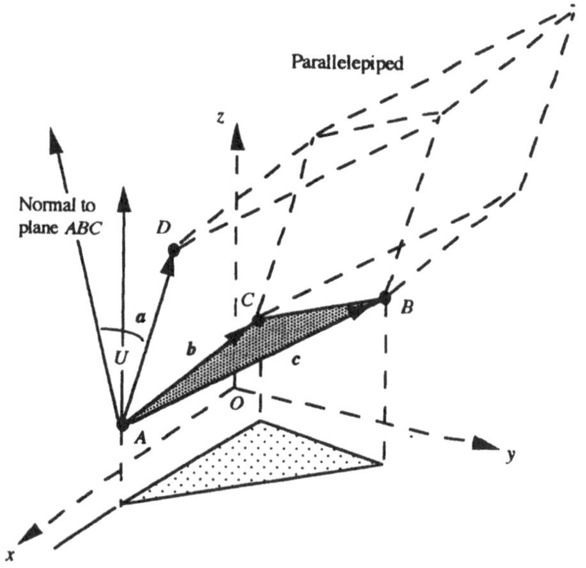

Figure 8.5

The area of the base parallelogram is given by the length of the vector product $\boldsymbol{b}\times\boldsymbol{c}$. Now the volume of the solid is the vertical height multiplied by the area of the base. If the vector $\boldsymbol{a}$ makes an angle U with the normal to the plane ABC,

$$\text{the vertical height} = |\boldsymbol{a}|\,\cos U$$

The volume of the solid is therefore

$$V = (|\boldsymbol{a}|\,\cos U)\,(|\boldsymbol{d}|) = |\boldsymbol{a}|\,|\boldsymbol{d}|\,\cos U$$

But U is the angle between $\boldsymbol{a}$ and $\boldsymbol{d}$ therefore V is in the form of a scalar product, and

$$V = \mathbf{a} \cdot \mathbf{d} = \mathbf{a} \cdot (\mathbf{b} \times \mathbf{c}) \tag{8.22}$$

Exercise 1 Find the volume of the parallelepiped whose edges are the vectors

	i	**j**	**k**
$AB = \boldsymbol{b}$	−10	3	−2
$AC = \boldsymbol{c}$	−8	−4	−1.5
$AD = \boldsymbol{a}$	−5	4	5

V = the value of the determinant

$$\begin{vmatrix} -10 & 3 & -2 \\ -8 & -4 & -1.5 \\ -5 & 4 & 5 \end{vmatrix} = 386.5 \text{ cubic units}$$

We have shown in Section 6.9 Exercise 5 that V is also six times the volume of the tetrahedron $ABCD$.

8.18 Volume in terms of position vectors

The volume of the parallelipiped may also be expressed in term of the position vectors of its apices ABC and D. If the position vectors for A, B, C and D are $\boldsymbol{p}$, $\boldsymbol{q}$, $\boldsymbol{r}$ and $\boldsymbol{s}$ say, then we have

$$AD = \boldsymbol{a} = \boldsymbol{s} - \boldsymbol{p} \quad AB = \boldsymbol{b} = \boldsymbol{q} - \boldsymbol{p} \quad AC = \boldsymbol{c} = \boldsymbol{r} - \boldsymbol{p}$$

The volume is then calculated as before from

$$V = \mathbf{a} \cdot \mathbf{d} = \mathbf{a} \cdot (\mathbf{b} \times \mathbf{c})$$

Exercise 1 Calculate the volume of the tetrahedron $ABCD$ from the coordinates

Point	x	y	z
A	11	6	2
B	1	9	0
C	3	2	0.5
D	6	10	7

Here

$$\boldsymbol{p} = 11\,\mathbf{i} + 6\,\mathbf{j} + 2\,\mathbf{k}$$
$$\boldsymbol{s} = 6\,\mathbf{i} + 10\,\mathbf{j} + 7\,\mathbf{k}$$
$$\boldsymbol{q} = \mathbf{i} + 9\,\mathbf{j} + 0\,\mathbf{k}$$
$$\boldsymbol{r} = 3\,\mathbf{i} + 2\,\mathbf{j} + 0.5\,\mathbf{k}$$

from which we have as before

	i	**j**	**k**
$AB = \boldsymbol{b}$	−10	3	−2
$AC = \boldsymbol{c}$	−8	−4	−1.5
$AD = \boldsymbol{a}$	−5	4	5

8.19 Coplanarity of vectors

If three vectors ***a***, ***b***, and ***c*** all lie in one plane (are coplanar) then the volume of the parallelepiped formed by them is zero, and therefore the *triple scalar product is zero* or

$$\mathbf{a}\,.\,(\mathbf{b}\times\mathbf{c}) = 0 \qquad (8.23)$$

Exercise 1 Show that the following vectors are coplanar

$$AB = \boldsymbol{b} = -\,10\,\mathbf{i} + 3\,\mathbf{j} - 2\,\mathbf{k}$$
$$AC = \boldsymbol{c} = -2\,\mathbf{i} + 7\,\mathbf{j} - 0.5\,\mathbf{k}$$
$$AD = \boldsymbol{a} = -8\,\mathbf{i} - 4\,\mathbf{j} - 1.5\,\mathbf{k}$$

Evaluating the determinant we have

$$\begin{aligned}\text{Det} &= -\,10(-7\times1.5 - 4\times0.5) - 3(2\times1.5 - 8\times0.5) - 2(2\times4 + 8\times7)\\ &= (-\,10\times-12.5) - (3\times-1) - (2\times64)\\ &= 125 + 3 - 128\\ &= 0\end{aligned}$$

Therefore the vectors are coplanar. This result has applications in photogrammetry.

SUMMARY OF KEY WORDS

scalar, vector, unit vector, free vector, position vector, orthogonal unit vectors, vector multiplication, scalar multiplication or 'dot product', vector multiplication or 'cross product', scalar triple product, parallelepiped

SUMMARY OF FORMULAE

Vectors $\quad \boldsymbol{a} = |\boldsymbol{a}|\bar{\boldsymbol{a}}$ (8.1)

Components

$$\boldsymbol{a} = \boldsymbol{a}_x + \boldsymbol{a}_y + \boldsymbol{a}_z$$

$$= |\boldsymbol{a}_x|\bar{\boldsymbol{a}}_x + |\boldsymbol{a}_y|\bar{\boldsymbol{a}}_y + |\boldsymbol{a}_z|\bar{\boldsymbol{a}}_z \qquad (8.2)$$

Unit vectors

$$\bar{\boldsymbol{a}}_x = \mathbf{i} \quad \bar{\boldsymbol{a}}_y = \mathbf{j} \quad \bar{\boldsymbol{a}}_z = \mathbf{k} \qquad (8.3)$$

$$a_x = |\boldsymbol{a}|\cos\alpha \quad a_y = |\boldsymbol{a}|\cos\beta \quad a_z = |\boldsymbol{a}|\cos\gamma \qquad (8.4)$$

$$\boldsymbol{a}_x = |\boldsymbol{a}|L\mathbf{i} \quad \boldsymbol{a}_y = |\boldsymbol{a}|M\mathbf{j} \quad \boldsymbol{a}_z = |\boldsymbol{a}|N\mathbf{k} \qquad (8.5)$$

Length

$$|\boldsymbol{a}| = \sqrt{(a_x a_x + a_y a_y + a_z a_z)} \qquad (8.6)$$

Dot product

$$\boldsymbol{a}\,.\,\boldsymbol{b} = |\boldsymbol{a}|\;|\boldsymbol{b}|\cos U \qquad (8.7)$$

Orthogonal unit vectors

$$\mathbf{i}\,.\,\mathbf{i} = 1 \quad \mathbf{j}\,.\,\mathbf{j} = 1 \quad \mathbf{k}\,.\,\mathbf{k} = 1 \qquad (8.8)$$

$$\mathbf{i}\,.\,\mathbf{j} = 0 \quad \mathbf{i}\,.\,\mathbf{k} = 0 \quad \mathbf{j}\,.\,\mathbf{k} = 0 \qquad (8.9)$$

$$\mathbf{j}\,.\,\mathbf{i} = 0 \quad \mathbf{k}\,.\,\mathbf{i} = 0 \quad \mathbf{k}\,.\,\mathbf{j} = 0 \qquad (8.10)$$

$$\mathbf{a}\,.\,\mathbf{b} = (a_x\,\mathbf{i} + a_y\,\mathbf{j} + a_z\,\mathbf{k})\,.\,(b_x\,\mathbf{i} + b_y\,\mathbf{j} + b_z\,\mathbf{k}) \qquad (8.11)$$

$$\mathbf{a}\,.\,\mathbf{b} = a_x b_x + a_y b_y + a_z b_z \qquad (8.12)$$

Angle between vectors

$$\cos A = \frac{\mathbf{p}\,.\,\mathbf{q}}{|\boldsymbol{p}||\boldsymbol{q}|} \qquad (8.14)$$

Cross product

$$\boldsymbol{a} \times \boldsymbol{b} = |\boldsymbol{a}|\;|\boldsymbol{b}|\sin U\;\bar{\boldsymbol{c}} \qquad (8.15)$$

$$\boldsymbol{b} \times \boldsymbol{a} = -\boldsymbol{a} \times \boldsymbol{b} \qquad (8.16)$$

Orthogonal unit vectors

$$\mathbf{i} \times \mathbf{i} = \mathbf{0} \quad \mathbf{j} \times \mathbf{j} = \mathbf{0} \quad \mathbf{k} \times \mathbf{k} = \mathbf{0} \qquad (8.17)$$

$$\mathbf{i} \times \mathbf{j} = \mathbf{k} \quad \mathbf{j} \times \mathbf{k} = \mathbf{i} \quad \mathbf{k} \times \mathbf{i} = \mathbf{j} \qquad (8.18)$$

$$\mathbf{j} \times \mathbf{i} = -\mathbf{k} \quad \mathbf{k} \times \mathbf{j} = -\mathbf{i} \quad \mathbf{i} \times \mathbf{k} = -\mathbf{j} \qquad (8.19)$$

Area of parallelogram

$$|\boldsymbol{a} \times \boldsymbol{b}| = |\boldsymbol{a}|\;|\boldsymbol{b}|\sin U \qquad (8.21)$$

Scalar triple product
Vol. of parallelepiped

$$V = \mathbf{a}\,.\,\mathbf{d} = \mathbf{a}\,.\,(\mathbf{b} \times \mathbf{c}) \qquad (8.22)$$

Chapter 9
Calculus

9.1 Calculus

The *differential and integral calculus* are very important in mathematics, both for their own sake, as in curve tracing, and as a means of deriving other results, such as the determination of areas and volumes. The differential calculus in particular has many applications to surveying and map making, particularly in the treatment of quality control and error propagation. Once the notation has been grasped, the beginner should find the subject quite straightforward, although the purpose of it all may appear to be something of a puzzle. It is only when the technique is applied to problems that the power of the calculus is made clear and the initial effort made to learn the rules is rewarded.

9.2 Functions

Consider the following expressions

$$y = x \quad y = x^2 \quad y = x^3$$

In each case we say that *y is a function of x*. By this we mean that if we give x a value, say $x = 2$, then y will have a corresponding value, in these cases $y = 2$, $y = 4$ and $y = 8$.

Without specifying what the function is we can also say

$$y = f(x)$$

This just states that *y is dependent upon x* in some way. In the three cases above we have specifically that

$$f_1(x) = x \quad f_2(x) = x^2 \quad f_3(x) = x^3$$

The notation $f_1(x)$ just means 'the first function of x', $f_2(x)$ the 'second function of x' and so on. Any letter can be used to mean a function of any *variable* such as x. For example we might put ' let P be a function of s', or

$$P = F(s) \quad \text{or} \quad P = \phi(s)$$

Generally the letters f and F are used to denote functions.

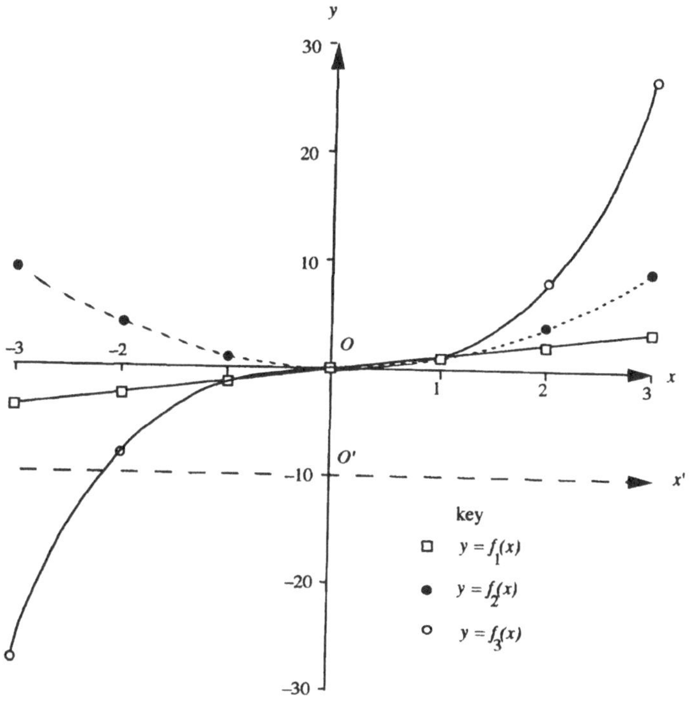

Figure 9.1

Figure 9.1 shows a sketch of the graphs of $f(x)$ in each of these three cases, for values of x between –3 and +3. The word 'sketch' is used to indicate that an accurate graph is not required. We first draw up a table of respective values.

Table 9.1

x	–3	–2	–1	0	1	2	3
Function							
$y = f_1(x) = x$	–3	–2	–1	0	1	2	3
$y = f_2(x) = x^2$	9	4	1	0	1	4	9
$y = f_3(x) = x^3$	–27	–8	–1	0	1	8	27

These functions are said to be *continuous* because you could draw them on a sheet of paper without lifting the pencil and they have no kinks.

Exercise 1 Verify from the above table that

$$f_1(3) = 3 \quad f_1(2) = 2 \quad f_2(3) = 9 \quad f_2(2) = 4 \quad f_3(-3) = -27 \quad f_3(2) = 8$$

The notation means that the value of f_1 when $x = 3$ is 3, the value of f_2 when $x = 3$ is 9 and so on.

Exercise 2 Show that

$$f_1(-2) + f_3(2) = -2 + 8 = 6$$

and that

$$f_1(-2).f_3(2) = -2 \times 8 = -16$$

Exercise 3 Plot corresponding values of y and x for the functions in Table 9.1. These are sketched in Figure 9.1. The following factors should be noted.

- All three graphs pass through the origin.
- Because all values of $f_2(x)$ are *positive*, $f_2(x)$ is called an *even function* of x.
- Because values of f_1 and f_3 can be *positive and negative* they are called *odd functions* of x.

Exercise 4 Sketch the graphs of the following functions

$$f_4(x) = x + 10 \quad f_5(x) = x^2 + 10 \quad f_6(x) = x^3 + 10$$

Note: It is is easier to draw a new $O'x'$ axis 10 units below the original Ox axis on the first graph than to replot all the graphs. This new axis is shown on Figure 9.1 as a broken line.

The graph of $y = f_4(x + 3)$ however would need to be redrawn because

$$y = f_4(x + 3) = (x + 3) + 10 = x + 13$$

when

$$x = 1 \quad y = 14$$

and

$$y = f_5(x + 3) = (x + 3)^2 + 10 = x^2 + 6x + 9 + 10$$

when

$$x = 1 \quad y = 26$$

In Section 9.4 we will show how to find the slope (gradient) of a curve at any point and so plot its shape more closely.

Exercise 5 Evaluate $y = F(x) = \sin x$, when $x = 90°$.

$$y = F(90°) = \sin 90° = 1$$

9.3 Limits of functions

We shall soon need to use the idea of a *limit of a function*. Consider the following function

$$F(x) = \frac{x^2 - 1}{x - 1} \qquad (9.1)$$

When

$$x = 1$$

$$F(x) = \frac{0}{0}$$

which has no meaning, and yet as x approaches closer and closer to 1, $F(x)$ does have a meaning. We can see this from the following table

x	0.9	0.99	0.999	1.001	1.01	1.1
$F(x)$	1.9	1.99	1.999	2.001	2.01	2.1

Other than when $x = 1$, $F(x)$ is nearly 2. We write this fact in the following way

$$\text{Lim}\frac{x^2-1}{x-1} \text{ as } x \to 1 = 2$$

In words this means 'the limit of $F(x)$ as x tends to 1 is 2'

Note: Often the limit can be determined by first recasting the function $F(x)$ in an appropriate form. In this case, we can factorise the numerator and simplify the function as follows

$$F(x) = \frac{x^2-1}{x-1} = \frac{(x-1)(x+1)}{x-1} = x+1$$

and therefore

$$F(x) = 2 \text{ when } x = 1$$

or more neatly

$$F(1) = 2$$

9.4 Tangents to curves

The tangent to a curve is the straight line which just touches the curve at some point. We will consider the function

$$y = f_2(x) = x^2$$

Exercise 1 Sketch the tangent to the curve at point P when $x = +1$. See Figure 9.2. When $x = 1$, $y = 1$. To find the tangent at the point P (1, 1) consider two points close to P on either side of P, for example at Q when $x = 1 - 0.1 = 0.9$ and at R where $x = 1 + 0.1 = 1.1$. The corresponding values of y are 0.81 and 1.21 giving

$$Q\ (0.9, 0.81) \quad \text{and} \quad R\ (1.1, 1.21)$$

The slope or gradient of QR is then

$$\frac{1.21-0.81}{1.1-0.9} = \frac{0.4}{0.2} = 2$$

This gradient is approximately that of the curve at P.

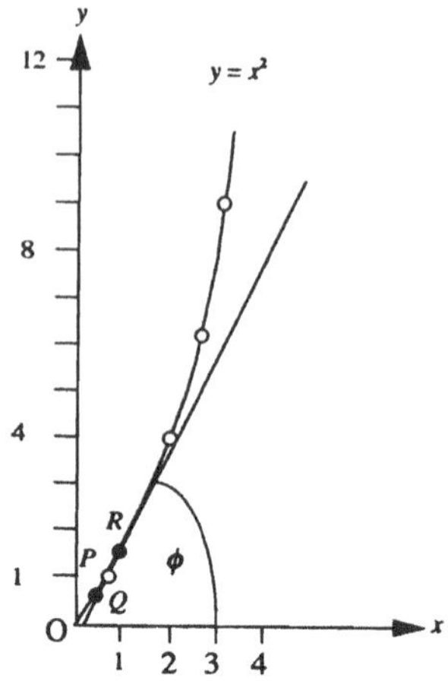

Figure 9.2

Exercise 2 In just the same way, sketch the tangent to the curve at the point (2, 4).
Here we find that the slope is

$$\frac{4.41-3.61}{2.1-1.9}=\frac{0.8}{0.2}=4$$

Thus the approximate gradient of the curve is 4 at the point (2, 4)

Exercise 3 Find the general expression for the gradient to the curve at any point $P(x, y)$. In this case we let the changes to x and y be small amounts called δx and δy. (In the numerical example above $\delta x = 0.1$). Note that we cannot separate the δ from the x. The notation 'δx' means 'a small change to x close to x'.

Because

$$y = f(x) = x^2 \tag{9.2}$$

if we alter x by δx, y is altered by δy thus we have

$$y + \delta y = (x + \delta x)^2 = (x + \delta x)(x + \delta x) = x^2 + 2x.\delta x + \delta x^2$$

And if we alter y by $-\delta y$ we have from (9.2)

$$y - \delta y = (x - \delta x)^2 = (x - \delta x)(x - \delta x) = x^2 - 2x.\delta x + \delta x^2$$

Thus the gradient of the tangent is approximately

$$\frac{y+\delta y-(y-\delta y)}{x+\delta x-(x-\delta x)}=\frac{2\delta y}{2\delta x}=\frac{\delta y}{\delta x}=\frac{(x+\delta x)^2-(x-\delta x)^2}{x+\delta x-(x-\delta x)}$$

$$=\frac{x^2+2x.\delta x+\delta x^2-\left(x^2-2x.\delta x+\delta x^2\right)}{2\delta x}=\frac{4x.\delta x}{2\delta x}=2x$$

Hence we have the important result that the gradient of the curve $y = f(x) = x^2$ is $2x$. In the case of this function, the ratio of the changes δy and δx is always equal to $2x$ even if they are quite large. Only when they are small can we say the tangent approximates to the curve at P. If we make them smaller and smaller approaching zero, their ratio still remains $2x$. We write this information in the following way

$$\text{Lim}\frac{\delta y}{\delta x} \text{ as } \delta x \to 0 \text{ is } 2x$$

and we write this special limiting case of $\delta y/\delta x$ as dy/dx.

To summarise, we can say that if

$$y = f(x) = x^2$$

then the gradient is given by

$$\frac{dy}{dx} = \text{Lim}\frac{\delta y}{\delta x} \text{ as } \delta x \to 0 = 2x$$

or even more shortly as

$$\frac{d\left(x^2\right)}{dx} = 2x \qquad (9.3)$$

This process is called *differentiation of y or $f(x)$ with respect to x*. It is often abbreviated to the form

$$f'(x) = \frac{dy}{dx}$$

Exercise 4 If $f(x) = x^2$, show that $f'(1) = 2$, and $f'(2) = 4$. Here $f'(x) = 2x$. The meaning of '$f'(1)$' is the value of $f'(x)$ when $x = 1$. These results were already found numerically. The angles which the tangents make with the x axis are respectively

$$\tan^{-1}(1) = 45° \quad \text{and} \quad \tan^{-1}(2) = 63.4°$$

Verify these results by drawing.

Exercise 5 Differentiate $y = f(x) = x^3$ from first principles. This means that we have to derive the result in exactly the same way as before. The process of selecting points close to P and equating the gradient reduces to

$$\frac{y + \delta y - (y - \delta y)}{x + \delta x - (x - \delta x)} = \frac{2\delta y}{2\delta x} = \frac{\delta y}{\delta x}$$

$$= \frac{(x + \delta x)^3 - (x - \delta x)^3}{x + \delta x - (x - \delta x)}$$

$$= \frac{x^3 + 3x^2.\delta x + 3x.\delta x^2 + \delta x^3 - \left(x^3 - 3x^2.\delta x + 3x.\delta x^2 - \delta x^3\right)}{2\delta x}$$

$$= \frac{6x^2.\delta x + 2\delta x^3}{2\delta x} = 3x^2 + \delta x^2$$

Thus, because

$$\delta x^2 \to 0$$

the gradient of the curve is obtained from

$$\frac{dy}{dx} = \text{Lim}\frac{\delta y}{\delta x} \text{ as } \delta x \to 0 = 3x^2$$

and we may write this as

$$\frac{d\left(x^3\right)}{dx} = 3x^2 \tag{9.4}$$

Exercise 6 If $f(x) = x^3$, verify that $f'(1) = 3$, and $f'(2) = 12$. The angles which the tangents make with the x axis are respectively

$$\tan^{-1}(3) = 71.6° \quad \text{and} \quad \tan^{-1}(12) = 85.2°$$

9.5 General case

The result for the general case where $y = x^n$ and n is an integer (whole number) is

$$\frac{dy}{dx} = nx^{n-1} \tag{9.5}$$

Exercise 1 Show that, if $f(x) = x^n$, and $n = 2$, $f'(2) = 4$, and $f'(3) = 6$.

$$f(x) = x^2 \quad f'(x) = 2x \quad \text{therefore} \quad f'(2) = 4, \text{ and } f'(3) = 6$$

Exercise 2 If again $f(x) = x^n$ and $n = 5$ show that $f'(2) = 80$

$$f'(x) = nx^{n-1} = 5x^4 \quad \text{and} \quad f'(2) = 5 \times 2^4 = 5 \times 2 \times 2 \times 2 \times 2 = 80$$

9.6 Proof of the general result

The proof of the general result follows the same method used for the previous functions

$$f_2(x) = x^2 \quad \text{and} \quad f_3(x) = x^3$$

We consider the gradient of the line between two points either side of $f(x)$ at $f(x + \delta x)$ and at $f(x - \delta x)$. Thus we have

$$\frac{\delta y}{\delta x} = \frac{(x+\delta x)^n - (x-\delta x)^n}{x+\delta x-(x-\delta x)}$$

The terms $(x + \delta x)^n$ and $(x - \delta x)^n$ are expanded by the binomial theorem (see Section 1.25) to give

$$\frac{\delta y}{\delta x} = \frac{x^n + nx^{n-1}\delta x + \ldots - \left(x^n - nx^{n-1}\delta x - \ldots\right)}{2\delta x}$$

$$= \frac{2nx^{n-1}\delta x}{2\delta x} \text{ plus terms in higher powers of } \delta x$$

Thus

$$\frac{dy}{dx} = \text{Lim}\frac{\delta y}{\delta x} \text{ as } \delta x \to 0 \;=\; nx^{n-1}$$

because δx and its higher powers tend to zero.

Note: If the function to be differentiated is multiplied by a constant A, i.e. if

$$y = F(x) \;= Ax^n$$

then

$$\frac{dy}{dx} = F'(x) = nAx^{n-1} \tag{9.6}$$

This may also be written in the form

$$\frac{d(F(x))}{dx} = F'(x)$$

Note: We shall discuss later the reverse process in which we know $F'(x)$ and wish to find $F(x)$. (See Section 9.23.)

9.7 Multiple problems

The differentiation of a multiple expression, or polynominal, is treated term by term as follows. If

$$y = F(x) = A + Bx + Cx^2 + Dx^3$$

$$\frac{dy}{dx} = F'(x) = B + 2Cx + 3Dx^2$$

because A is a constant

$$\frac{dA}{dx} = 0$$

Exercise 1 Differentiate the following expression with respect to x

$$y \;= 3 + 4x - 5x^2 + 6x^3$$

$$\frac{dy}{dx} = F'(x) = 4 - 10x + 18x^2$$

Exercise 2 Sketch the curve $y = 3 + 4x - 5x^2 + 6x^3$ in the vicinity of the point ($x = 1, y = 8$) and verify by drawing that its gradient is 12 at this point.

Exercise 3 Sketch the graph of the function $F'(x) = 4 - 10x + 18x^2$ and demonstrate that its gradient at the point where $x = 1$ is 26.

9.8 Second and higher differentials

It is quite possible to differentiate the function $F'(x) = 4 - 10x + 18x^2$ with respect to x. Thus we write

$$\frac{d(F'(x))}{dx} = F''(x) = \frac{d(4-10x+18x^2)}{dx} = -10 + (2)(18)x = -10 + 36x$$

The notation for this *second differential* of the original function is $F''(x)$. An alternative notation is to write

$$\frac{d^2y}{dx^2}$$

We can go on differentiating indefinitely or until a constant is obtained. For example the third differential of y with respect to x is written

$$F'''(x) \quad \text{or} \quad \frac{d^3y}{dx^3}$$

and the general case of the nth differential of y with respect to x is written

$$F^n(x) \quad \text{or} \quad \frac{d^ny}{dx^n}$$

Exercise 1 Find the third differential of y with respect to x of the function

$$y = 3x^5$$

and show that this equals 720 when $x = 2$.

$$F'(x) = 5 \times 3x^4$$

$$F''(x) = 4 \times 5 \times 3x^3$$

$$F'''(x) = 3 \times 4 \times 5 \times 3x^2 = 180x^2$$

When $x = 2$

$$F'''(x) = 180x^2 = 180(4) = 720$$

Note: The differentiations have to be carried out *before* the value of the function is substituted.

9.9 Maclaurin's theorem

In the above exercises we have been building up knowledge of the differentiation of polynomials representing curves or other useful functions. Two very important theorems were developed by two 18th century mathematicians, Maclaurin and Taylor. First we deal with *Maclaurin's theorem*. Many (though not all) curves can be expressed as a polynomial of the form

$$y = F(x) = A + Bx + Cx^2 + Dx^3 + Ex^4 + \ldots + Px^n$$

Differentiating with respect to x we have

$$F'(x) = B + 2Cx + 3Dx^2 + 4Ex^3 + \dots + nPx^{n-1}$$

Suppose we now differentiate $F'(x)$ with respect to x, we write this as

$$F''(x) = 2C + 2 \times 3Dx + \dots + (n-1)nPx^{n-2}$$

and if we again differentiate with respect to x we obtain

$$F'''(x) = 2 \times 3D + \dots + (n-2)(n-1)n\,Px^{n-3}$$

Examining the values of $F(x)$, $F'(x)$, ... when $x = 0$, we can see at once that

$$F(0) = A,\ F'(0) = B,\ F''(0) = 2C,\ F'''(0) = 2 \times 3D,\ F''''(0) = 2 \times 3 \times 4E$$

and so on. Therefore we can express the original polynomial

$$y = F(x) = A + Bx + Cx^2 + Dx^3 + Ex^4 + \dots + Px^n$$

in the form

$$y = F(x)$$

$$= F(0) + F'(0)x + \frac{F''(0)}{1\times 2}x^2 + \frac{F'''(0)}{1\times 2\times 3}x^3 + \frac{F''''(0)}{1\times 2\times 3\times 4}x^4 + \dots \frac{F^n(0)}{1\times 2\dots(n-1)n}x^n$$

and remembering that 3! (factorial 3) = 1 × 2 × 3, we have

$$y = F(x)$$

$$= F(0) + F'(0)x + \frac{F''(0)}{2!}x^2 + \frac{F'''(0)}{3!}x^3 + \frac{F''''(0)}{4!}x^4 + \dots \frac{F^n(0)}{n!}x^n \qquad (9.7)$$

This expression (9.7) is Maclaurin's theorem.

Exercise 1 There will be an important exercise using Maclaurin's theorem after we have obtained the derivatives of $\sin x$ and $\cos x$.

9.10 Differentiation of trigonometrical functions $\sin x$ and $\cos x$

To find the derivative of $\sin x$ with respect to x, we must first consider the important limit

$$\text{as } x \to 0,\ \text{Lim}\ \frac{\sin x}{x} = 1 \qquad (9.8)$$

If we just write

$$\text{as } x \to 0,\ \text{Lim}\ \frac{0}{0}$$

this has no meaning. So we must consider the matter more carefully by a geometrical method.

Consider Figure 9.3 in which AC is the arc of a circle, centre O and radii $OA = OC = R$. AB is a tangent to the circle at A. CD is perpendicular to OA. The angle subtended at the centre of the circle is x radians. (See Section 2.13.)

By inspection

CD is less than arc AC, and arc AC is less than AB

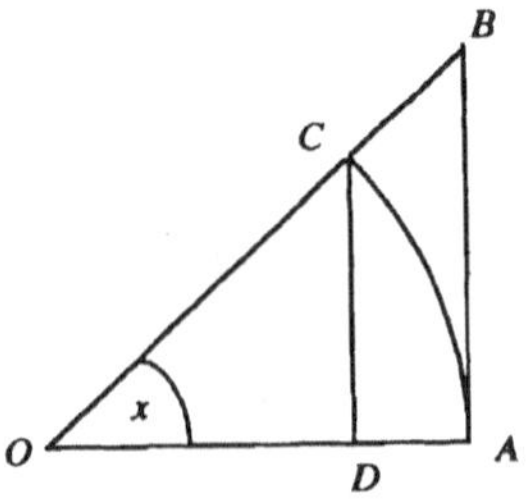

Figure 9.3

Expressed more neatly

$$CD < \text{arc}\,AC < AB$$

The sign '<' means 'less than', and note we omit the trivial case when $x = 0$.

Dividing throughout by the radius R

$$\frac{CD}{R} < \frac{\text{arc}\,AC}{R} < \frac{AB}{R}$$

$$\frac{CD}{OC} < \frac{\text{arc}\,AC}{R} < \frac{AB}{OA}$$

$$\sin x < x < \tan x$$

where x is in radians.

Exercise 1 Using your calculator demonstrate that $\sin x < x < \tan x$ for angles less than 90°.

If $x = 10°$ $\quad \sin x = 0.1736 \quad x = \frac{10\pi}{180} = 0.1745 \quad \tan x = 0.1763$

If $x = 70°$ $\quad \sin x = 0.9397 \quad x = \frac{70\pi}{180} = 1.2217 \quad \tan x = 2.7475$

If $x = 1°$ $\quad \sin x = 0.017\,452 \quad x = \frac{1\pi}{180} = 0.017\,453 \quad \tan x = 0.017455$

Consider the expression again

$$\sin x < x < \tan x$$

$$\sin x < x < \frac{\sin x}{\cos x}$$

Dividing throughout by $\sin x$ gives

$$1 < \frac{x}{\sin x} < \frac{1}{\cos x}$$

inverting gives

$$1 > \frac{\sin x}{x} > \frac{\cos x}{1}$$

The sign '>' means 'greater than'.

Now $\cos x = 1$ if $x = 0$, thus as x tends to zero $\sin x/x$ is squeezed between 1 and 1 or this means that

$$\text{as } x \to 0, \text{Lim} \frac{\sin x}{x} = 1$$

Exercise 2 Given that

$$1 < 2 < 3$$

Show that by inverting

$$\frac{1}{1} > \frac{1}{2} > \frac{1}{3}$$

9.11 Derivative of sin *x*

Let

$$y = F(x) = \sin x$$

then, from 3.14

$$y + \delta y = F(x + \delta x) = \sin(x + \delta x) = \sin x \cos \delta x + \cos x \sin \delta x$$
$$y - \delta y = F(x - \delta x) = \sin(x - \delta x) = \sin x \cos \delta x - \cos x \sin \delta x$$

Subtracting gives

$$2\delta y = 2\cos x \sin \delta x$$

and dividing by $2\delta x$

$$\frac{\delta y}{\delta x} = \cos x \frac{\sin \delta x}{\delta x}$$

then

$$\text{as } \delta x \to 0 \text{ Lim} \frac{\delta y}{\delta x} = \frac{dy}{dx} = \cos x$$

because

$$\text{as } x \to 0 \text{ Lim} \frac{\sin \delta x}{\delta x} = 1$$

Thus we have the interesting and important result that

$$\frac{d(\sin x)}{dx} = \cos x \qquad (9.9)$$

9.12 Derivative of cos *x*

We follow the same procedure as for $\sin x$. Suppose

$$y = F(x) = \cos x$$

then, remembering formulae (3.13) and (3.15), we have

$$y + \delta y = F(x + \delta x) = \cos(x + \delta x) = \cos x \cos \delta x - \sin x \sin \delta x$$
$$y - \delta y = F(x - \delta x) = \cos(x - \delta x) = \cos x \cos \delta x + \sin x \sin \delta x$$

Subtracting gives

$$2\delta y = -2\sin x \sin \delta x$$

and dividing by δx

$$\frac{\delta y}{\delta x} = -\sin x \frac{\sin \delta x}{\delta x}$$

then

$$\text{as } \delta x \to 0 \text{ Lim} \frac{\delta y}{\delta x} = \frac{dy}{dx} = -\sin x$$

because

$$\text{as } x \to 0 \text{ Lim} \frac{\sin \delta x}{\delta x} = 1$$

Thus we have another interesting result that

$$\frac{d(\cos x)}{dx} = -\sin x \qquad (9.10)$$

Note the change of sign.

Exercise 1 Derive a series for $\sin x$ using Maclaurin's theorem (9.7). Let

$$y = F(x) = \sin x$$

$$= F(0) + F'(0)x + \frac{F''(0)}{2!}x^2 + \frac{F'''(0)}{3!}x^3 + \frac{F''''(0)}{4!}x^4 + \ldots \frac{F^n(0)}{n!}x^n \quad (9.11)$$

The successive differentials, applying (9.9) and (9.10), are

$$F'(x) = \cos x \quad F''(x) = -\sin x \quad F'''(x) = -\cos x \quad F''''(x) = \sin x$$

evaluating these functions for $x = 0$ we have

$$F(0) = \sin 0 = 0 \quad \text{and}$$

$$F'(0) = \cos 0 = 1 \quad F''(0) = -\sin 0 = 0$$

$$F'''(0) = -\cos 0 = -1 \quad F''''(0) = \sin 0 = 0$$

Substituting in (9.11) we obtain the series for $\sin x$

$$\sin x = x - \frac{x^3}{3!} + \frac{x^5}{5!} - \frac{x^7}{7!} - \ldots \qquad (9.12)$$

Remember that x is in radians. Note also that alternate terms are positive and negative, and that only odd powers of x are present.

Exercise 2 Calculate sin 10° to three significant figures. (Remember to convert x to radians before substituting in the series.) Check your result with a calculator.

Note: An electronic calculator does not use this series to calculate the sine of an angle because it takes too long by this method. A more efficient algorithm using a curve fitting routine is used instead.

Exercise 3 Show by applying Maclaurin's theorem that the series for $\cos x$ is

$$\cos x = 1 - \frac{x^2}{2!} + \frac{x^4}{4!} - \frac{x^6}{6!} + \ldots \tag{9.13}$$

Remember that x is in radians. Note also that alternate terms are positive and negative, and that only even powers of x are present.

Exercise 4 Verify from the series that the differential of $\cos x$ with respect to x is $-\sin x$.

$$\frac{d(\cos x)}{dx} = \frac{d}{dx}\left(1 - \frac{x^2}{2!} + \frac{x^4}{4!} - \frac{x^6}{6!} + \ldots\right)$$

$$= 0 - 2\frac{x^1}{2!} + 4\frac{x^3}{4!} - 6\frac{x^5}{6!} + \ldots$$

$$= 0 - x + \frac{x^3}{3!} - \frac{x^5}{5!} + \ldots$$

$$= -\sin x$$

9.13 The differentials of e^x and $\ln x$

In Chapter 1 some discussion of 'e' the base of natural logarithms was given. It was stated that

$$e^x = 1 + x + \frac{x^2}{2!} + \frac{x^3}{3!} + \ldots \frac{x^n}{n!} \tag{9.14}$$

and that e is obtained when $x = 1$ namely

$$e = 1 + 1 + \frac{1^2}{2!} + \frac{1^3}{3!} + \ldots \frac{1^n}{n!}$$

$$e = 1 + 1 + \frac{1}{2!} + \frac{1}{3!} + \ldots \frac{1}{n!} \tag{9.15}$$

$$= 2.7182$$

Note that e is an irrational number, it cannot be expressed as a fraction. Napier chose e as the base for his logarithms to enable him to calculate the values of logarithms. If we now differentiate e^x with respect to x we obtain the interesting and important result that

$$\frac{d(e^x)}{dx} = \frac{d}{dx}\left(1 + x + \frac{x^2}{2!} + \frac{x^3}{3!} + \ldots \frac{x^n}{n!}\right)$$

$$= 1 + x + \frac{x^2}{2!} + \frac{x^3}{3!} + \ldots \frac{x^{n-1}}{n-1!}$$

which tends to e^x as $n \to \infty$

Therefore

$$\frac{\mathrm{d}(\mathrm{e}^x)}{\mathrm{d}x} = \mathrm{e}^x \tag{9.16}$$

Now consider

$$\frac{\mathrm{d}(\ln x)}{\mathrm{d}x}$$

Let

$$y = \ln x$$

then by definition of a logarithm

$$\mathrm{e}^y = x$$

By (9.16)

$$\frac{\mathrm{d}x}{\mathrm{d}y} = \frac{\mathrm{d}(\mathrm{e}^y)}{\mathrm{d}y} = \mathrm{e}^y$$

and therefore

$$\frac{\mathrm{d}y}{\mathrm{d}x} = \frac{1}{\mathrm{e}^y} = \frac{1}{x}$$

therefore

$$\frac{\mathrm{d}(\ln x)}{\mathrm{d}x} = \frac{1}{x} \tag{9.17}$$

Exercise 1 Show by Maclaurin's theorem that

$$\ln(1+x) = x - \frac{x^2}{2} + \frac{x^3}{3} - \frac{x^4}{4} \ldots \tag{9.18}$$

Let

$$y = f(x) = \ln(1+x)$$

The successive differentials are

$$f'(x) = \frac{1}{1+x} \qquad f''(x) = -\frac{1}{(1+x)^2}$$

$$f'''(x) = +\frac{2}{(1+x)^3} \qquad f''''(x) = -\frac{6}{(1+x)^4}$$

and putting $x = 0$ in these differentials we have

$$f'(0) = \frac{1}{1+0} = 1 \qquad f''(0) = \frac{1}{(1+0)^2} = -1$$

$$f'''(0) = +\frac{2}{(1+0)^3} = 2 \qquad f''''(0) = -\frac{6}{(1+0)^4} = -6$$

$$y = f(x) = \ln(1+x)$$

$$= f(0) + f'(0)x + \frac{f''(0)}{2!}x^2 + \frac{f'''(0)}{4!}x^4 + \ldots \frac{f^n(0)}{n!}x^n$$

Remembering that ln (1) = 0 because $e^0 = 1$ by definition

$$\ln(1+x) = 0 + x - \frac{1}{2!}x^2 + \frac{2}{3!}x^3 - \frac{6}{4!}x^4 + \ldots$$

because 3! = 3 × 2 etc.

$$\ln(1+x) = x - \frac{x^2}{2} + \frac{x^3}{3} - \frac{x^4}{4} + \ldots$$

Note that there are no factorials in the denominators of (9.18). If x is greater than +1or less than −1 the expression (9.18) never converges to a practical value. If x lies in the range −1 to +1 it does converge.

Exercise 2 Calculate ln 1.5 from the first six terms of (9.25). Here $x = 0.5$ and

$$\begin{aligned}\ln(1.5) &= 0.5 - \frac{0.25}{2} + \frac{0.125}{3} - \frac{0.0625}{4} + \frac{0.03125}{5} - \frac{0.01562}{6} \\ &= 0.5 - 0.125 + 0.0417 - 0.0156 + 0.00625 - 0.00312 \\ &= 0.4042\end{aligned}$$

A calculator gives ln 1.5 = 0.4055.

Note: In practice Napier calculated logarithms by a more effective method which we need not describe here.

9.14 Taylor's series

A slight modification to Maclaurin's series gives the very useful *Taylor's series*. In Figure 9.4 two axes systems are shown, one (x, y) with origin at O and the other (x', y') with the origin at Q.

If y is a function of x or

$$y = F(x)$$

then

$$x' = x - a$$
$$y' = y - F(a)$$

Now let y be a continuous function or polynomial

$$y = F(x) = A + Bx + Cx^2 + Dx^3 + Ex^4 + \ldots + Px^n$$

We can find the value of $F(x)$ at the point where $x = a$ from

$$F(a) = A + Ba + Ca^2 + Da^3 + Ea^4 + \ldots + Pa^n$$

Suppose we now want to find the value of y at a point R close to Q where $x' = h$ say. Consider the origin of coordinates at point Q $(a, F(a))$ of Figure 9.4. By Maclaurin's theorem we can put

$$y' = F(0) + \frac{F'(0)}{1!}h + \frac{F''(0)}{2!}h^2 + \frac{F'''(0)}{3!}h^3 + \frac{F''''(0)}{4!}h^4 + \ldots \frac{F^n(0)}{n!}h^n$$

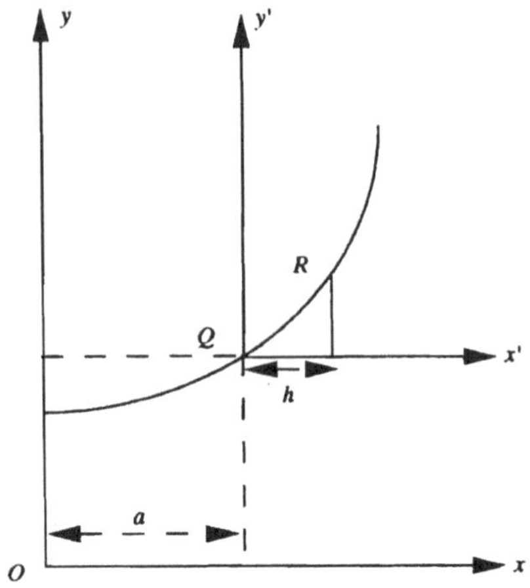

Figure 9.4

or in terms of the original axes

$$y = F(a) + \frac{F'(a)}{1!}h + \frac{F''(a)}{2!}h^2 + \frac{F'''(a)}{3!}h^3 + \frac{F''''(a)}{4!}h^4 + \dots \frac{F^n(a)}{n!}h^n \quad (9.20)$$

This is Taylor's series for one variable.

Exercise 1 Show that sin 10° 15' = 0.177944 given that sin 10° = 0.173 648. Here $F(x) = \sin x$ and $a = 10°$. Then $F(a) = F(10°) = 0.1736$. We first convert 15' to radians to obtain h. Therefore

$$h = \frac{15\pi}{60 \times 180} \text{ radians } = 0.004\,363$$

$$y = F(a) + \frac{F'(a)}{1!}h + \frac{F''(a)}{2!}h^2 + \frac{F'''(a)}{3!}h^3 + \frac{F''''(a)}{4!}h^4 + \dots \frac{F^n(a)}{n!}h^n$$

$$= \sin 10° + \frac{\cos 10°}{1!}h + \frac{-\sin 10°}{2!}h^2 + \frac{-\cos 10°}{3!}h^3 + \frac{\sin 10°}{4!}h^4 + \dots \quad (9.21)$$

Now $\cos^2 x = 1 - \sin^2 x$ thus cos 10° = 0.984 808

Substituting in (9.21) we have

$$\sin x = 0.173\,648 + \frac{0.984\,808}{1}0.004\,363 + \frac{-0.1736}{2}0.004\,363^2$$

$$+ \frac{-0.9848}{6}0.004\,363^3 + \frac{0.1736}{24}0.004\,363^4 + \dots$$

$$= 0.173\,648 + 0.004\,297 - 0.000\,002 - \text{negligible terms } = 0.177\,943$$

Note: Because h is small it is sufficient to use approximate values for $F''(a)$ etc.

9.15 Differentiation of a product

Suppose y is the product of two variables u and v each of which is a function of x, and we wish to differentiate y with respect to x. For example we might have

$$y = \sin x \cos x$$

If we let

$$\sin x = u \quad \text{and} \quad \cos x = v$$

then

$$y = uv$$

We shall prove that

$$\frac{dy}{dx} = u\frac{dv}{dx} + v\frac{du}{dx} \tag{9.22}$$

Consider small changes to y, u, and v, then we have

$$y + \delta y = (u + \delta u)(v + \delta v)$$

$$y - \delta y = (u - \delta u)(v - \delta v)$$

Subtracting gives

$$y + \delta y - (y - \delta y) = (u + \delta u)(v + \delta v) - (u - \delta u)(v - \delta v)$$

$$2\delta y = uv + u\delta v + v\delta u + \delta u\delta v - (uv - u\delta v - v\delta u + \delta u\delta v)$$

$$= 2u\delta v + 2v\delta u$$

Dividing through by $2\delta x$ we have

$$\frac{\delta y}{\delta x} = u\frac{\delta v}{\delta x} + v\frac{\delta u}{\delta x}$$

then

$$\text{Lim}\ \frac{\delta y}{\delta x} \text{ as } \delta x \to 0 = \frac{dy}{dx}$$

$$\text{Lim}\ \frac{\delta u}{\delta x} \text{ as } \delta x \to 0 = \frac{du}{dx}$$

$$\text{Lim}\ \frac{\delta v}{\delta x} \text{ as } \delta x \to 0 = \frac{dv}{dx}$$

therefore

$$\frac{dy}{dx} = u\frac{dv}{dx} + v\frac{du}{dx}$$

Exercise 1 Derive, by the product rule, the differential of $\sin x \cos x$ with respect to x. Let

$$y = \sin x \cos x,\ u = \sin x \text{ and } v = \cos x$$

then

$$y = uv$$

$$\frac{du}{dx} = \cos x \quad \text{and} \quad \frac{dv}{dx} = -\sin x$$

so
$$\frac{dy}{dx} = u\frac{dv}{dx} + v\frac{du}{dx} = \sin x(-\sin x) + \cos x \cos x = \cos^2 x - \sin^2 x$$

9.16 Differentiation of a function of a function

We now consider the case when y is a double function of x such as

$$y = \sin^2 x = (\sin x)^2$$

We say 'double function' because there are two operations involved; first to find $\sin x$ then to square it. To differentiate such a function of a function we proceed as follows.

Let $\sin x = u$; then

$$y = u^2$$

$$\frac{dy}{du} = 2u \quad \text{and} \quad \frac{du}{dx} = \cos x$$

Now
$$\frac{dy}{dx} = \frac{dy}{du}\frac{du}{dx}$$

therefore
$$\frac{dy}{dx} = 2\sin x \cos x$$

Exercise 1 Differentiate $y = \sin 2x$ with respect to x. Let

$$2x = u \text{ then } \frac{du}{dx} = 2$$

$$y = \sin u \text{ then } \frac{dy}{du} = \cos u$$

therefore
$$\frac{dy}{dx} = \frac{dy}{du}\frac{du}{dx} = 2\cos 2x$$

Note: After some practice the actual substitution of u is not required. To differentiate $\sin 2x$ with respect to x we first differentiate $\sin 2x$ with respect to $2x$ and obtain $\cos 2x$; then multiply it by the differential of $2x$ with respect to x i.e. by 2, obtaining the result $2 \cos 2x$ straight away.

Exercise 2 Differentiate $y = \cos 2x$ with respect to x.
First differentiate $\cos 2x$ with respect to $2x$ and obtain $-\sin 2x$, then differentiate $2x$ with respect to x to obtain 2, and multiply the two results, to obtain the final answer of

$$\frac{d(\cos 2x)}{dx} = -2\sin 2x$$

Exercise 3 Differentiate $y = \cos^2 2x$ with respect to x.

$$y = \cos^2 2x = (\cos 2x)^2$$

First differentiate y with respect to $\cos 2x$ to obtain $2 \cos 2x$, then differentiate $\cos 2x$ with respect to $2x$ and obtain $-\sin 2x$, and differentiate $2x$ with respect to x to obtain 2. Multiply these three results together to give the final result

$$\frac{\mathrm{d}\left(\cos^2 2x\right)}{\mathrm{d}x} = -4\sin 2x \cos 2x$$

What we have done is to avoid making the following substitutions.
Let

$$\cos 2x = v$$

then

$$y = v^2 \quad \text{and} \quad \frac{\mathrm{d}y}{\mathrm{d}v} = 2v$$

Next let

$$2x = u$$

then

$$v = \cos u \quad \text{and} \quad \frac{\mathrm{d}v}{\mathrm{d}u} = -\sin u \quad \text{and} \quad \frac{\mathrm{d}u}{\mathrm{d}x} = 2$$

$$\frac{\mathrm{d}y}{\mathrm{d}x} = \frac{\mathrm{d}y}{\mathrm{d}v}\,\frac{\mathrm{d}v}{\mathrm{d}u}\,\frac{\mathrm{d}u}{\mathrm{d}x} = 2v(-\sin u)2 = -4\sin 2x \cos 2x$$

Exercise 4 Derive the differential, with respect to x, of $\tan x$ by the product and chain rules. Let

$$y = \tan x = \frac{\sin x}{\cos x} = \sin x(\cos x)^{-1}$$

Let

$$u = \sin x \quad \text{and} \quad v = (\cos x)^{-1}$$

then

$$\frac{\mathrm{d}u}{\mathrm{d}x} = \cos x \text{ and } \frac{\mathrm{d}v}{\mathrm{d}x} = -1(\cos x)^{-2}(-\sin x) = \sin x(\cos x)^{-2}$$

then

$$\frac{\mathrm{d}y}{\mathrm{d}x} = u\frac{\mathrm{d}v}{\mathrm{d}x} + v\frac{\mathrm{d}u}{\mathrm{d}x} = \sin^2 x(\cos x)^{-2} + (\cos x)^{-1}\cos x = 1 + \tan^2 x = \sec^2 x$$

$$\frac{\mathrm{d}(\tan x)}{\mathrm{d}x} = \sec^2 x \qquad (9.23)$$

9.17 Quotient rule for differentiation

Consider the case where y, u and v are functions of x and

$$y = \frac{u}{v}$$

Here, y is a function of the division process, it is the *quotient* of u (the *numerator*) by v (the *denominator*). To differentiate y with respect to x we merely use the product and chain rules, for

$$y = \frac{u}{v} = u.v^{-1}$$

$$\frac{\mathrm{d}y}{\mathrm{d}x} = u\frac{\mathrm{d}\left(v^{-1}\right)}{\mathrm{d}x} + \left(v^{-1}\right)\frac{\mathrm{d}u}{\mathrm{d}x} = u\left(-1v^{-2}\right)\frac{\mathrm{d}v}{\mathrm{d}x} + \left(v^{-1}\right)\frac{\mathrm{d}u}{\mathrm{d}x}$$

$$= -\frac{u}{v^2}\frac{\mathrm{d}v}{\mathrm{d}x} + \frac{1}{v}\frac{\mathrm{d}u}{\mathrm{d}x}$$

$$\frac{\mathrm{d}y}{\mathrm{d}x} = \frac{v\dfrac{\mathrm{d}u}{\mathrm{d}x} - u\dfrac{\mathrm{d}v}{\mathrm{d}x}}{\mathrm{v}^2} \qquad (9.24)$$

This formula is often used to differentiate quotients, but I prefer to use only the product rule in the same way as above.

Exercise 1 Differentiate tan x with respect to x by the quotient rule.

Let

$$y = \tan x = \frac{\sin x}{\cos x} = \frac{u}{v}$$

$$\frac{\mathrm{d}u}{\mathrm{d}x} = \cos x \text{ and } \frac{\mathrm{d}v}{\mathrm{d}x} = -\sin x$$

therefore

$$\frac{\mathrm{d}y}{\mathrm{d}x} = \frac{v\cos x - u(-\sin x)}{\mathrm{v}^2} = \frac{\cos^2 x + \sin^2 x}{\cos^2 x} = \frac{1}{\cos^2 x} = \sec^2 x$$

Exercise 2 Show, by Maclaurin's theorem, that the series for sec x is

$$\sec x = 1 + \frac{1}{2}x^2 + \frac{5}{24}x^4 + \frac{61}{720}x^6 + \ldots$$

Note: This important series is used in deriving formulae for the Transverse Mercator projection. There are some quite heavy differentiations to work out to obtain this result. We will give working for the first three terms only to save space. Remember to use results previously found as the differentations proceed

Let

$$F(x) = \sec x = (\cos x)^{-1} \tag{9.25}$$

Differentiating $F(x)$ we have

$$F'(x) = -1(\cos x)^{-2}(-\sin x) = \frac{\sin x}{\cos^2 x} = \tan x \sec x \tag{9.26}$$

Differentiating $F'(x)$ we have, using (9.22)

$$F''(x) = \tan x \tan x \sec x + \sec x \sec^2 x = \tan^2 x \sec x + \sec^3 x \tag{9.27}$$

Differentiating $F''(x)$ we have

$$\begin{aligned} F'''(x) &= \sec x.2\tan x \sec^2 x + \tan^2 x(\tan x \sec x) + 3\sec^2 x(\tan x \sec x) \\ &= 3\sec^3 x \tan x + 2\sec^3 x \tan x + \tan^3 x \sec x \\ &= 5\sec^3 x \tan x + \tan^3 x \sec x \end{aligned} \tag{9.28}$$

Differentiating $F'''(x)$ we have

$$\begin{aligned} F''''(x) &= 5\tan x \frac{d}{dx}\left(\sec^3 x\right) + 5\sec^3 x.\sec^2 x \\ &\quad + \sec x\left(3\tan^2 x \sec^2 x\right) + \tan^3 x \frac{d}{dx}(\sec x) \end{aligned} \tag{9.29}$$

When $x = 0$, since $\tan(0) = 0$ and $\sec(0) = 1$, the differentials expressed in (9.24) to (9.27) are respectively

$$F'(0) = 0 \quad F''(0) = 1 \quad F'''(0) = 0 \quad F''''(0) = 5$$

Substituting these values in the Maclaurin series (9.7) gives

$$\sec x = 1 + \frac{1}{2}x^2 + \frac{5}{24}x^4 + \ldots \tag{9.30}$$

or more accurately, if more terms are taken

$$\sec x = 1 + \frac{1}{2}x^2 + \frac{5}{24}x^4 + \frac{61}{720}x^6 + \ldots \tag{9.31}$$

9.18 Partial differentiation

Consider the equation of a circle of radius R in the x/y plane.

$$x^2 + y^2 = R^2$$

We wish to see how small changes in these parameters affect each other. We see that

$$(x + \delta x)^2 + (y + \delta y)^2 = (R + \delta R)^2$$

$$x^2 + 2x\delta x + \delta x^2 + y^2 + 2y\delta y + \delta y^2 = R^2 + 2R\delta R + \delta R^2$$

$$2x\delta x + 2y\delta y = 2R\delta R + (\delta R^2 - \delta x^2 - \delta y^2)$$
$$x\delta x + y\delta y = R\delta R + e$$

If the changes (δx, δy and δR) are small, e is small because it involves higher powers of small terms, then

$$x\delta x + y\delta y \fallingdotseq R\delta R \qquad (9.32)$$

Exercise 1 If $R = 5$, $x = 3$, and $y = 4$, find the approximate change in R as a result of changing x and y by $+0.01$.

Here
$$\delta x = \delta y = 0.01$$
then
$$R\delta R = x\delta x + y\delta y$$
$$5\delta R = 3(0.01) + 4(0.01) = 0.03 + 0.04 = 0.07$$
$$\delta R = \frac{0.07}{5} = 0.014$$

Note: We can obtain the result of the changes from
$$R = \sqrt{(3.01^2 + 4.01^2)} = 5.014$$

The advantage of the differential method is that it involves only multiplication, and avoids the more complex processes of squaring and square rooting. It enables us to reduce complex mathematical formulae to the much simpler linear equations. This advantage is particularly useful in statistics and error theory.

9.19 Total differential

We express the process of *partial differentiation* more formally by the following procedure. Because

$$x^2 + y^2 - R^2 = 0$$

we can write

$$F(x, y, R) = 0$$

The differential of $F(x, y, R)$ with respect to x only, treating all other variables as constant, is written in the form

$$\frac{\partial F}{\partial x}$$

In this case

$$\frac{\partial F}{\partial x} = 2x$$

The curly tailed 'd' or ∂ is called *del*. This is called the partial differentiation of F with respect to x only. The other partial differentials are

$$\frac{\partial F}{\partial y} = 2y \quad \text{and} \quad \frac{\partial F}{\partial R} = -2R$$

We can formally obtain the results: if

$$F(x, y, R) = 0$$

then

$$\frac{\partial F}{\partial x}\delta x + \frac{\partial F}{\partial y}\delta y + \frac{\partial F}{\partial R}\delta R = 0 \qquad (9.33)$$

This is called the total differential of $F(x, y, R)$. The partial differentials are sometimes written in the form

$$\frac{\partial F}{\partial x} = F'_x \quad \frac{\partial F}{\partial y} = F'_y \quad \frac{\partial F}{\partial R} = F'_R$$

Note: This result is only valid if the three variables are linked only by this function. For example if x and y are linked by another function such as $y = \sin x$ the partial differentials with respect to x and y are also linked. Equations formed by total differentials are much used in error estimation problems in surveying, cartography and geodesy.

Exercise 1 In triangle ABC the side a is calculated from side b, and angles A and B by the formula

$$a = \frac{b \sin A}{\sin B} \qquad (9.34)$$

Derive an expression to be used in calculation of the effect on the computed side a of small errors δb in b, δA in A and δB in B. We rewrite the expression as

$$a \sin B - b \sin A = 0$$

Thus we have

$$F(a, b, A, B) = 0$$

The total differential of F is

$$\frac{\partial F}{\partial a}\delta a + \frac{\partial F}{\partial b}\delta b + \frac{\partial F}{\partial A}\delta A + \frac{\partial F}{\partial B}\delta B = 0$$

The required expression is

$$\delta a = -\frac{\dfrac{\partial F}{\partial b}\delta b + \dfrac{\partial F}{\partial A}\delta A + \dfrac{\partial F}{\partial B}\delta B}{\dfrac{\partial F}{\partial a}}$$

Evaluating the differentials we have

$$\frac{\partial F}{\partial a} = \sin B, \quad \frac{\partial F}{\partial b} = -\text{sinA}, \quad \frac{\partial F}{\partial A} = -b \cos A, \quad \frac{\partial F}{\partial B} = a \cos B$$

thus

$$\delta a = \frac{\sin A.\delta b + b\cos A.\delta A - a\cos B.\delta B}{\sin B}$$

$$= \frac{\sin A.\delta b}{\sin B} + \frac{b\cos A.\delta A}{\sin B} - \frac{a\cos B.\delta B}{\sin B}$$

$$= \frac{a.\delta b}{b} + \frac{a\cos A.\delta A}{\sin A} - \frac{a\cos B.\delta B}{\sin B}$$

therefore

$$\frac{\delta a}{a} = \frac{\delta b}{b} + \cot A\delta A - \cot B\delta B \tag{9.35}$$

Note that δA and δB are in radians.

9.20 Logarithmic partial differentiation

It is interesting to derive the result of (9.35) more easily by first taking logarithms of the monomial function (9.34)

$$a = \frac{b\sin A}{\sin B}$$

The function is called *monomial* because there is only one term on each side of the equation. It is therefore possible to take logarithms of both sides, giving

$$\ln a = \ln b + \ln \sin A - \ln \sin B$$

then partial differentiation gives

$$\frac{1}{a}\delta a = \frac{1}{b}\delta b + \frac{\cos A}{\sin A}\delta A - \frac{\cos B}{\sin B}\delta B$$

$$\frac{\delta a}{a} = \frac{\delta b}{b} + \cot A\delta A - \cot B\delta B$$

Many monomial functions which arise in map making are much more easily differentiated by this logarithmic method.

Exercise 1 In triangle *ABC* the angles *A* and *B* are each 70° and the side $b = 100$ metres. What is the effect on the computed side *a* of the following changes: +1' in angle *A*, –1' in angle B, and –0.05 metre in side *b*? It is instructive to derive the result in two ways:

(1) by recalculation of the triangle after making the changes, and
(2) by the total differential formula.

Method (1) The new values are

$$A = 70°\,01' \quad B = 69°\,59' \quad \text{and} \quad b = 99.95 \text{ m}$$

$$a = \frac{b\sin A}{\sin B} = \frac{99.95 \times 0.939\,79}{0.939\,59}$$

$$= 99.971$$

Method (2) Remembering that there are approximately 3438 minutes of arc in a radian, we have

$$\frac{\delta a}{a} = \frac{\delta b}{b} + \cot A \delta A - \cot B \delta B$$

$$\frac{\delta a}{100} = \frac{\delta b}{100} + 0.3640\delta A - 0.3640\delta B$$

$$\frac{\delta a}{100} = \frac{-0.05}{100} + \frac{0.3640}{3438} + \frac{0.3640}{3438}$$

$$\begin{aligned}\delta a &= -0.05 + 0.0106 + 0.0106\\ &= -0.029\end{aligned}$$

This gives the new value of a as $100 - 0.029 = 99.971$ as before.

Exercise 2 Find the total differential of the function $F(S, x, y, z)$ where

$$F(S, x, y, z) = x^2 + y^2 + z^2 - s^2 = 0$$

Logarithmic differentiation is not possible because the function is not monomial.

$$\frac{\partial F}{\partial s}\delta s + \frac{\partial F}{\partial x}\delta x + \frac{\partial F}{\partial y}\delta y + \frac{\partial F}{\partial z}\delta z = 0$$

therefore

$$-2s\delta s + 2x\delta x + 2y\delta y + 2z\delta z = 0$$

$$-s\delta s + x\delta x + y\delta y + z\delta z = 0$$

$$-\delta s + \frac{x}{s}\delta x + \frac{y}{s}\delta y + \frac{z}{s}\delta z = 0$$

$$-\delta s + L\delta x + M\delta y + N\delta z = 0 \qquad (9.36)$$

where L, M and N are the direction cosines of s. (See Section 5.8.)

9.21 Curvature

Consider the ellipse shown in Figure 9.5. Two circles of different sizes are also shown touching the ellipse. Their centres are O_1 and O_2 and radii R_1 and R_2. These circles are called *osculating circles* because they *kiss* the ellipse.

At the points of contact, the circles and the ellipse have the same *radius of curvature*. The radius of curvature of the ellipse varies from a minimum to a maximum as we move round the curve. The ellipse therefore does not have constant curvature. In geodesy we are interested in the curvature of an ellipse at any point on its arc. Engineering surveying also requires knowledge of the curvature of arcs.

We define the curvature of an arc at a point P to be the reciprocal of the radius of the osculating circle at P.

$$\text{curvature} = \frac{1}{R}$$

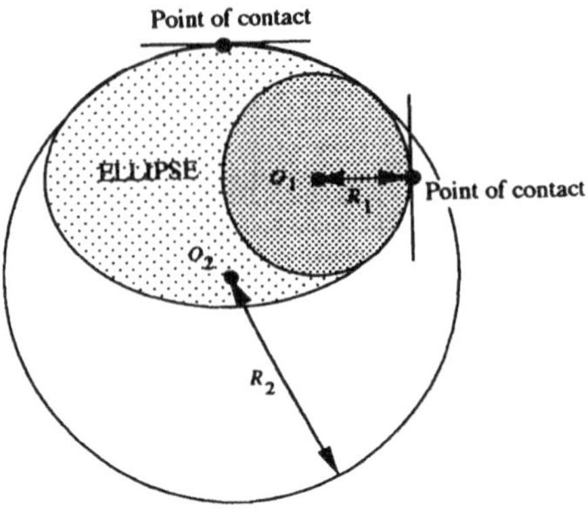

Figure 9.5

A curve which lies in a plane can be expressed as

$$y = f(x)$$

The radius of curvature of an arc of this curve is given by

$$R = \frac{\left[1+\left(\dfrac{\mathrm{d}y}{\mathrm{d}x}\right)^2\right]^{3/2}}{\dfrac{\mathrm{d}^2y}{\mathrm{d}x^2}} \qquad (9.37)$$

The differentials are evaluated for the curve at the point of interest. To derive this expression, consider Figure 9.6 showing a curve in a plane coordinate system. PS is the tangent to the curve at point P where the curvature is required. This tangent makes an angle A with the positive direction of the Ox axis. The centre of curvature is at U. As P moves round the curve to Q the tangent changes by an angle δA. Because the radii are perpendicular to these tangents, the angle at the centre U also changes by δA. The arc δs of the osculating circle subtended by δA is given by

$$\delta s = R\delta A$$

therefore

$$\delta A = \frac{\delta s}{R}$$

but

$$\delta s^2 = \delta x^2 + \delta y^2$$

so

$$\frac{\delta s^2}{\delta x^2} = \frac{\delta x^2}{\delta x^2} + \frac{\delta y^2}{\delta x^2} = 1 + \left(\frac{\delta y}{\delta x}\right)^2$$

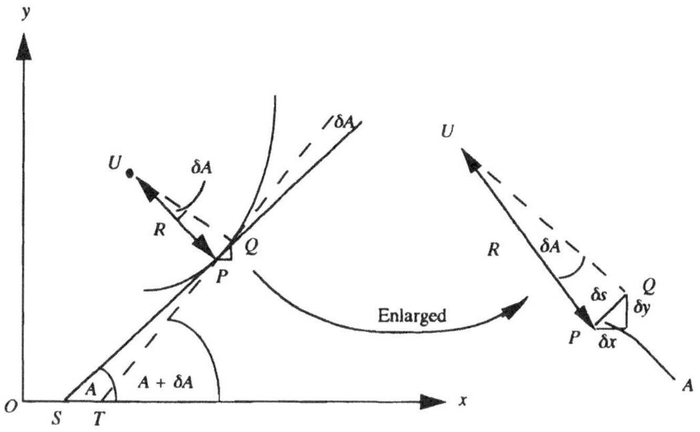

Figure 9.6

and

$$\frac{\delta s}{\delta x} = \left[1 + \left(\frac{\delta y}{\delta x}\right)^2\right]^{1/2}$$

but

$$\frac{\delta A}{\delta x} = \frac{1}{R}\frac{\delta s}{\delta x}$$

$$= \frac{1}{R}\left[1 + \left(\frac{\delta y}{\delta x}\right)^2\right]^{1/2}$$

Proceeding to the limits as δx tends to zero gives

$$\frac{dA}{dx} = \frac{1}{R}\left[1 + \left(\frac{dy}{dx}\right)^2\right]^{1/2}$$

but

$$\frac{dy}{dx} = \tan A$$

therefore

$$\frac{d^2y}{dx^2} = \frac{d(\tan A)}{dx} = \frac{d(\tan A)}{dA}\frac{dA}{dx}$$

$$= \sec^2 A \frac{dA}{dx}$$

But

$$\sec^2 A = 1 + \tan^2 A$$

therefore

$$\frac{d^2y}{dx^2} = \left(1 + \tan^2 A\right)\frac{dA}{dx}$$

$$=\left[1+\left(\frac{dy}{dx}\right)^2\right]\frac{1}{R}\left[1+\left(\frac{dy}{dx}\right)^2\right]^{1/2}$$

$$=\frac{1}{R}\left[1+\left(\frac{dy}{dx}\right)^2\right]^{3/2}$$

Finally

$$R=\frac{\left[1+\left(\frac{dy}{dx}\right)^2\right]^{3/2}}{\frac{d^2y}{dx^2}}$$

Exercise 1 Calculate the radius of curvature of the point P (2, 2) on the curve whose equation is

$$y = 4x^2$$

Now

$$\frac{dy}{dx}=8x \quad \frac{d^2y}{dx^2}=8$$

At P

$$\frac{dy}{dx}=16$$

$$R=\frac{\left[1+(16)^2\right]^{3/2}}{8^2}=\frac{257^{3/2}}{64}$$

$$=64.37$$

In Section 10.10 the expression for the radius of curvature of an ellipse is derived.

9.22 Integral calculus

The *integral calculus* is very important in cartography and geodesy, both for its own sake, as in map projection theory, and as a means of deriving other results, such as the determination of areas and volumes. In the section on differentiation we discussed functions such as the following expressions

$$y = x \quad y = x^2 \quad y = x^3$$

In each case we say that *y is a function of x* or

$$y = f(x)$$

More specifically we can write

$$y_1 = f_1(x) = x \quad y_2 = f_2(x) = x^2 \quad y_3 = f_3(x) = x^3 \tag{9.38}$$

This just states that *y is dependent upon x* in some way. The differentiation of $f(x)$ with respect to x was expressed in the form

$$\frac{dy}{dx} = f'(x)$$

and the result for the general case where $y = x^n$ and n is an integer is

$$\frac{dy}{dx} = nx^{n-1}$$

Thus in each case of (9.38)

$$f_1'(x) = \frac{dy}{dx} = 1 \qquad f_2'(x) = \frac{dy}{dx} = 2x \qquad f_3'(x) = \frac{dy}{dx} = 3x^2$$

Now consider the reverse problem. Given that

$$f_1'(x) = \frac{dy}{dx} = 1$$

what is $f(x)$? We know one answer to be

$$f(x) = x$$

Is this the only answer? If a constant is added to $f(x)$ the differential remains the same, because a constant cannot be changed or in this case if

$$f(x) = x + 3$$

$$f_1'(x) = \frac{dy}{dx} = 1$$

Thus if

$$f_1'(x) = \frac{dy}{dx} = 1$$

then we must say that

$$f'(x) = x + k$$

where k is any constant.

Exercise 1 Given that $f'(x) = 2x$ what is $f(x)$? The result is

$$f(x) = x^2 + k$$

Exercise 2 Given that $f'(x) = 3x^2$ what is $f(x)$? The result is

$$f(x) = x^3 + k$$

General rule for integration

By inference from these three results, we can see that the general rule, where n is an integer, is that where

$$f'(x) = x^n$$

$$f(x) = \frac{1}{n+1} x^{n+1} + k$$

This is valid except for the case in which the integer $n = -1$.

Exercise 3 If $f'(x) = 1$ find $f(x)$. Here $n = 0$ therefore

$$f(x) = \frac{1}{n+1}x^{n+1} + k = x + k$$

Exercise 4 If $f'(x) = x$ find $f(x)$, and test the result by differentiation. Here $n = 1$ therefore

$$f(x) = \frac{1}{n+1}x^{n+1} + k = \tfrac{1}{2}x^2 + k$$

Differentiating with respect to x

$$f'(x) = \frac{\mathrm{d}}{\mathrm{d}x}\left(\tfrac{1}{2}x^2 + k\right) = 2.\tfrac{1}{2}x + 0 = x$$

9.23 The notation for an integral

The expression 'If $f'(x) = x$ find $f(x)$' is cumbersome, so a word '*integration*' is used to describe the process. Thus instead of writing

'If $f'(x) = x$ find $f(x)$'

we say

'*Integrate* $f'(x) = x$ with respect to x'

or simply

'*Integrate* $f'(x)$'

A even neater way is to use the symbol $\int$, for an elongated S, to denote integration. For example

$$\int f'(x)\,\mathrm{d}x = f(x) + k$$

Thus

$$\int \mathrm{d}x = x + k$$

and

$$\int x\,\mathrm{d}x = \tfrac{1}{2}x^2 + k$$

and

$$\int x^2\,\mathrm{d}x = \tfrac{1}{3}x^3 + k$$

$$\int x^n \mathrm{d}x = \frac{1}{n+1}x^{n+1} + k \tag{9.39}$$

These results are known as *indefinite integrals* because the constant k in each case is unknown.

Exercise 1 Write down the indefinite integrals of the following functions

(1) $\cos x$ (2) $\sin x$ (3) $\sec^2 x$

(1) $\int \cos x\,\mathrm{d}x = \sin x + \mathrm{k}$
(2) $\int \sin x\,\mathrm{d}x = -\cos x + \mathrm{k}$
(3) $\int \sec^2 x = \tan x + \mathrm{k}$

These results can be proved by differentiating them with respect to x. See Sections 9.9, 9.10 and 9.16.

Note: The integration of many functions is a difficult process often requiring clever substitutions. Again some important formulae, such as that of the ellipse, are not integrable in a strict mathematical sense. The reader is referred to advanced texts for more information about integration.

Exercise 2 Verify by differentiation that

$$\int \sec x \, dx = \ln \tan\left(\frac{\pi}{4} + \frac{x}{2}\right) + k \tag{9.40}$$

Let

$$y = \tan\left(\frac{\pi}{4} + \frac{x}{2}\right) \quad \text{and} \quad u = \left(\frac{\pi}{4} + \frac{x}{2}\right)$$

then

$$y = \tan u$$

$$\begin{aligned} \mathrm{d}\frac{(\ln y)}{\mathrm{d}x} &= \mathrm{d}\frac{(\ln y)}{\mathrm{d}y}\frac{\mathrm{d}y}{\mathrm{d}x} = \frac{1}{y}.\frac{\mathrm{d}y}{\mathrm{d}x} \\ &= \frac{1}{y}.\frac{\mathrm{d}y}{\mathrm{d}u}.\frac{\mathrm{d}u}{\mathrm{d}x} = \frac{\sec^2 u}{\tan u}.\frac{1}{2} \\ &= \frac{\cos u}{2\sin u.\cos^2 u} \\ &= \frac{1}{2\sin u.\cos u} = \frac{1}{\sin 2u} \end{aligned}$$

Now

$$2u = 2\left(\frac{\pi}{4} + \frac{x}{2}\right) = x + \frac{\pi}{2}$$

$$\therefore \sin 2u = \cos x$$

therefore

$$\sec x = \frac{1}{\sin 2u} = \mathrm{d}\frac{(\ln y)}{\mathrm{d}x}$$

Summarising, we have

$$\mathrm{d}\frac{\ln \tan\left(\frac{\pi}{4} + \frac{x}{2}\right)}{\mathrm{d}x} = \sec x$$

or

$$\int \sec x \, dx = \ln \tan\left(\frac{\pi}{4} + \frac{x}{2}\right) + k$$

where k is an unknown constant. This integral is basic to the theory of the Transverse Mercator projection of the sphere.

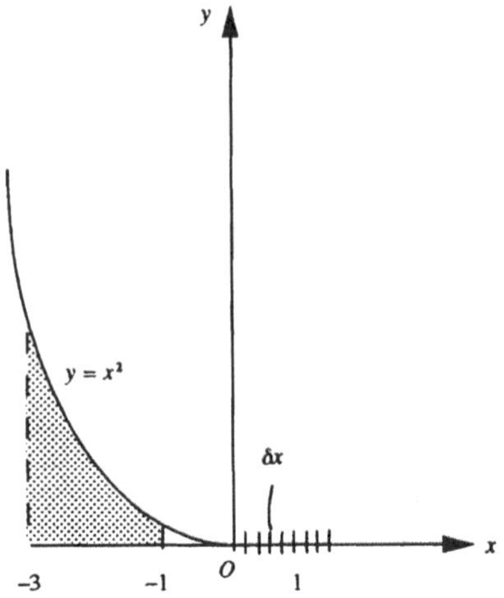

Figure 9.7

9.24 Geometrical interpretation of integration

Consider the x axis of Figure 9.7. A value x could be divided into many little equa parts δx. Then we could reverse the process and say that x is the sum of all th small parts δx, or

$$x = \delta x + \delta x + \delta x + \delta x + \ldots$$

If we think of the δxs becoming as small as we like, x is still their sum and we sa that

$$x = \int \mathrm{d}x \qquad (9.41)$$

The integration symbol $\int$ means 'sum up the parts' dx. The function resultin from the integration procedure is called the *integral.* The function integrated i called the *integrand.*

9.25 Integration limits

To be more specific, we have to define the ends of the line x. If one end lies at th origin where $x = 0$, and the other is at a point where $x = 2$, we write the integra as

$$\int_0^2 \mathrm{d}x$$

This means, integrate, or sum up the parts, starting at 0 and ending at 2. Th next stage is to carry out the integration part and place it in square brackets wit the limits placed outside. In this case

$$\int \mathrm{d}x = x$$

$$\int_0^2 \mathrm{d}x = [x]_0^2$$

We now assign the values of $x = 2$ and $x = 0$ to the integrand

$$[x]_0^2 = [2-0] = 2$$

The whole process is written

$$\int_0^2 dx = [x]_0^2 = [2-0] = 2$$

Note that the value of the integral at the lower limit is subtracted from the value of the integral at the upper limit. The use of these limits gives a *definite integral* without the arbitrary constant k.

Exercise 1 Write down the notation for the integration of dx between the range $x = 5$ to $x = 28$, and obtain the result of this calculation.

$$\int_5^{28} dx = [x]_5^{28} = [28-5] = 23$$

Exercise 2 Write down the notation for the integration of x dx between the range $x = 5$ to $x = 28$, and obtain the result of this calculation.

$$\int_5^{28} x\,dx = \left[\tfrac{1}{2}x^2\right]_5^{28} = \tfrac{1}{2}\left[28^2 - 5^2\right] = \tfrac{1}{2}[784-25] = 379.5$$

Exercise 3 Write down the notation for the integration of x^2 dx between the range $x = -4$ to $x = 2$, and obtain the result of this calculation.

$$\int_{-4}^{2} x^2\,dx = \left[\tfrac{1}{3}x^3\right]_{-4}^{2} = \tfrac{1}{3}\left[2^3 - (-4)^3\right] = \tfrac{1}{3}[8+64] = 24$$

9.26 Areas by Integration

Consider the graph in Figure 9.7 of the function

$$y = f(x) = x^2$$

Consider the area, δA, of a small strip between the curve and the x axis. The width of the strip is δx and the mid ordinate is y. Then the area of the strip is given by

$$\delta A = y\,\delta x$$

The area A, bounded by two ordinates y_1 and y_2 is the sum of many strips, that is

$$A = \int_{x_1}^{x_2} \delta A = \int_{x_1}^{x_2} y\delta x = \int_{x_1}^{x_2} x^2\delta x = \left[\tfrac{1}{3}x^3\right]_{x_1}^{x_2} = \left[\tfrac{1}{3}x_2^{\,3} - \tfrac{1}{3}x_1^{\,3}\right]$$

Exercise 1 Given that

$$y = f(x) = x^2$$

find the area between the curve and the x axis bounded by ordinates at $x = 1$ and $x = 3$. Applying the formula we have

$$A = \left[\tfrac{1}{3}x_2^{\,3} - \tfrac{1}{3}x_1^{\,3}\right] = \tfrac{1}{3}[27-1] = 8.67 \text{ sq. units}$$

Exercise 2 Given that

$$y = f(x) = x^2$$

find the area between the curve and the x axis bounded by ordinates at $x = -3$ and $x = -1$. Applying the formula we have

$$A = \left[\tfrac{1}{3}x_2^{\,3} - \tfrac{1}{3}x_1^{\,3}\right] = \tfrac{1}{3}[-27 + 1] = -8.67 \text{ sq. units}$$

Note: A negative value for the area indicates that it lies to the left of the origin. If we integrate from $x = -3$ to $x = +3$ the area is calculated as zero by the formula, because negative and positive areas balance. Care with signs is therefore needed in solving area problems. In this case it is better to treat the problem in its two halves either side of the origin.

9.27 Simpson's rule for approximate integration

We can recalculate the above area using the formula

$$A = \tfrac{1}{6}(y_1 + 4y_m + y_2)(x_2 - x_1) \qquad (9.42)$$

where y_1 and y_2 are the ordinates corresponding to x_1 and x_2, and y_m corresponds to the ordinate at x_m where

$$x_m = \tfrac{1}{2}(x_1 + x_2)$$

The formula (9.42) is called *Simpson's rule* for approximate integration. It gives the exact result if $f(x)$ involves no more than squares of x, and a good approximation in other cases provided the range $(x_2 - x_1)$ is kept small.

Exercise 1 Consider the above case in which

$$y = f(x) = x^2$$

Then

$$y_1 = x_1^2 \quad y_2 = x_2^2 \quad y_m = x_m^2 = \tfrac{1}{4}(x_1 + x_2)^2$$

From Simpson's rule we have

$$A = \tfrac{1}{6}(y_1 + 4y_m + y_2)(x_2 - x_1)$$
$$= \tfrac{1}{6}\left(x_1^2 + 4\tfrac{1}{4}(x_1 + x_2)^2 + x_2^2\right)(x_2 - x_1)$$

which multiplied out becomes

$$A = \tfrac{1}{3}\left[x_2^{\,3} - x_1^{\,3}\right]$$

which is the formula derived by integration.

Exercise 2 Show that if $y = f(x) = 2x + 3x^2$, the areas under the curve derived by integration and by Simpson's rules are the same, i.e. $x_2^2 + x_2^3 - x_1^2 - x_1^3$. This is left as an exercise for the reader, who should follow the method used for the above case where $y = x^2$.

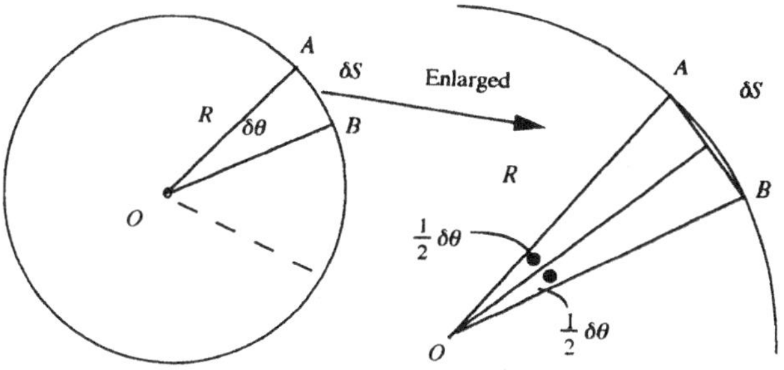

Figure 9.8

9.28 Properties of the circle

Consider a circle of radius R shown in Figure 9.8. The radian is defined as that angle subtended at the centre of the circle by an arc S equal to R. Thus there are 2π radians to a whole circle. When $S = C$, the circumference of the circle, $C = 2\pi R$. Let a small element of arc be δS which subtends a small angle $\delta\theta$ at the centre of the circle. Then if $\delta\theta$ is in radians

$$\delta S = R\delta\theta$$

The total arc length round the circle is the circumference C therefore

$$C = \int_0^S \delta S = \int_0^{2\pi} R\delta\theta = R[\theta]_0^{2\pi} = R[2\pi - 0] = 2\pi R$$

Now consider the small sector of the circle ABO. The small angle

$$AOb = \delta\theta$$

and the area of the small triangle AOB is given by

$$\delta A = R\cos\tfrac{1}{2}\delta\theta . R\sin\tfrac{1}{2}\delta\theta = \tfrac{1}{2}R^2\sin\delta\theta$$

$$= \tfrac{1}{2}R^2\frac{\sin\delta\theta}{\delta\theta}.\delta\theta$$

If we make $\delta\theta$ and δA smaller and smaller until they are the infinitesimally small ($d\theta$ and dA respectively) we have

$$dA = \tfrac{1}{2}R^2 d\theta$$

because the ratio $\dfrac{\sin\delta\theta}{\delta\theta}$ tends to 1 (See (9.8)).

The area of the whole circle is therefore given by

$$A = \tfrac{1}{2}\int_0^{2\pi} R^2\delta\theta = \tfrac{1}{2}R^2[\theta]_0^{2\pi} = \tfrac{1}{2}R^2[2\pi - 0] = \pi R^2$$

Exercise 1 Show that the area A_S of a sector of the circle defined by its angle θ is given by

$$A_S = \tfrac{1}{2}R^2\theta \qquad (9.43)$$

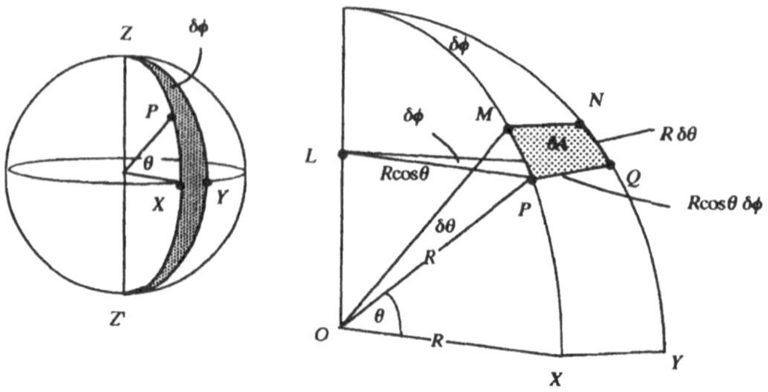

Figure 9.9

$$A_S = \tfrac{1}{2}\int_0^{\theta} R^2\delta\theta = \tfrac{1}{2}R^2[\theta]_0^{\theta} = \tfrac{1}{2}R^2\theta$$

9.29 Properties of the sphere

Consider a sphere of radius R shown in Figure 9.9. We are going to prove by integration that the area of the curved surface of the sphere is given by

$$A = 4\,\pi\,R^2 \tag{9.44}$$

and that its volume is given by

$$V = \frac{4}{3}\pi R^3 \tag{9.45}$$

The shaded portion $ZXZ'Y$ of the surface, shown in the small diagram to the left, is called a *lune*. The surface of the sphere can be divided into small patches such as

$$\delta A = PMNQ$$

If these patches are summed over the whole sphere the suface area is obtained. Let the small angles

$$QLP = \delta\phi \quad \text{and} \quad MOP = \delta\theta$$

The area of the patch is given approximately by

$$\delta A = PQ \times PM$$

If the patch is infinitesimally small the approximation is negligible. Now since

$$LP = R\cos\theta \quad \text{and} \quad OP = R \text{ we have}$$

$$\delta A = PQ \times PM = R\cos\theta\,\delta\phi\,R\,\delta\theta$$

To obtain the area of the lune shown in the small diagram on the left of Figure 9.9 we sum up the patches δA from

$$\theta = -\tfrac{1}{2}\pi \quad \text{to} \quad \theta = +\tfrac{1}{2}\pi$$

For the whole lune $\delta\phi$ is constant and the area of the lune is

$$R^2\delta\phi\int_{-\frac{1}{2}\pi}^{\frac{1}{2}\pi}\cos\theta\,\delta\theta$$

$$= R^2\delta\phi[\sin\theta]_{-\frac{1}{2}\pi}^{\frac{1}{2}\pi} = R^2[1+1] = 2R^2\delta\phi$$

To obtain the total surface area of the sphere we integrate again with respect to $\delta\phi$ over a whole circle. Thus the surface area of the sphere is

$$A = 2R^2\int_0^{2\pi}\delta\phi$$

$$A = 2R^2[\phi]_0^{2\pi} = 2R^2[2\pi - 0] = 4\pi R^2 \tag{9.46}$$

To obtain the volume of the sphere all we need to do is to sum up the volume of *spherical shells* like the layers of an onion. The thickness of each of these shells is dR. Starting from O, each shell is dR thick, so the volume will be given by

$$V = 4\pi\int_0^R R^2\delta R = 4\pi\left[\tfrac{1}{3}R^3\right]_0^R = 4\pi\left[\tfrac{1}{3}R^3 - 0\right]$$

therefore

$$V = \tfrac{4}{3}\pi R^3$$

9.30 The normal distribution function

As an example from statistics which combines curve tracing and the application of both differentiation and integration we consider the normal distribution curve. We do not attempt to derive this curve in this book, but merely use it as a useful example to illustrate the application of the calculus. *Beginners who find this too difficult should just ignore this section until they have studied the basic statistics required for an understanding of the problem.*

In error theory and statistical analysis the bell shaped curve, of Figure 9.10, which represents the frequency (y) of residuals (x) about a mean derived from randomly observed variables, may be expressed by the probability density function or PDF given by the formula

$$f(x) = y = \frac{h}{\sqrt{\pi}}\mathrm{e}^{-h^2x^2} \tag{9.47}$$

Such statistical or stochastic mathematical models are derived from probability theory. See for example an advanced statistical textbook such as Yule and Kendal, 1950, *An Introduction to the Theory of Statistics* to which the reader may refer. However, to give some understanding of this most important distribution, we will trace the function and show that a bell shaped curve results. Consider the formula (9.47).

When

$$x = 0 \quad y_m = \frac{h}{\sqrt{\pi}}$$

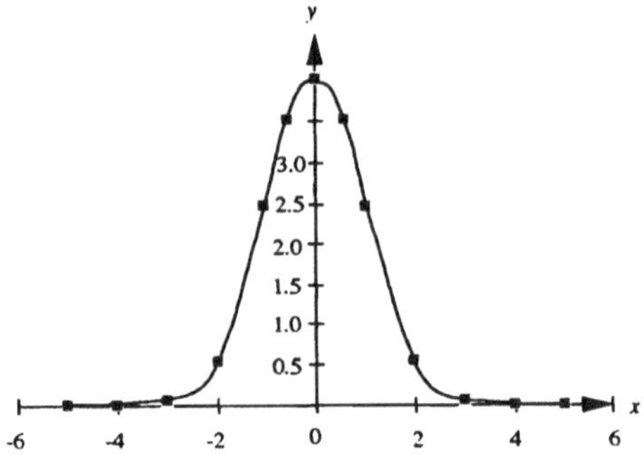

Figure 9.10

giving the maximum height of the curve y_m. The quantity h is known as the index of precision. Again, differentiating we obtain

$$\frac{dy}{dx} = -2h^2xy \qquad (9.48)$$

therefore

$$\frac{dy}{dx} = 0 \text{ when } x = 0 \text{ and when } y = 0$$

In the second case $y = 0$ when the curve touches the x axis at a very great (or infinite) distance away on either side of zero. The symbol for this infinite distance is a figure-of-eight on its side or $\pm\infty$. The curve is parallel to the x axis at these points.

Differentiating again we have

$$\frac{d^2y}{dx^2} = -2h^2y\left(1 - 2h^2x^2\right) \qquad (9.48)$$

therefore

$$\frac{d^2y}{dx^2} = 0$$

when

$$x = \pm\frac{1}{h\sqrt{2}} \qquad (9.50)$$

When the change of gradient is zero, the curve has a *point of inflexion* and the gradient reaches a maximum value on either side of the y axis. A rough sketch of the curve can now be made using all the information about gradients and special values just derived above. The curve clearly follows the bell shaped pattern shown in Figure 9.10. The index of precision h is related to the statistical quantity (variance) by the expression

$$s = \pm\frac{1}{h\sqrt{2}} \qquad (9.51)$$

By definition, the population variance (s) is given by

$$s^2 = \frac{\text{sum of squares of residuals}}{\text{total number of residuals}}$$

or in mathematical terms

$$s^2 = \frac{\int_{-\infty}^{+\infty} x^2 y \, dx}{\int_{-\infty}^{+\infty} y} dx \tag{9.52}$$

It is to be remembered that there are y values of each residual x. But

$$\int_{-\infty}^{+\infty} y \, dx \tag{9.53}$$

is the probability of all residuals being selected, namely a certainty, therefore

$$\int_{-\infty}^{+\infty} y \, dx = 1 \tag{9.54}$$

therefore

$$s^2 = \int_{-\infty}^{+\infty} x^2 y \, dx \tag{9.55}$$

Again

$$\frac{d^2 y}{dx^2} = -2h^2 y(1 - 2h^2 x^2)$$

therefore, integrating all terms, we have

$$\int_{-\infty}^{+\infty} \frac{d^2 y}{dx^2} dx = 4h^4 \int_{-\infty}^{+\infty} x^2 y \, dx - 2\int_{-\infty}^{+\infty} h^2 y \, dx \tag{9.56}$$

Remembering that the gradient of the curve is zero at infinity, taking the left hand side of (9.56), we have

$$\int_{-\infty}^{+\infty} \frac{d^2 y}{dx^2} dx = \left[\frac{dy}{dx} - \frac{dy}{dx}\right]_{-\infty}^{-\infty} = [0 - 0] = 0$$

Taking the right hand side of (9.56), remembering (9.55) and (9.53), we have

$$4h^4 \int_{-\infty}^{+\infty} x^2 y \, dx - 2\int_{-\infty}^{+\infty} h^2 y \, dx = 4h^4 s^2 - 2h^2$$

and therefore

$$0 = 4h^4 s^2 - 2h^2$$

giving the result

$$s = \pm \frac{1}{h\sqrt{2}}$$

and

$$h = \pm \frac{1}{s\sqrt{2}}$$

Thus the original equation (9.47) may also be written

$$f(x) = y = \frac{1}{s\sqrt{2\pi}} e^{-x^2/2s^2} \qquad (9.57)$$

By putting $s = 1$ this expression is often *standardised* to be

$$f(x) = y = \frac{1}{\sqrt{2\pi}} e^{-x^2/2} \qquad (9.58)$$

Exercise 1 Verify that the values of $10y$ correspond to residuals x of the following table, assuming that a standardised normal distribution is present. Sketch the curve.

x	–5	–4	–3	–2	–1	–0.5	–0.25	–0.125	0
$10y$	0.00	0.00	0.04	0.54	2.42	3.52	3.87	3.96	3.99

x	0.125	0.25	0.5	1	2	3	4	5
$10y$	3.96	3.87	3.52	2.42	0.54	0.04	0.00	0.00

SUMMARY OF KEY WORDS

differential and integral calculus, function, continuous, even function, odd function, slope (gradient), limit of a function, differentiation of y or f(x) with respect to x, second differential, Maclaurin's theorem, derivative of sin *x*, *derivative of* cos *x*, *Taylor's series, partial differentiation, osculating circles, radius of curvature, integral calculus, indefinite integrals, integrand, definite integral, Simpson's rule, point of inflexion, normal distribution curve*

SUMMARY OF FORMULAE

If $y = x^n$ $\qquad \dfrac{dy}{dx} = nx^{n-1}$ (9.5)

If $y = F(x) = Ax^n$ $\qquad \dfrac{dy}{dx} = F'(x) = nAx^{n-1}$ (9.6)

Maclaurin's theorem $\qquad y = F(x)$

$$= F(0) + F'(0)x + \frac{F''(0)}{2!}x^2 + \frac{F'''(0)}{3!}x^3 + \frac{F''''(0)}{4!}x^4 + \ldots \frac{F^n(0)}{n!}x^n \quad (9.7)$$

$$\frac{d(\sin x)}{dx} = \cos x \quad (9.9)$$

$$\frac{d(\cos x)}{dx} = -\sin x \quad (9.10)$$

$$\sin x = x - \frac{x^3}{3!} + \frac{x^5}{5!} - \frac{x^7}{7!} - \ldots \quad (9.12)$$

$$\cos x = 1 - \frac{x^2}{2!} + \frac{x^4}{4!} - \frac{x^6}{6!} + \ldots \quad (9.13)$$

$$\frac{d(e^x)}{dx} = e^x \quad (9.16)$$

$$\frac{d(\ln x)}{dx} = \frac{1}{x} \quad (9.17)$$

$$\ln(1+x) = x - \frac{x^2}{2} + \frac{x^3}{3} - \frac{x^4}{4} \ldots \quad (9.18)$$

Taylor's theorem

$$y = F(a) + \frac{F'(a)}{1!}h + \frac{F''(a)}{2!}h^2 + \frac{F'''(a)}{3!}h^3 + \frac{F''''(a)}{4!}h^4 + \ldots \frac{F^n(a)}{n!}h^n \quad (9.20)$$

Product rule if $y = uv$ $\qquad \dfrac{dy}{dx} = u\dfrac{dv}{dx} + v\dfrac{du}{dx}$ (9.22)

$$\frac{d(\tan x)}{dx} = \sec^2 x \quad (9.23)$$

Quotient rule if $y = \frac{u}{v}$ $$\frac{dy}{dx} = \frac{v\frac{du}{dx} - u\frac{dv}{dx}}{v^2} \qquad (9.24)$$

$$\sec x = 1 + \frac{1}{2}x^2 + \frac{5}{24}x^4 + \frac{61}{720}x^6 + \ldots \qquad (9.31)$$

Partial differentiation

If $F(x, y, R) = 0$ $$\frac{\partial F}{\partial x}\delta x + \frac{\partial F}{\partial y}\delta y + \frac{\partial F}{\partial R}\delta R = 0 \qquad (9.33)$$

If $a = \frac{b \sin A}{\sin B}$ $$\frac{\delta a}{a} = \frac{\delta b}{b} + \cot A \delta A - \cot B \delta B \qquad (9.35)$$

Curvature

$$R = \frac{\left[1 + \left(\frac{dy}{dx}\right)^2\right]^{3/2}}{\frac{d^2y}{dx^2}} \qquad (9.37)$$

Integration $$\int x^n dx = \frac{1}{n+1}x^{n+1} + k \qquad (9.39)$$

$$\int \sec x \, dx = \ln \tan\left(\frac{\pi}{4} + \frac{x}{2}\right) + k \qquad (9.40)$$

$$x = \int dx \qquad (9.41)$$

Simpson's Rule $$A = \tfrac{1}{6}(y_1 + 4y_m + y_2)(x_2 - x_1) \qquad (9.42)$$

Circle. Area of sector $$A_S = \tfrac{1}{2}R^2\theta \qquad (9.43)$$

Sphere. Area of surface $$A = 4\pi R^2 \qquad (9.44)$$

Sphere. Volume $$V = \frac{4}{3}\pi R^3 \qquad (9.45)$$

Normal distribution $$f(x) = y = \frac{h}{\sqrt{\pi}}e^{-h^2x^2} \qquad (9.47)$$

Chapter 10
Conic Sections

10.1 Conic sections

The *conic sections* are the well-known curves

circle, parabola, ellipse and hyperbola

These curves are surprisingly useful in the applied sciences, such as optics and engineering, and in geometry, geodesy, cartography and statistics. Their mathematics is elegant and complex. From within the vast subject of *conic sections* we select only a few basic ideas and develop those properties immediately valuable to the map maker.

10.2 Sections of a cone

As their name suggests, these curves are the various sections described in planes which cut a cone. We will consider only a right circular cone, which has a circle as its base and the axis normal to the base passing through its centre. In mathematics, a double cone, shaped like an hour glass or egg-timer, is usually considered.

In Figure 10.1 we see a right circular cone touching a sphere along a small circle B_1B_2Q. A small circle of a sphere does not pass through its centre. All points marked by a black dot lie in the plane of the paper. This plane is illustrated separately in Figure 10.2 which is a section containing the axis of the cone and a diameter of the sphere. The points marked in Figure 10.1 by an open circle do not lie in this plane.

The plane of the small circle B_1B_2QC is orthogonal to the plane of the paper. Another plane A_1A_2D, also orthogonal to the plane of the paper, touches the sphere at T and makes an angle V with the cone axis. We are interested in the curve drawn out by the cone in the plane A_1A_2D. Let P be any point on this curve. Once we have selected the sphere, the point T and the line CD are fixed relative to the cone. P can vary round the curve. PN is drawn parallel to the axis of the cone and meets the plane B_1B_2QC in point N. Angle $NPQ = U$. ND is drawn parallel to the line B_1B_2C. CD is orthogonal to the plane of the paper. Therefore the angle $NPD = V$. Also

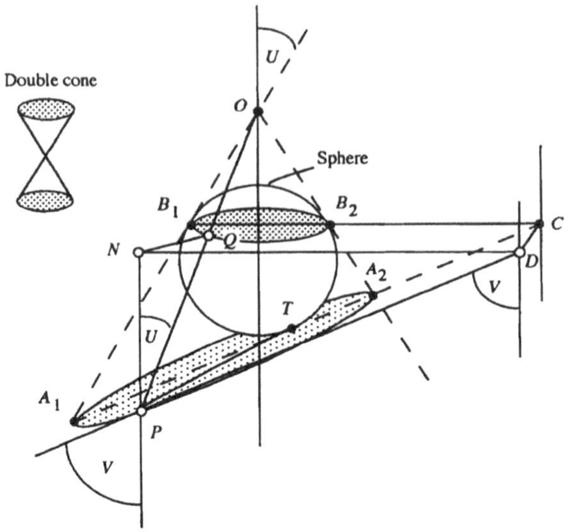

Figure 10.1

$$PN = PQ \cos U \quad \text{and} \quad PN = PD \cos V$$

Because PQ and PT are both tangents to the sphere

$$PQ = PT$$

therefore

$$PT \cos U = PQ \cos U = PN = PD \cos V$$

$$\frac{PT}{PD} = \cos V \sec U \qquad (10.1)$$

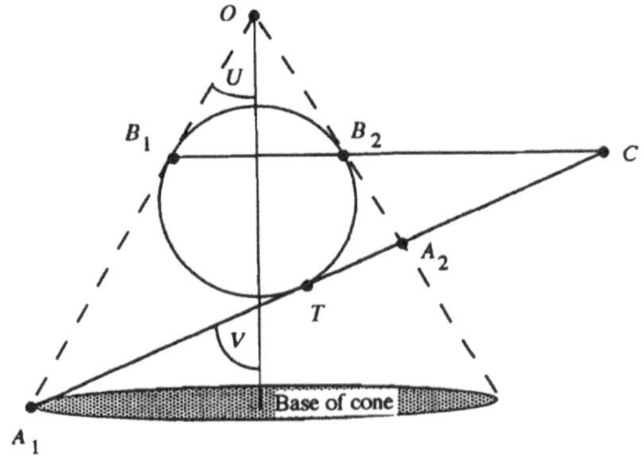

Figure 10.2

As P varies round the curve, this ratio is always satisfied: it is called the *eccentricity* (e) of the curve. Therefore

$$PT = e\,PD \tag{10.2}$$

where $e = \cos V \sec U$.

Curves defined by this ratio are called *conic sections* or 'conics'. The fixed point T is called the *focus* of the conic, and the fixed line CD the *directrix* of the conic.

The angle U defines the shape of the cone. Of specific interest are

(1) when $U = 0$ we obtain a *cylinder* touching the sphere, and

(2) when $U = 90°$ the cone degenerates into a plane touching the sphere.

Given a cone defined by U, the angle V determines which type of curve is generated.

Case (1) Circle

If $V = 90°$, $\cos V = 0$, and $e = 0$, the curve A_1A_2P becomes the circle B_1B_2Q.

Case (2) Parabola

If $V = U$, $e = 1$, the curve A_1A_2P cuts the cone only once to become a *parabola*.

Case (3) Ellipse

If $V > U$, $e < 1$, the curve A_1A_2P is closed on itself, and cuts the cone only once to become an *ellipse*.

Case (4) Hyperbola

If $V < U$, $e > 1$, the curve A_1A_2P cuts the *extended cone* twice to become an *hyperbola* with two separate parts called *sheets*.

10. 3 Circle

We have already considered many properties of the circle in other parts of this book, especially in Section 2.8 and 4.10. For convenience a summary of these and some other properties useful to a surveyor are given here.

First we will consider geometrical properties. A circle is a plane curve defined by its centre O and radius R. All points on its *circumference* are a distance R from the centre. Thus, in Figure 10.3,

$$AO = BO = CO = R$$

Chords of the circle are lines joining two points on its circumference, such as AB and AC. A chord passing through the centre is called a *diameter.*

In Figures 10.3 and 10.4, various equal angles are marked by dots and a square and others by combinations of these symbols. The reader new to this topic should draw these figures and verify these angular relationships by measurement. Perhaps the most important relationship is that shown in Figure 10.3 where

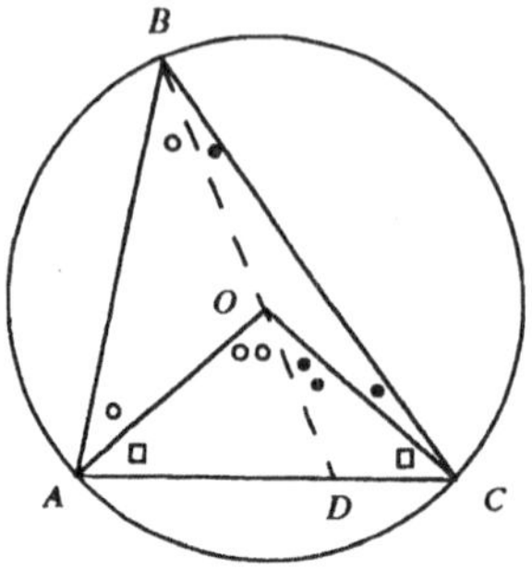

Figure 10.3

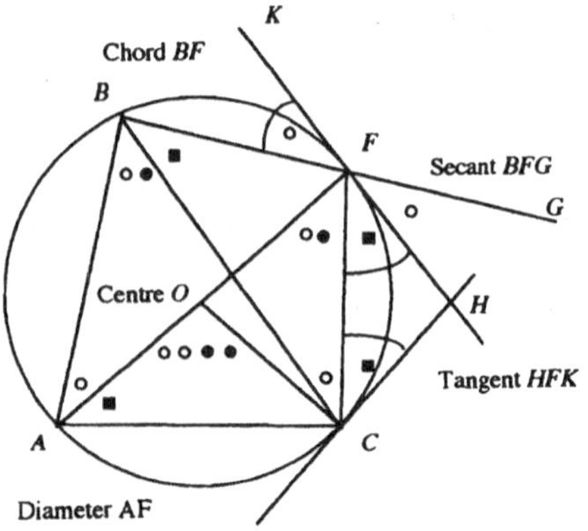

Figure 10.4

$$\text{angle } AOC = 2\times \text{ angle } ABC$$

The sides OA, OB and OC are all radii of the circle and equal to R. Therefore triangles ABO, ACO and CBO are *isosceles* (they have two equal sides). Therefore angles opposite equal sides in these three triangle are equal. They are marked by dot and square symbols. Because

$$\angle AOD = 2\angle ABO \text{ and } \angle COD = 2\angle CBO$$

$$\angle AOC = 2\angle ABC \tag{10.3}$$

This important result is much used in surveying especially in setting out engineering curves. The line *KFH* which touches the circle at *F* is called a *tangent* to the circle. It is perpendicular to the diameter *AF*. A very useful property is that the angle subtended by a diameter at the circumference is a right angle. For example angle $ABF = 90°$

To prove this, applying (10.3) we have,

$$\angle AOF = 2\angle ABF$$

But

$$\angle AOF = 180°$$

therefore

$$\angle ABF = 90° \tag{10.4}$$

Another very useful property in setting out curves is that

$$\angle KFB = \angle BAF$$

This follows from the fact that

$$\angle BAF + \angle BFA = 180° - \angle ABF = 90°$$

Also

$$\angle KFB + \angle BFA = 90°$$

therefore

$$\angle KFB = \angle BAF \tag{10.5}$$

Consider also the second tangent HC touching the circle at C. Then

$$\angle HCF = \angle CAF \text{ and } \angle HFC = \angle CAF$$

therefore

$$\angle HCF = \angle HFC$$

Thus the triangle HCF has two sides equal opposite these angles. That is, the two tangents HF and HC drawn from a point to a circle are equal. This is yet another important property of the circle.

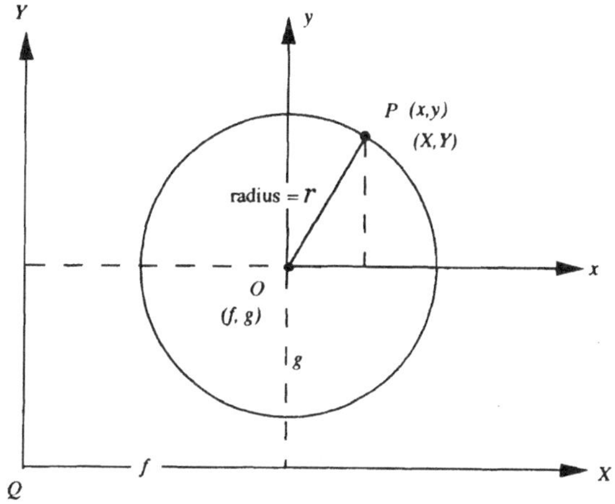

Figure 10.5

We now turn our attention to the equation of the circle. Refer to Figure 10.5 showing a circle centre O and radius r. P is any point (x, y) on the circle; therefore by Pythagoras's theorem

$$x^2 + y^2 = r^2 \qquad (10.6)$$

This is the equation of a circle based on the origin at its centre. See also Section 4.10. If we translate the origin of coordinates to some other point Q $(-a, -b)$ not at its centre, the new coordinates of P are (X, Y) given by

$$x = X - a \quad \text{and } y = Y - b$$

and the equation (10.6) becomes

$$(X - a)^2 + (Y - b)^2 = r^2$$

$$X^2 + Y^2 - 2aX - 2bY + a^2 + b^2 = r^2$$

$$X^2 + Y^2 - 2aX - 2bY + a^2 + b^2 - r^2 = 0 \qquad (10.7)$$

If we put

$$c = a^2 + b^2 - r^2 \text{ and } f = -a \text{ and } g = -b$$

equation (10.7) becomes

$$X^2 + Y^2 + 2fX + 2gY + c = 0 \qquad (10.8)$$

This is the general equation of a circle in terms of its plane coordinates in a two-dimensional system. (See also Section 5.20 for a consideration of the circle in a three dimensional system). Notice that the quantity $\sqrt{c}$ is the length of the tangent from Q to the circle.

Exercise 1 Refer to Figure 10.6. Find the equation of the circle passing through the points ABC whose coordinates are shown in the diagram.

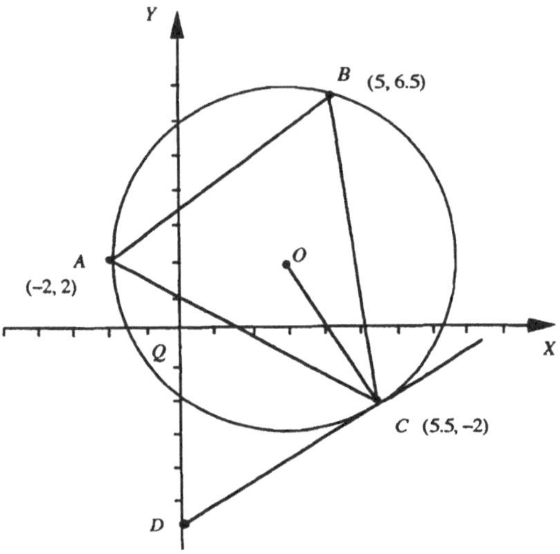

Figure 10.6

Let the equation of the circle be

$$X^2 + Y^2 + 2fX + 2gY + c = 0 \qquad (10.8)$$

$$+ 2fX + 2gY + c = -X^2 - Y^2 \qquad (10.9)$$

Substituting in (10.9) for the X and Y values of A, we obtain one equation in f, g and c

$$+2f(-2) + 2g(2) + c = -4 - 4$$

$$-4f + 4g + c = -8 \qquad (10.10)$$

Similarly for B and C we obtain

$$+10f + 13g + c = -67.25 \qquad (10.11)$$

$$+11f - 4g + c = -34.25 \qquad (10.12)$$

Solving equations (10.10), (10.11) and (10.12) gives

$$f = -2.876 \quad g = -2.110 \quad c = -11.061$$

Therefore the equation of the circle is

$$X^2 + Y^2 - 5.752\,X - 4.220\,Y - 11.061 = 0 \qquad (10.13)$$

Exercise 2 Verify equation (10.13) by showing that the original coordinates of A, B and C satisfy it. The working for point A is

$$4 + 4 - 5.752\,(-2) - 4.220\,(2) - 11.061 = 8 + 11.504 - 8.44 - 11.061 = 0$$

Exercise 3 Find the coordinates of the centre of the circle and its radius. The coordinates of the centre O are given directly as

$$-f = 2.876 \qquad -g = 2.110$$

and the radius from

$$r = \sqrt{(g^2 + f^2 - c)} = 4.877$$

Exercise 4 Find the equation of the tangent to the circle at C.
The formula for the gradient to the circle is found by differentiating the equation of the circle with respect to X, i.e. from

$$\frac{\mathrm{d}}{\mathrm{d}X}\left(+Y^2 + 2gY + c = -X^2 - 2fX\right)$$

$$2Y\frac{\mathrm{d}Y}{\mathrm{d}X} + 2g\frac{\mathrm{d}Y}{\mathrm{d}X} + 0 = -2X - 2f$$

therefore

$$\frac{\mathrm{d}Y}{\mathrm{d}X} = -\frac{X+f}{Y+g}$$

Substituting the values for the problem we have

$$\frac{\mathrm{d}Y}{\mathrm{d}X} = -\frac{5.5 - 2.876}{-2 - 2.110} = \frac{-2.624}{-4.110} = 0.6384$$

Expressing the equation of the tangent in the gradient form we have

$$Y = mX + d$$

where d is the intercept on the Y axis; the equation is

$$Y = 0.6384X + d$$

Substituting the values for C gives d to be

$$d = Y - 0.6384X = -2 - 0.6384\,(5.5) = -5.511$$

and the equation of the tangent is

$$Y = 0.6384X - 5.511$$

Exercise 5 It is left as an exercise for the reader to calculate the following angles from respective bearings

$$\angle ACD = 60.63° \quad \angle CBA = 60.63° \quad \angle COA = 121.26°$$

This verifies that the angle between a tangent and chord equals the angle in the alternate segment of the circle and is equal to twice the angle subtended at the centre of the circle. See (10.3) and (10.5).

10.4 Parabola

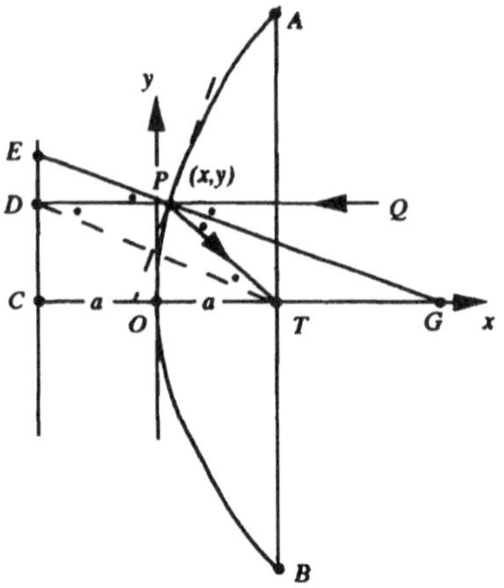

Figure 10.7

Figure 10.7 shows a parabola with *focus* T and *directrix* CD, defined by the relationship, $e = 1$, therefore

$$PD = PT$$

The line CT is chosen as x axis, and the coordinate origin to be at O where the curve cuts this axis. The Oy axis is perpendicular to Ox. Since O is a point on the curve,

$$OC = OT = a$$

Let P be the point (x, y) then at once we have coordinates

$$D(-a, y) \text{ and } T(a, 0)$$

also

$$PD^2 = PT^2$$
$$(x + a)^2 = (a - x)^2 + y^2$$
$$x^2 + 2ax + a^2 = x^2 - 2ax + a^2 + y^2$$

therefore

$$y^2 = 4ax \tag{10.14}$$

This is the equation of a parabola. It is symmetrical about the x axis and takes the form shown in the Figure 10.7. It is often used as a vertical curve in road design. Perhaps its most important property is that a line such as DP, parallel to the x axis, is equally inclined to the normal at P as is the line PT which passes through the focus T. In other words, the normal PE bisects the angle QPT. This enables the parabola to be used in optics as a means of bringing parallel light, or radio waves, to a focus. Or conversely to create a parallel beam by emission from a focus. To prove this property, differentiating (10.14) we have

$$2y\frac{dy}{dx} = 4a$$
$$\frac{dy}{dx} = \frac{2a}{y}$$

This is the gradient of the tangent at P. The normal to the curve at P therefore has the gradient

$$-\frac{y}{2a}$$

But this is also the gradient of DT. Therefore the normal at P is parallel to DT. In triangle DPT, $DP = TP$, therefore the angles at D and T, marked by dots, are equal: and from the parallel lines, the angles QPG and TPG, again marked by dots, are also equal. Finally at A and B, $x = a$ therefore $y = \pm 2a$. Therefore

$$AB = 4a$$

10.5 Ellipse

The ellipse is by far the most important conic section for the map maker. It is the basis of the ellipsoid of revolution used as a standard figure to represent the Earth; it is the theoretical path taken by an earth-orbiting satellite, and by the Earth itself round the Sun. It also features in engineering, surveying, systems of distance measurement, and in error theory. This is by no means a complete list.

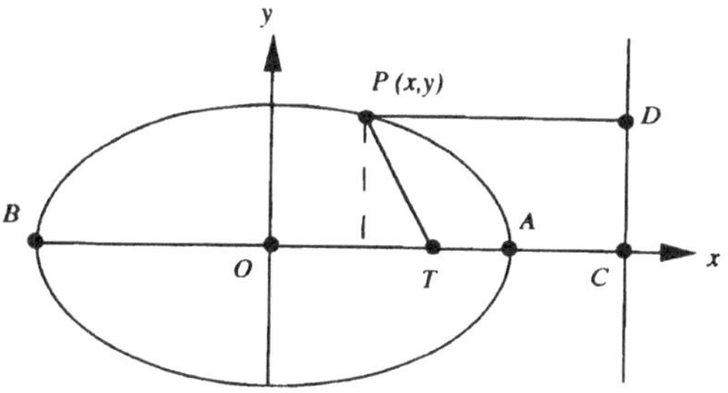

Figure 10.8

10.6 Equation of an ellipse

In Figure 10.8, T is the fixed focus and CD is the directrix of the curve. Any point P is located so that $PT = e\,PD$ where $e < 1$. When P is at A and at B, the ratio still holds therefore

$$TA = e\,AC \quad \text{and} \quad TB = e\,BC$$

Let $AB = 2a$, then

$$2a = AB = TB + TA = e(BC + AC) = e(AB + 2.AC) = e(2a + 2.AC)$$

$$a = ea + e.AC$$

$$AC = \frac{a(1-e)}{e}$$

$$OC = a + AC = a + \frac{a(1-e)}{e} = \frac{ae + a - ae}{e} = \frac{a}{e} \qquad (10.15)$$

$$TB = a + OT \quad \text{and} \quad TA = a - OT$$

therefore

$$2.OT = TB - TA = e(BC - AC) = 2ae$$

$$OT = ae \qquad (10.16)$$

Now

$$PT^2 = (OT - x)^2 + y^2 = (ae - x)^2 + y^2$$

Also

$$e^2PD^2 = e^2(OC - x)^2 = e^2\left(\frac{a}{e} - x\right)^2 = (a - ex)^2 \qquad (10.17)$$

But by definition of the ellipse

$$PT^2 = e^2PD^2$$

therefore

$$(ae - x)^2 + y^2 = (a - ex)^2$$

$$a^2e^2 + x^2 - 2aex + y^2 = x^2e^2 + a^2 - 2aex$$

so

$$x^2(1 - e^2) + y^2 = a^2(1 - e^2)$$

and finally

$$\frac{x^2}{a^2} + \frac{y^2}{a^2(1-e^2)} = 1 \qquad (10.18)$$

This is the equation of an ellipse in terms of its two parameters, the semi-major axis and the eccentricity. By inspection, it can be seen that the curve is symmetrical about both axes. Also it could have been defined from another focus at G and another directrix to the left of the curve (not shown in the figure). If $e = 0$ the equation becomes the equation of a circle of radius a, which circumscribes the ellipse. This circle is called the *auxiliary circle* to the ellipse. See Figure 10.9.

When P is at E, $x = 0$ and if we put $y = b$, we have

$$y^2 = b^2 = a^2(1 - e^2)$$

Hence we may write the equation of the ellipse in the commonly used form

$$\frac{x^2}{a^2} + \frac{y^2}{b^2} = 1 \qquad (10.19)$$

The ellipse can also be thought of as circle deformed uniformly in one direction. Consider the point Q on the auxiliary circle in Figure 10.9. It has the same abscissa (x) as P. From the equation of the auxiliary circle

$$HQ^2 = y^2 = a^2 - OH^2$$

and from the equation of the ellipse

$$\frac{OH^2}{a^2} + \frac{HP^2}{b^2} = 1$$

$$OH^2 = a^2 - a^2 \frac{HP^2}{b^2}$$

therefore

$$HQ^2 = a^2 - a^2 + a^2 \frac{HP^2}{b^2}$$

and

$$\frac{HP}{HQ} = \frac{b}{a} \tag{10.20}$$

This is true for all positions of P and Q on the ellipse and auxiliary circle. Thus the ellipse is a circle uniformly contracted in the ratio b/a.

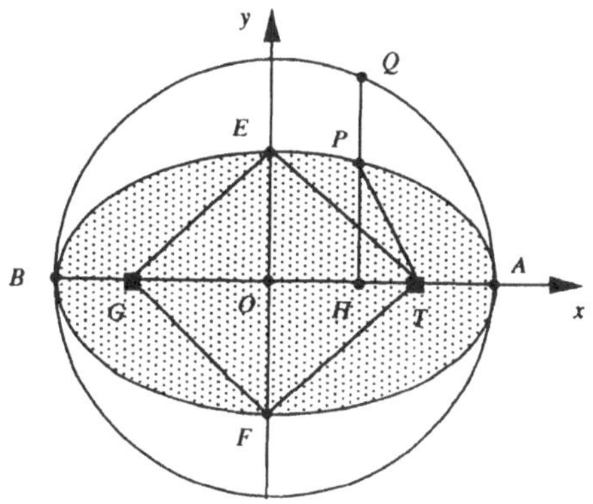

Figure 10.9

10.7 Area of an ellipse

Following equation (10.20), because all ordinates of the circle are contracted into the ellipse, their total sum is also contracted. Since the total sum of all ordinates is the area within the curve we have that

$$\frac{\text{area of the ellipse}}{\text{area of the auxiliary circle}} = \frac{b}{a}$$

$$\frac{\text{area of the ellipse}}{\pi a^2} = \frac{b}{a}$$

$$\text{area of the ellipse} = \pi ab \tag{10.21}$$

10.8 Sum of focal distances

From (10.2) and figure 10.8 we have

$$PT^2 = e^2PD^2$$

and from (10.17)

$$= (a - ex)^2$$

$$PT = a - ex \qquad (10.22)$$

The distance from P to the second focus is PG, $OG = OT = ae$, therefore

$$PG^2 = (ae + x)^2 + y^2$$

$$PG^2 = (ae + x)^2 + \left[a^2(1 - e^2) - x^2(1 - e^2)\right]$$

$$= a^2e^2 + 2aex + x^2 + \left[a^2(1 - e^2) - x^2(1 - e^2)\right]$$

$$= a^2e^2 + 2aex + x^2 + a^2 - a^2e^2 - x^2 + x^2e^2$$

$$= a^2 + 2aex + x^2e^2 = (a + ex)^2$$

therefore

$$PG = a + ex \qquad (10.23)$$

Immediately from (10.22) and (10.23) we have

$$PT + PG = a - ex + a + ex = 2a \qquad (10.24)$$

This is a most important result which, among other things, enables an ellipse to be drawn on a map or set out in a field.

Exercise 1 A cylindrical pipe of diameter 200 mm meets a wall at a point P at angle of 30°. Mark out the shape on the wall to be cut to enable the pipe to pass through. (A similar calculation is required for the other side of the wall at another point Q.) The reader is encouraged to drawn this ellipse on a piece of paper, from the following instructions.

The shape required is an ellipse with the following parameters.

$$b = 100 \text{ mm} \quad \text{and } a = 100 \operatorname{cosec} 30° = 100 \times 2 = 200 \text{ mm}$$

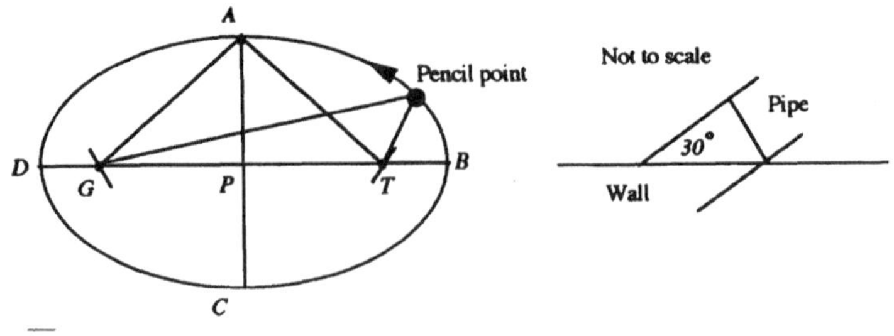

Figure 10.10

The axes are drawn through P as shown in Figure 10.10 and a and b measured off to give points A, B, C and D on the ellipse. From A and from C arcs are drawn of length a to cut the x axis in the foci at T and G. Pins are placed at these foci. Each end of a string of length $2a$ is attached to each pin. If a pencil is moved round keeping the string taut as shown, an ellipse is drawn out as required.

10.9 Freedom equations of the ellipse

It is important to be able to express the coordinates of any point $P(x, y)$ in terms of a useful independent parameter. Equations in this parameter are called *freedom* equations. The parameter most useful to the map maker is the angle U of Figure 10.11. This is the angle between the normal PC to the ellipse and the x axis. In geodetic problems this angle is the latitude of a place on the earth's surface. The line OQ passes from the origin perpendicular to the tangent at P meeting it in Q. The length, $OQ = p$, is called the *pedal distance* of P with respect to the ellipse. If we put

$$x = ka^2 \cos U \quad \text{and} \quad y = kb^2 \sin U$$

then from the equation of the ellipse

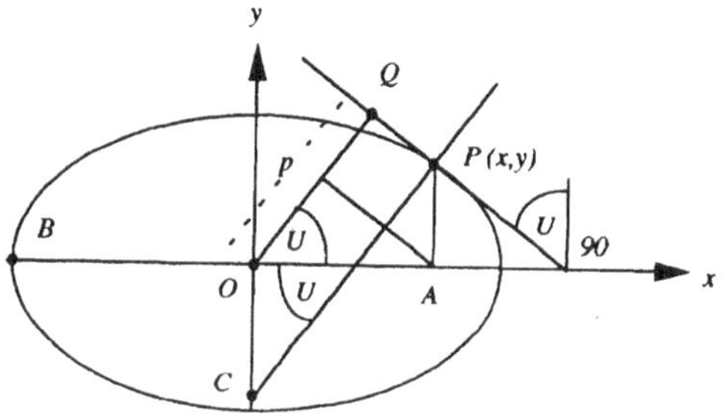

Figure 10.11

$$\frac{x^2}{a^2} + \frac{y^2}{b^2} = 1 \tag{10.25}$$

we have

$$\frac{\left(ka^2 \cos U\right)^2}{a^2} + \frac{\left(kb^2 \sin U\right)^2}{b^2} = 1 \tag{10.26}$$

which gives

$$\frac{1}{k^2} = a^2 \cos^2 U + b^2 \sin^2 U$$

or

$$\frac{1}{k^2} = a^2 \cos^2 U + a^2\left(1 - e^2\right)\sin^2 U = a^2\left(1 - e^2 \sin^2 U\right)$$

By inspection of Figure 10.11 we see that

$$p = x\cos U + y\sin U$$

and remembering that

$$x = ka^2\cos U \quad \text{and} \quad y = kb^2\sin U$$

we have

$$p = ka^2\cos^2 U + kb^2\sin^2 U$$

therefore

$$\frac{p}{k} = \frac{1}{k^2}$$

therefore

$$k = \frac{1}{p}$$

therefore

$$p^2 = a^2\cos^2 U + b^2\sin^2 U = a^2(1 - e^2\sin^2 U) \tag{10.27}$$

10.10 Radius of curvature of the ellipse

The *radius of curvature of the ellipse* is required for geodetic calculations and map projections. From (9.37) we have the radius of curvature R given by

$$R = \frac{\left[1+\left(\frac{dy}{dx}\right)^2\right]^{3/2}}{\frac{d^2y}{dx^2}} \tag{9.37}$$

Differentiating the equation for the ellipse gives

$$2\frac{x}{a^2} + 2\frac{y}{b^2}\frac{dy}{dx} = 0$$

$$\frac{dy}{dx} = -\frac{b^2x}{a^2y} \tag{10.28}$$

Therefore

$$1+\left(\frac{dy}{dx}\right)^2 = 1+\left(\frac{-b^2x}{a^2y}\right)^2 = 1+\frac{b^4x^2}{a^4y^2} = \frac{a^4y^2+b^4x^2}{a^4y^2}$$

but from

$$x = ka^2\cos U \quad \text{and} \quad y = kb^2\sin U$$

$$a^4y^2 + b^4x^2 = a^4k^2b^4\sin^2 U + b^4k^2a^4\cos^2 U = a^4k^2b^4(\sin^2 U + \cos^2 U) = a^4k^2b^4$$

therefore

$$1+\left(\frac{dy}{dx}\right)^2 = \frac{a^4k^2b^4}{a^4y^2} = \frac{k^2b^4}{y^2} \tag{10.29}$$

Now differentiating (10.28), we have

$$\frac{d^2y}{dx^2} = -\frac{d}{dx}\left(\frac{b^2x}{a^2y}\right) = -\frac{b^2}{a^2}\frac{d}{dx}\left(xy^{-1}\right) = -\frac{b^2}{a^2}\left(y^{-1} - y^{-2}\frac{dy}{dx}\right)$$

$$= -\frac{b^2}{a^2}\left(y^{-1} + y^{-2}\frac{b^2x}{a^2y}\right) = -\frac{b^2}{a^2}\left(\frac{a^2y^2 + b^2x^2}{a^2y^3}\right)$$

But from (10.19)

$$a^2y^2 + b^2x^2 = a^2b^2$$

herefore

$$\frac{d^2y}{dx^2} = -\frac{b^2a^2b^4}{a^4y^3} = -\frac{b^4}{a^2y^2} \tag{10.30}$$

Now substituting from (10.29) and (10.30) in (9.37), we have

$$R = \frac{\left(\frac{k^2b^4}{y^2}\right)^{3/2}}{-\frac{b^4}{a^2y^3}} = -\frac{k^3b^6}{y^3}.\frac{a^2y^3}{b^4} = -k^3a^2b^2$$

r because $k = \frac{1}{p}$

$$R = -\frac{a^2b^2}{p^3} \tag{10.31}$$

This is a useful form of R. The negative sign merely means that the ellipse is oncave towards the origin at its centre. To recast (10.31) into the more traditionl orm, we have

$$p^2 = a^2\left(1 - e^2\sin^2 U\right) \quad \text{and} \quad b^2 = a^2\left(1 - e^2\right)$$

herefore

$$R = -\frac{a^4\left(1 - e^2\right)}{a^3\left(1 - e^2\sin^2 U\right)^{3/2}}$$

$$|R| = \frac{a\left(1 - e^2\right)}{\left(1 - e^2\sin^2 U\right)^{3/2}} \tag{10.32}$$

10.11 Length of the normal to an ellipse

Also of great importance in geodesy is the length of PC. Now

$$PC = x\sec U = ka^2\cos U\sec U = ka^2 = \frac{a^2}{p} \tag{10.33}$$

10.12 Length of an arc of the ellipse

In map projections, the length of an elliptical arc is required. If R is the radius of curvature at point P and dS is a small arc , then

$$dS = R\,dU$$

$$S = \int R\,dU = \int \frac{a(1-e^2)}{\left(1-e^2 \sin^2 U\right)^{3/2}} dU \qquad (10.34)$$

There is no closed solution to this *elliptic integral.* It has to be converted into a series in terms of U, integrated term by term, with the results added. The number of terms used in the series depends on the accuracy needed. The whole process is tedious and not very illuminating.

Another very practical way to evaluate the integral is to do so numerically. A value of R is calculated for each of a very large number of angles U, spaced over the range of interest, by selected amounts dU. The successive values of $dS = R dU$ are added to give the required result. The accuracy of the calculation depends on how small we select dU to be.This computational process is ideally suited for a computer.

Exercise 1 The range sum from two fixed shore points P and Q, 10 km apart, is measured by an electronic system to be 12 km. Sketch the ellipse on which the ship lies, at a scale of 1 cm = 1 km.

Let the equation of the ellipse relative to the standard origin at the mid point of the base line be

$$\frac{x^2}{a^2} + \frac{y^2}{b^2} = 1$$

From the range sum we have

$$2a = 12 \quad a = 6 \quad ae = 5$$

$$e = 5/6 = 0.8333$$

$$b = a\sqrt{0.30556} = 3.317$$

Thus the equation of the ellipse is

$$\frac{x^2}{36} + \frac{y^2}{11} = 1$$

$$11x^2 + 36y^2 = 396$$

Drawing up a table of values we have

x	1	2	3	4	5	6
y	3.27	3.13	2.87	2.47	1.83	0

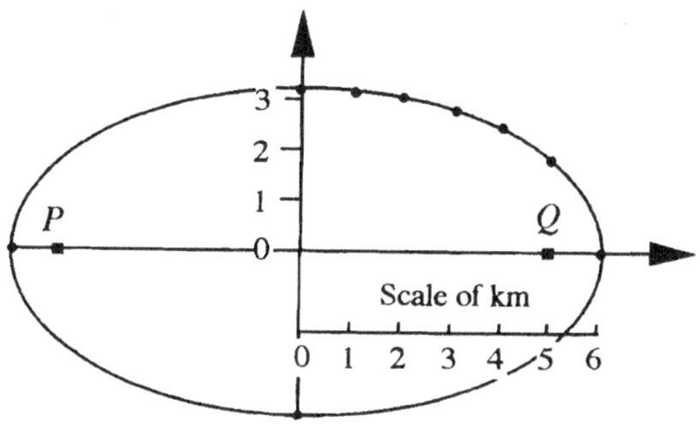

Figure 10.12

Hence we can sketch the ellipse in Figure 10.12.
Note: This sketch can be used later in conjuction with the sketch of a similar hyperbola to locate the ship. See Exercise 2 of Section 10.14.

10.13 The hyperbola

The last conic section to be considered is the hyperbola. This curve is drawn out in the intersecting plane when $V < U$ (see Figure 10.2) and $e > 1$. There are two parts of the curve where the plane cuts the double or extended cone. Figure 10.13 shows the principal section through the cone axis. In each part of the cone there is a sphere which the intersecting plane touches in points T and T_1. Associated with these spheres are two directrices. There are also two points on each directrix at C and C_1. The points of interest on the cone and curve are A and A_1. By inspection of Figure 10.13 it appears that the following relationships hold

$$TA = T_1A_1 \quad \text{and} \quad CA = C_1A_1$$

We now prove this to be the case. Remember that tangents to a circle from a common point are equal.

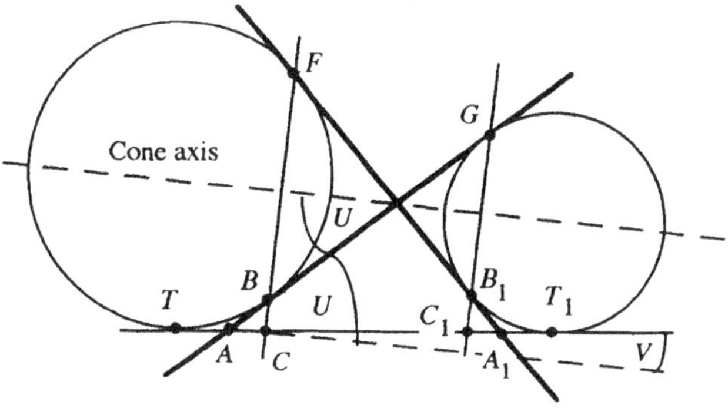

Figure 10.13

From Figure 10.13 the following stages are evident

$$BG = FB_1$$
$$AG - AB = FA_1 - B_1A_1$$
$$\therefore AT_1 - TA = TA_1 - T_1A_1$$
$$AA_1 + T_1A_1 - TA = TA + AA_1 - T_1A_1$$
$$2T_1A_1 = 2TA$$

therefore

$$T_1A_1 = TA \tag{10.35}$$

Considering similar triangles, we have

$$\frac{AC}{CC_1} = \frac{AB}{BG} \quad \text{and} \quad \frac{A_1C_1}{CC_1} = \frac{A_1B_1}{B_1F}$$

but

$$AB = TA = T_1A_1 = A_1B_1 \quad \text{and} \quad BG = B_1F$$

therefore

$$\frac{AC}{CC_1} = \frac{AB}{BG} \quad \text{and} \quad \frac{A_1C_1}{CC_1} = \frac{AB}{BG}$$

therefore

$$\mathrm{AC} = A_1C_1 \tag{10.36}$$

Again from (10.1)

$$AC_1 \cos V \sec U = AG = AT_1$$

therefore

$$AT_1 = AC_1 \cos V \sec U = e\,AC_1 \tag{10.37}$$

Note: This relationship could be considered to follow immediately from the definition of the conic using the second focus and directrix. However, experience shows that it needs to be established separately as above.

From these three results (10.35), (10.36) and (10.37) we can establish the equation of the hyperbola in the traditonal form

$$\frac{x^2}{a^2} - \frac{y^2}{b^2} = 1 \tag{10.38}$$

Figure 10.14 shows the plane containing the conic section and the line of Figure 10.13, $TACC_1A_1T_1$. The directrices with respective foci T and T_1 are CD and C_1D_1. Let O be the mid point of AA_1 and let

$$AA_1 = 2a$$

Also

$$AT = e\,AC \text{ and from (10.37) } AT_1 = e\,AC_1$$

thus

$$AT_1 - AT = e\,(AC_1 - AC)$$

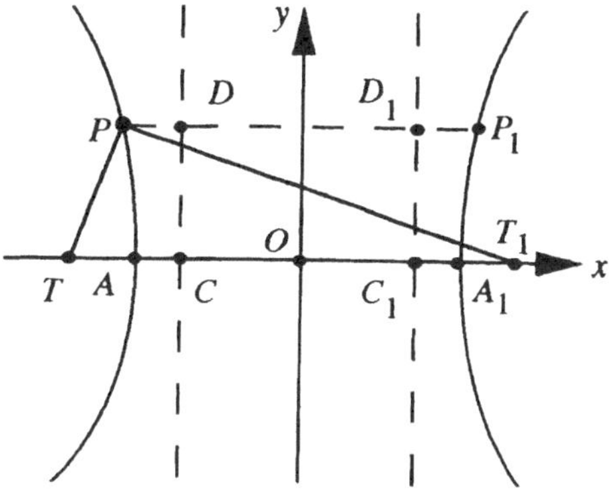

Figure 10.14

$$\therefore AT_1 - A_1T_1 = 2a = eCC_1 = 2e.OC \quad (AT = A_1T_1 \text{ from 10.35})$$

$$\therefore OC = \frac{a}{e} \tag{10.39}$$

$$AC = a - \frac{a}{e}$$

$$AT = ae - a = a(e-1)$$

thus

$$OT = a + AT = ae \tag{10.40}$$

By definition, for any point P (x, y) on the conic,

$$PT = e\,PD$$

therefore

$$PT^2 = e^2PD^2$$

$$(OT - x)^2 + y^2 = e^2(x - OC)^2$$

$$\therefore (ae - x)^2 + y^2 = e^2\left(x - \frac{a}{e}\right)^2 = (ex - a)^2 \quad \text{(from 10.39 and 10.40)}$$

$$a^2e^2 - 2aex + x^2 + y^2 = x^2e^2 - 2aex + a^2$$

$$x^2(e^2 - 1) - y^2 = a^2(e^2 - 1)$$

Finally, putting

$$b^2 = a^2(e^2 - 1)$$

we have

$$\frac{x^2}{a^2} - \frac{y^2}{b^2} = 1 \tag{10.38}$$

10.14 Range difference property

For the map maker, perhaps the most important property of the hyperbola is that the *range difference* from the two foci to a point on the curve is constant. Consider the range difference

$$PT_1 - PT = ePD_1 - ePD = e\left[\left(x + \frac{a}{e}\right) - \left(x - \frac{a}{e}\right)\right]$$

therefore

$$PT_1 - PT = 2a \qquad (10.41)$$

Range difference systems, such as in hydrographic position fixing or satellite Doppler systems, define hyperbolae. These in turn are used to find an observer's position.

Exercise 1 The range difference from two fixed shore points P and Q, 10 km apart, is measured by an electronic system to be 8 km. Sketch the hyperbola on which the ship lies, at a scale of 1 cm = 1 km.

Let the equation of the hyperbola relative to the standard origin at the mid point of the base line be

$$\frac{x^2}{a^2} - \frac{y^2}{b^2} = 1 \qquad (10.38)$$

From the range difference we have

$$2a = 8$$

and when

$$y = 0 \quad x = \pm a = \pm 4$$

The base line is 10 km long so

$$AT = 1 = a(e - 1) = 4\,(e - 1)$$

$$e = 1.25$$

$$b^2 = 4^2(1.25^2 - 1) = 16 \times 0.5625 = 9$$

The equation of the hyperbola is

$$\frac{x^2}{16} - \frac{y^2}{9} = 1$$

$$9x^2 - 16y^2 = 144 \qquad (10.42)$$

when

$$x = ae = 5 \quad y = \pm a\,(e^2 - 1) = \pm 4 \times 0.5625 = \pm 2.25$$

Corresponding selected values are

x	4	4.5	5	6	7	8	9
y	0	1.55	2.25	3.35	4.31	5.20	6.05

Hence we can sketch the two branches or *sheets* of the curve which is symmetrical about both axes.

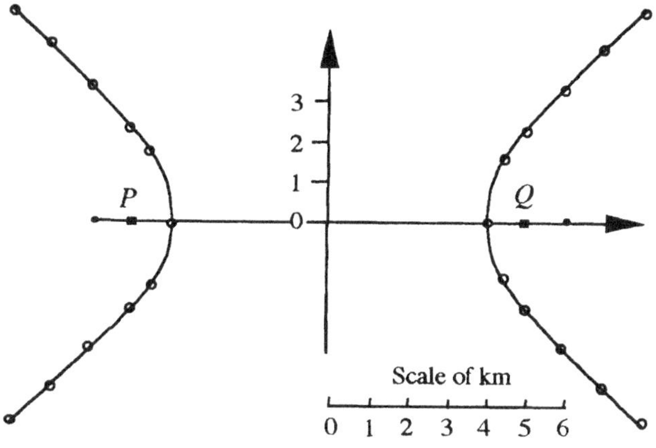

Figure 10.15

Exercise 2 If we now combine the two sketches, of the hyperbola (Figure 10.15) and the ellipse (Figure 10.12) into Figure 10.16, we produce part of a *lattice chart* showing the four possible positions of the ship. It is usually possible to select the correct solution using other information such as one point is the only one where water can be located! From the graphs drawn to scale we find the position of the ship to be (4.75, 2.05). These values do not satisfy the equations exactly. A better solution (4.8, 2.0) is obtained by Newton's method in Section 12.16.

Traditionally, mariners worked from *lattice charts* to obtain solutions to navigation problems. Today, solutions are obtained by purely analytical methods solving for the intersection of the conics by computer software.

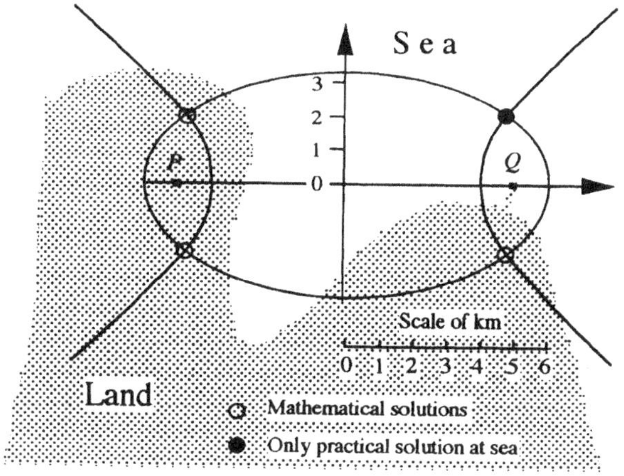

Figure 10.16

SUMMARY OF KEY WORDS

conic sections, circle, isosceles, parabola, ellipse, hyperbola, eccentricity, focus, directrix, radius of curvature of the ellipse, auxiliary circle to the ellipse, pedal distance, freedom equation, radius of curvature of the ellipse, elliptic integral, range sum, range difference, lattice charts

SUMMARY OF FORMULAE

Circle $$X^2 + Y^2 + 2fX + 2gY + c = 0 \quad (10.8)$$

Parabola $$y^2 = 4ax \quad (10.14)$$

Ellipse $$\frac{x^2}{a^2} + \frac{y^2}{b^2} = 1 \quad (10.19)$$

$$b^2 = a^2(1 - e^2)$$

$$\text{Area of the ellipse} = \pi ab \quad (10.21)$$

Pedal distance $$p^2 = a^2\,(1 - e^2 \sin^2 U) \quad (10.27)$$

$$R = \frac{a(1 - e^2)}{(1 - e^2 \sin^2 U)^{3/2}} \quad (10.32)$$

Length of normal to ellipse $$PC = \frac{a^2}{P} \quad (10.33)$$

Hyperbola $$\frac{x^2}{a^2} - \frac{y^2}{b^2} = 1 \quad (10.38)$$

Chapter 11
Spherical Trigonometry

11.1 Introduction

The study of the trigonometrical formulae connecting parts of a sphere is called *spherical trigonometry.* No new mathematical functions are involved: we are simply using standard trigonometrical functions applied to a sphere. Although most problems to do with spheres can be solved without the use of this branch of mathematics, it is sometimes more convenient to do so with it. Map makers mainly use spherical trigonometry to solve problems in field astronomy, navigation and map projections, although it also has applications to surveying instrumentation.

It is helpful if the reader has available a spherical ball, such as a tennis ball or table tennis ball, for inspection as we define terms and consider ideas.

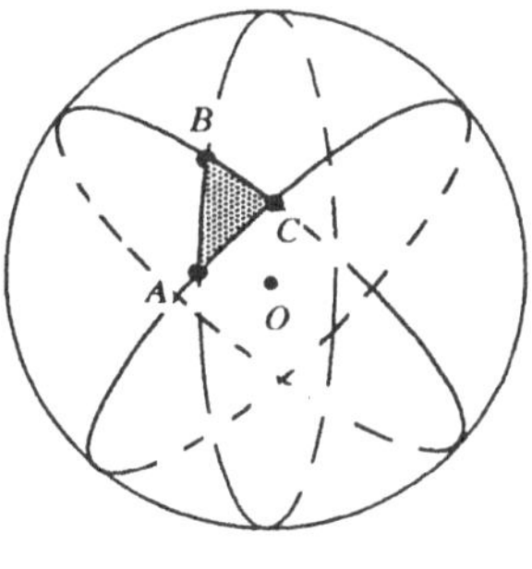

Figure 11.1

11.2 The sphere

A sphere can be defined in the following way with reference to Figure 11.1. Consider O, a fixed point in space and from it sweep out a surface by rotating a line of fixed length R. This surface forms a sphere, of radius R, centred on the fixed point. Any plane cutting the sphere does so in a circle, and, if the plane also passes through the centre of the sphere, the circle formed is the largest possible.

It is therefore called a *great circle* of the sphere. Clearly the radius of a great circle is also R. All circles smaller than a great circle are called *small circles* of the sphere. *In spherical trigonometry we are concerned solely with great circles.*

11.3 The spherical triangle

A spherical triangle ABC is formed by three great circles of a sphere as shown in Figure 11.1. A good way to visualise this is to draw great circles on a table-tennis ball. To do this, place the ball in an egg cup, hold a pencil point against the ball and rotate the ball carefully creating a great circle on its surface.

11.4 Model of spherical triangle

To assist the reader the following model of a spherical triangle should be constructed from a sheet of paper A4 size. With a centre O and radius about 15 cm describe the arc of a circle C_1BACB_1 as shown in Figure 11.2. Mark off, in the order indicated, the angles

$$a = 30° \quad c = 50° \quad b = 40° \quad a = 30°$$

Notice that these angles are denoted by lower case letters. Mark the points C_1, B, A, C and B_1 on the circumference as shown. Draw straight lines CFD and C_1ED perpendicular to OA and OB respectively to meet at D. Draw FHG perpendicular to OB, and HD perpendicular to GF, to complete the rectangle $GEDH$. Note that

$$\text{angle } HFD = c$$

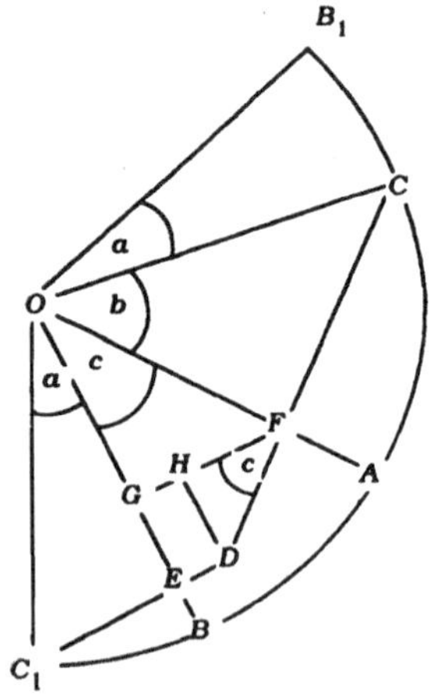

Figure 11.2

Carefully cut out the diagram. Fold it inwards along the radii and construct the spherical triangle ABC with the two faces OC_1B and OCB_1 overlapping. Pin these together. In the folded position, a shape like that illustrated in Figure 11.3 is obtained. Hold the model in this position to compare it with the drawing. The angles at the vertices of the spherical triangle are A, B and C, written in upper case letters. These are the angles between the tangents to the sphere at a vertex. (You should verify this by looking inwards along a radius of the model.)

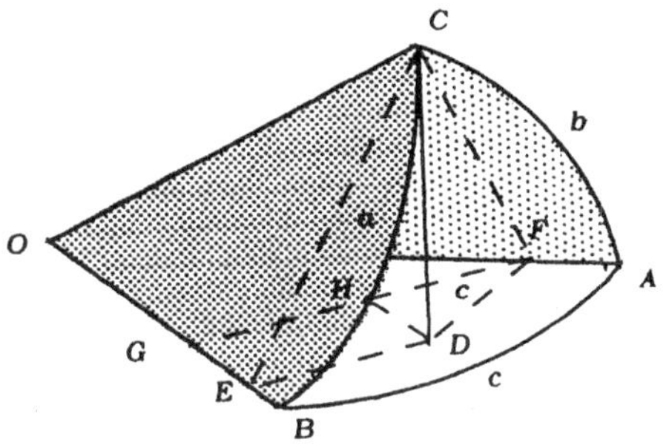

Figure 11.3

The sides of the triangle are Ra, Rb, and Rc respectively where R is the radius of the sphere and the small angles are converted to radians. Because the relative relationships of the sides and angles do not depend on the size of the sphere, it is usual to put $R = 1$ and deal only with the angles subtended at the centre. It is customary still to refer to these as 'sides' even although they are actually *angles*.

11.5 Spherical trigonometrical formulae

As with plane trigonometry, there is one basic formula in spherical trigonometry, the *cosine formula*, from which the others may be derived. It is

$$\cos a = \cos b \cos c + \sin b \sin c \cos A \qquad (11.1)$$

However, the proof given here lends itself to derive two other important formulae directly and with little extra effort. These are the *sine formula*

$$\frac{\sin a}{\sin A} = \frac{\sin b}{\sin B} = \frac{\sin c}{\sin C} \qquad (11.2)$$

and the *cotangent formula*

$$\cos A \cos c = \cot b \sin c - \sin A \cot B \qquad (11.3)$$

11.6 Proof of the cosine formula

By inspection of Figure 11.3, we have the following relationships

$$\text{angle } CFD = A \quad \text{and} \quad \text{angle } CED = B$$

CD is perpendicular to the plane OBA

Consider triangles OCF, CDF and FDH in turn to give

From Figure 11.2 $\quad CF = OC \sin b = R \sin b \quad$ (11.4)

From Figure 11.3 $\quad FD = CF \cos A = R \sin b \cos A \quad$ (11.5)

From Figure 11.2 $\quad HD = FD \sin c = R \sin b \cos A \sin c \quad$ (11.6)

From triangles OCE, OCF and OFG in turn we have also

$$OE = R \cos a \tag{11.7}$$

$$OF = R \cos b \tag{11.8}$$

$$OG = OF \cos c = R \cos b \cos c \tag{11.9}$$

$$HD = GE = OE - OG = R \cos a - R \cos b \cos c \tag{11.10}$$

Equating values of HD from (11.6) and (11.10), R cancels out, giving the cosine formula:

$$\cos a = \cos b \cos c + \sin b \sin c \cos A \tag{11.1}$$

This is the most useful formula of spherical trigonometry. It relates two sides and their included angle to the side opposite. There is no sign ambiguity in a between the first and second quadrants, because $\cos a$ is positive in the first and negative in the second quadrant.

Exercise 1 Write down the two other cases of the cosine formula relating the angles B and C to the 'sides'. These are

$$\cos b = \cos a \cos c + \sin a \sin c \cos B$$

$$\cos c = \cos a \cos b + \sin a \sin b \cos C$$

Exercise 2 Calculate the angles of the spherical model in which

$$a = 30° \quad b = 40° \quad c = 50°$$

We will give the working for A only. From (11.1)

$$\cos A = \frac{\cos a - \cos b \cos c}{\sin b \sin c}$$

$$= \frac{0.8660 - 0.7660 \times 0.6428}{0.6428 \times 0.7660}$$

$$= \frac{0.8660 - 0.4924}{0.4924}$$

$$= \frac{0.3736}{0.4924} = 0.7587$$

$$A = 40.65°$$

Verify this result by direct measurement on the paper model. It is left to the reader to check that the other two results are

$$B = 56.86° \quad \text{and} \quad C = 93.68°$$

Note: The sum of the angles of a spherical triangle do not add up to 180°. In this case the sum is 191.19°. The excess of this sum over 180° is called the *spherical excess e* of the triangle.Or we put

$$A + B + C - 180° = e = 11.19°$$

11.7 Proof of the sine formula

In triangles OC_1E, OCF of Figure 11.2, and triangles CED and CFD of Figure 11.3, we have respectively

$C_1E = R \sin a \quad CF = R \sin b \quad CD = R \sin a \sin B \quad CD = R \sin b \sin A$

Equating the two expressions for CD gives the sine rule

$$\frac{\sin a}{\sin A} = \frac{\sin b}{\sin B}$$

Similarly, by dropping a perpendicular from B to plane OAC, we could show that

$$\frac{\sin a}{\sin A} = \frac{\sin c}{\sin C}$$

therefore

$$\frac{\sin a}{\sin A} = \frac{\sin b}{\sin B} = \frac{\sin c}{\sin C} \tag{11.2}$$

This formula links *sides* with their opposite *angles*. Although a convenient formula, care has to be taken with signs, because the sine is positive in both the first and second quadrants. For this reason, an apparently more complicated alternative formula is often preferred.

Exercise 1 Use the sine rule to check the calculations of the angles A, B and C of Exercise 2 of Section 11.6. The respective 'sides' and angles are

a	30°	A	40.65°	$\sin a/\sin A = 0.7675$
b	40°	B	56.86°	$\sin b/\sin B = 0.7676$
c	50°	C	93.68°	$\sin c/\sin C = 0.7676$

The minor discrepancy is due to rounding errors in the arithmetic.

11.8 Proof of the cotangent formula

Consider triangles HFD and OCF of Figure 11.2 and triangle CDF of Figure 11.3 to give

$$HF = DF \cos c = R \sin b \cos A \cos c$$

From triangles OFG, OCF, CED, and OCE we have

$$HF = GF - GH = GF - ED = R \cos b \sin c - R \sin a \cos B$$

Equating values of HF gives

$$\sin b \cos A \cos c = \cos b \sin c - \sin a \cos B$$

Dividing by $\sin b$ gives

$$\cos A \cos c = \cot b \sin c - \frac{\sin a \cos B}{\sin b}$$

$$\cos A \cos c = \cot b \sin c - \sin A \cot B \qquad (11.3)$$

This cotangent formula relates four adjacent parts of the triangle.

Exercise 1 Use the cotangent formula to check the calculations of the angles A, B and C of Exercise 2 of Section 11.6. The data are

a	30°	A	40.65°
b	40°	B	56.86°
c	50°	C	93.68°

$$\cos A \cos c = \cot b \sin c - \sin A \cot B \qquad (11.3)$$

We shall calculate B from

$$\begin{aligned}
\cot B &= \frac{\cot b \sin c - \cos A \cos c}{\sin A} \\
&= \frac{\cot 40° \sin 50° - \cos 40.65° \cos 50°}{\sin 40.65°} \\
&= \frac{1.1918 \times 0.7660 - 0.7587 \times 0.6428}{0.6514} \\
&= \frac{0.9129 - 0.4877}{0.6514} \\
&= \frac{0.4252}{0.6514} \\
&= 0.6527 \\
B &= 56.86°
\end{aligned}$$

11.9 Spherical excess

The amount by which the sum of the angles of a spherical triangle exceeds 180° is called the spherical excess e: in other words

$$A + B + C = 180° + e \qquad (11.11)$$

On a sphere of radius R, the spherical excess in seconds of arc of a triangle, whose area is Δ, is given by

$$e'' = \frac{\Delta}{R} \times 206\,265 \qquad (11.12)$$

Thus a terrestrial triangle, such as an equilateral triangle of side 20 km, whose area is about 180 km^2 has a spherical excess of one second.

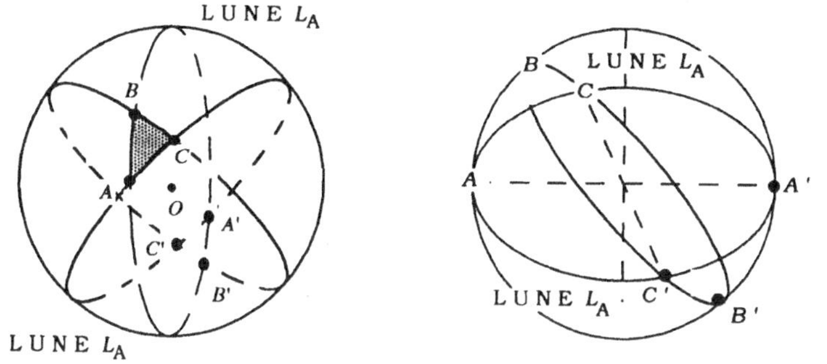

Figure 11.4

To prove equation (11.12) refer to Figure 11.4 which show two aspects of the same diagram. It will be seen that a second spherical triangle $A'B'C'$ is shown on the opposite side of the sphere from ABC and that AOA', BOB' and COC' are diameters of the sphere. The triangle $A'B'C'$ is identical to ABC and has the same area. Suppose the sphere is cut into four pieces, like the segments of an orange or tangerine, along the planes ACC' and ABB'. The two smaller segments formed by the angle BAC form the moon of A. The surface of this moon is called a *lune* L_A. The beginner may care to use a table tennis ball or orange as a model on which to draw lunes to follow the arguments used in the derivation. The surface area of a complete sphere can be thought of as the area swept out by a great circle as it rotates through 180°, or π radians, about a diameter. For any intermediate angle A the surface area of the *lune* swept out, L_A, is proportional to **A**, thus we have

$$\frac{A}{\pi} = \frac{L_A}{4\pi R^2}$$

Remembering, from (9.44) that the surface area of a sphere of radius R is

$$4\pi R^2$$

The formula for the spherical excess follows by considering the surface areas of the lunes L_A, L_B and L_C formed by the three angles A, B and C. Adding these areas we have

$$\frac{A}{\pi} + \frac{B}{\pi} + \frac{C}{\pi} = \frac{L_A}{4\pi R^2} + \frac{L_B}{4\pi R^2} + \frac{L_C}{4\pi R^2}$$

$$A + B + C = \frac{1}{4R^2}(L_A + L_B + L_C)$$

Let the area of the triangle ABC and of $A'B'C'$ be Δ. If we add the areas of the lunes, the area of triangle ABC is covered six times, two of which are needed to obtain the curved surface area of the sphere, four others are redundant. Thus

$$L_A + L_B + L_C = \text{the surface area of the sphere} + 4\Delta$$
$$= 4\pi R^2 + 4\Delta$$

therefore

$$A + B + C = \frac{1}{4R^2}\left(L_A + L_B + L_C\right)$$
$$= \frac{1}{4R^2}\left(4\pi R^2 + 4\Delta\right)$$
$$= \pi + \frac{\Delta}{R^2}$$

and we have finally

$$e = \frac{\Delta}{R^2} \qquad (11.13)$$

where e is in radians.

Exercise 1 Calculate the area of the triangle ABC of the model if $R = 120$ mm. From (11.11) the spherical excess is

$$11.19° = 0.1953 \text{ radians}$$

therefore

$$\Delta = eR^2 = 0.1953 \times 120 \times 120 \text{ mm}^2$$
$$= 2812.32 \text{ mm}^2$$

11.10 Navigation and spherical trigonometry

An aircraft is to fly from London to New York at a speed of 800 km per hour. How long will the flight take, and in what direction will the plane leave London? The radius of the Earth is 6378 km and the latitudes and longitudes of the cities are

City	Latitude North	Longitude West
London	51.30°	00.10°
New York	40.40°	73.50°

Refer to Figure 11.5. In this case we select the point A to be at the Earth's North Pole, B is at London and C is at New York on a spherical Earth. The arc $AB = c$ is the angular distance from London to the pole, the complement of its latitude. Therefore

$$c° = 90 - 51.30 = 38.70$$

and similarly the arc $CA = b$, the complement of the latitude of New York, there fore

$$b° = 90 - 40.40 = 49.60$$

The angle at the North Pole is the difference in longitude between the two cities, that is

$$A° = 73.50 - 00.10 = 73.40$$

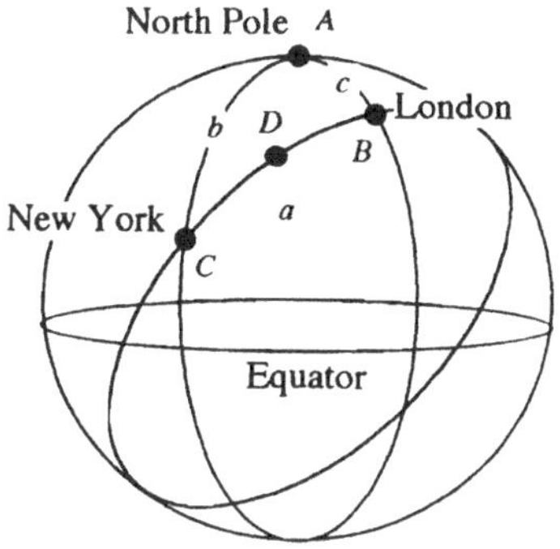

Figure 11.5

To find the inter-city distance we calculate the arc BC by the cosine formula

$$\begin{aligned}\cos a &= \cos b \cos c + \sin b \sin c \cos A \\ &= 0.6481 \times 0.7804 + 0.7615 \times 0.6252 \times 0.2857 \\ &= 0.5058 + 0.1360 \\ &= 0.6418 \\ a &= 50.07° = 0.8740 \text{ rad}\end{aligned}$$

The linear distance is

$$6378 \times 0.8740 = 5574 \text{ km}$$

Since the speed of the aircraft is 800 km per hour the flight time is 6.97 hours.

11.11 Direction at take-off

The direction at take-off is the angle B given again by (11.3) or by

$$\begin{aligned}\sin B &= \frac{\sin A \sin b}{\sin a} \\ &= \frac{0.9583 \times 0.7615}{0.7668} \\ B &= 72.11°\end{aligned}$$

The direction from north in which the plane takes off is 287.88°.

Note: As the plane flies along the great circle BC the pilot has to alter the plane's direction relative to north from time to time. Spherical triangles are solved at these times to give the new direction.

Exercise 1 Find the direction to fly when the plane is halfway between the two cities. In this case the cotangent formula is most appropriate because four adjacent parts of the triangle are involved. The angle $ADB = D$ is required. It is given by

$$\cot D = \frac{\sin\frac{1}{2}a\cot c - \cos B\cos\frac{1}{2}a}{\sin B}$$

$$c = 38.70^\circ \quad B = 72.11^\circ \quad a = 50.07^\circ,\ \tfrac{1}{2}a = 25.035^\circ$$

$$\begin{aligned}\cot D &= \frac{0.4232 \times 1.2482 - 0.3072 \times 0.9060}{0.9516} \\ &= 0.2627 \\ D &= 75.28^\circ\end{aligned}$$

The required forward flying direction is now 255.28°.

11.12 Map projections of a sphere

Map projections were formerly chosen for their ease of computation and drawing. Computer methods free the user from such restrictions. We shall demonstrate procedures with an example of the *oblique stereographic projection*, which has the property of being *conformal* (shows small shapes correctly) and is therefore suitable for topographical mapping. Although it also happens to be easy to draw by graphical methods, we will demonstrate the computational procedures here. The position of a point on the surface of a sphere is defined by its latitude (the angular distance from the equator towards the North Pole or South Pole) and its longitude (the angular distance round the equator from a standard meridian, such as the meridian of Greenwich). Whilst the origin of latitude has some natural basis, the origin of longitude is purely arbitrary and was selected for historical reasons. In mapping, any point can be chosen as the 'pole' and any line for reference.

11.13 Oblique coordinate system

In the case of an oblique projection, the first stage is to select the geographical position of the *centre of the map* and the *reference line* from which the spherical arc distances and directions to the points on the graticule are computed by spherical trigonometry.We have selected an example from a popular atlas. Refer to Figure 11.6. The reference line is the meridian through the centre of projection at the point Q whose geographical coordinates are

Latitude $\phi = 40^\circ$ N   Longitude $U = 85^\circ$ East of Greenwich

The points of the graticule to be plotted are A to J (see Figure 11.6) at the intersections of parallels 40°, 60° and 80°, with the meridians 20° West, 0° and 20° East of Greenwich.The land mass covered is Western and central Europe.

Consider a typical point F (60° N, 20° E) The first stage is to change the origin

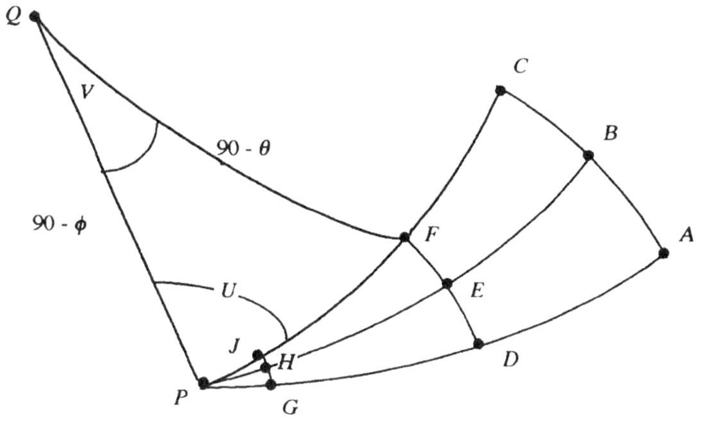

Figure 11.6

of longitude to the point Q, thus the new longitude of F becomes 65° W of Q. The other longitudes become 85° W and 105° W of Q.

We have to calculate the arc distance QF and the direction V to obtain the coordinates of F referred to the new origin at Q. Applying the cosine formulae (11.1) to the spherical triangle PFQ gives

$$\cos(90-\theta_F)=\cos(90-\phi_F)\cos(90-\phi_Q)+\sin(90-\phi_F)\sin(90-\phi_Q)\cos U_F \qquad (11.14)$$

Exercise 1 Given the coordinates of F (60°N, 65°W from Q), and Q (40°N, 0) calculate $(90-\theta_F)$.

$$\cos(90-\theta_F) = \cos(30°)\cos(50°) + \sin(30°)\sin(50°)\cos(65°)$$

therefore

$$(90-\theta_F) = 44.066°$$

To find V we again use the cosine rule to give

$$\cos(90-\phi_F)=\cos(90-\theta_F)\cos(90-\phi_Q)+\sin(90-\theta_F)\sin(90-\phi_Q)\cos V_F \qquad (11.15)$$

$$\cos V_F = \frac{\cos(90-\phi_F)-\cos(90-\theta_F)\cos(90-\phi_Q)}{\sin(90-\theta_F)\sin(90-\phi_Q)}$$

$$= \frac{\cos(30°)-\cos(44.066°)\cos(50°)}{\sin(44.066°)\sin(50°)}$$

$$V_F = 40.660°$$

Exercise 2 Verify that the spherical coordinates in degrees of points A to J referred to Q are as follows

Table 11.1

	$p = 90 - \theta°$	$V°$
A	74.85	50.05
B	62.33	59.50
C	48.61	67.73
D	62.77	32.90
E	53.84	38.09
F	44.07	40.66
G	53.23	12.09
H	49.86	13.08
J	46.43	12.54
Pole	50.00	00.00

11.14 Stereographic projection of a sphere

Consider a typical point such as F on a generating sphere of radius R representing the Earth. See Figure 11.7. The plane of the map touches this sphere at Q. In the stereographic projection the point of projection is chosen to be at S diammetrically opposite to Q. The line through S and F meets the map plane in F'.

To calculate QF'

The angular arc distance $p_F = 90 - \theta_F = 44.07°$ has already been calculated by spherical trigonometry. We obtain the length of $QF' = r_F$ from

$$QF' = r_{F'} = 2R \tan \tfrac{1}{2} p_F \qquad (11.16)$$

This follows because

$$\text{angle } QSF = \tfrac{1}{2} p_F$$

Exercise 1 Show that $QF' = 56.8$ mm using $R = 70.2$ mm, and $p_F = 90 - \theta_F = 44.07°$.

$$QF' = r_{F'} = 2R \tan \tfrac{1}{2} p_F = 56.82 \text{ mm}$$

11.15 Azimuthal property of the stereographic projection

The stereographic projection is one example of projections in which we choose to draw the angles through the origin at Q correctly preserved on the map, i.e. the same on the sphere as on the map. Thus we draw the lines radiating from Q at the computed angles V. This means the the *azimuths (true directions)* at Q are correct. If we select the y axis of the map to pass through the North Pole N, and the x axis pointing to the west, the final cartesian coordinates of the projected point F are

$$x = r\sin V \quad \text{and} \quad y = r\cos V \qquad (11.17)$$

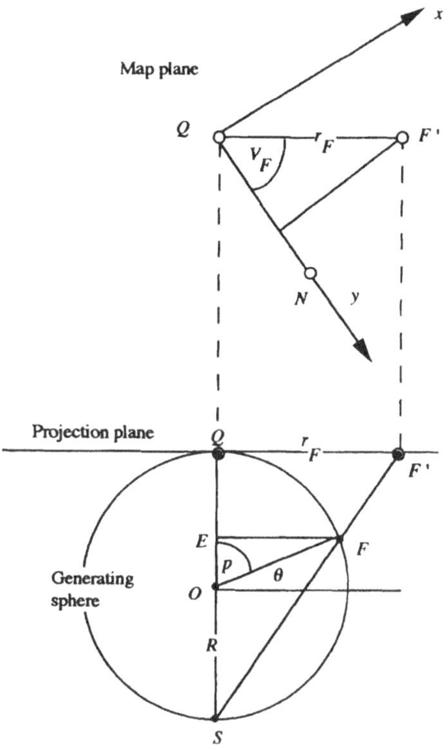

Figure 11.7

Exercise 1 Verify that the cartesian coordinates in mm of all points *A* to *J* are as follows

Table 11.2

	r	x	y
A	107.4	82.4	69.0
B	84.9	73.2	43.1
C	63.4	58.7	24.0
D	85.7	46.5	71.9
E	71.3	44.0	56.1
F	56.8	37.0	43.1
G	70.4	14.7	68.8
H	65.3	14.8	63.6
J	60.2	13.1	58.8
Pole	65.5	0	65.5

Figure 11.6 shows these points plotted at a suitable scale and the graticule sketched in.

Exercise 2 Plot Figure 11.6 from the data of Table 11.2 and verify that the lines of the sketched graticule are all circles, and that they meet at right angles. (It is a property of the stereographic projection of the sphere that *all* circles on the surface of the sphere appear as circles on the map.)

11.16 Conformality of the stereographic projection

The scale factor K of a map relates the map distance to the corresponding ground distance by

$$K = \frac{\text{map distance}}{\text{ground distance}} = \frac{\text{map distance}}{\text{spherical distance}}$$

Because of the distortion introduced when converting the three-dimensional surface of the sphere to a two-dimensional plane, the scale factor varies from point to point over the map. If the scale factor at a point *does not vary with direction* the shapes of small land parcels are correctly preserved but at a different size. Such a projection is called *conformal* or *orthomorphic*. For example, the theodolite circle on which horizontal angles are observed, maps into another circle on these projections. At finite distances over 100 metres or so the accuracy of angle measurement cannot be preserved and distortion occurs. For further information on this topic see Allan (1997). An examination of the scale factors at a point, in two orthogonal directions, reveals whether a projection is conformal or not.

It is convenient to select the two directions of the spherical coordinates for an examination of scale factors.

11.17 Scale factor in direction *QF*

Refer to Figure 11.8. In the vicinity of F a small angular change to p of δp will cause a corresponding change to the spherical arc of δs given by

$$\delta s = R\,\delta p \tag{11.18}$$

This is the change in ground distance caused by a small change in the angle p. The corresponding change to the map distance in the direction QF' in the vicinity of the projected point F' is

$$\delta r = \frac{\mathrm{d}r}{\mathrm{d}p}\delta p \tag{11.19}$$

$$= \frac{\mathrm{d}}{\mathrm{d}p}\left(2R\tan\tfrac{1}{2}p\right)\delta p$$

$$= 2R\sec^2\tfrac{1}{2}p.\left(\tfrac{1}{2}\right)\delta p$$

$$= \frac{R}{\cos^2\tfrac{1}{2}p}\delta p \tag{11.20}$$

The scale factor along QF' is therefore

$$K_1 = \frac{\delta r}{\delta s} = \frac{R}{R\cos^2\tfrac{1}{2}p} = \sec^2\tfrac{1}{2}p \tag{11.21}$$

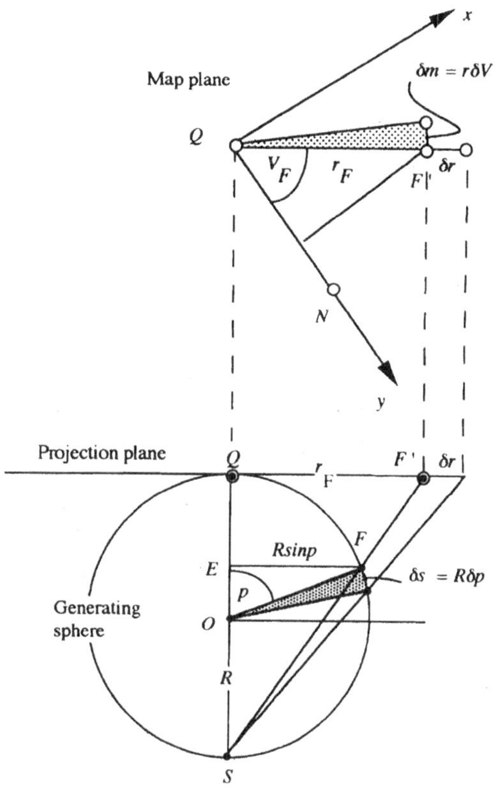

Figure 11.8

11.18 Scale factor in direction orthogonal to *QF*

A small arc of the sphere orthogonal to the plane of Figure 11.8 in the vicinity of *F* is given by

$$\delta S = R \sin p \, \delta V \tag{11.22}$$

This is because the effective radius is the length *EF* of Figure 11.8. Because of the azimuthal property of the projection, the corresponding map distance is

$$\delta m = r \, \delta V \tag{11.23}$$

Thus the scale factor orthogonal to *QF'* is given by

$$K_2 = \frac{\delta m}{\delta S} = \frac{r\delta V}{R \sin p \delta V} \tag{11.24}$$

$$= \frac{2\tan\frac{1}{2}p}{\sin p} = \frac{2\sin\frac{1}{2}p}{\cos\frac{1}{2}p} \cdot \frac{1}{2\sin\frac{1}{2}p\cos\frac{1}{2}p}$$

therefore

$$K_2 = \sec^2 \tfrac{1}{2} p \qquad (11.25)$$

and

$$K_1 = K_2$$

The point scale factors in two orthogonal directions are equal, therefore the pro-jection is conformal. Inspection of the plotted graticule shows that its lines meet at right angles as they do on the sphere.

SUMMARY OF KEY WORDS

spherical trigonometry, great circle, small circles, 'sides', angles, cosine formula, sine formula, cotangent formula, spherical excess, lune, map projections, oblique stereographic projection, conformal, azimuths (true directions), scale factor, orthomorphic

SUMMARY OF FORMULAE

$$\cos a = \cos b \cos c + \sin b \sin c \cos A \qquad (11.1)$$

$$\frac{\sin a}{\sin A} = \frac{\sin b}{\sin B} = \frac{\sin c}{\sin C} \qquad (11.2)$$

$$\cos A \cos c = \cot b \sin c - \sin A \cot B \qquad (11.3)$$

Spherical excess

$$A + B + C - 180° = e \qquad (11.11)$$

$$e'' = \frac{\Delta}{R^2} \times 206\,265 \qquad (11.12)$$

Lunes

$$e = \frac{\Delta}{R^2} \qquad (11.13)$$

Stereographic projection

$$r_F = 2R \tan \tfrac{1}{2} p_F \qquad (11.16)$$

$$K_1 = K_2 = \sec^2 \tfrac{1}{2} p \qquad (11.25)$$

Chapter 12
Solution of Equations

12.1 Introduction

Many problems in mathematics involve the solution of equations of various forms. Typical examples are

$$ax + by + cz + d = 0 \quad \text{(A)}$$
$$ax^2 + bx + c = 0 \quad \text{(B)}$$
$$ax^3 + bx^2 + cx + d = 0 \quad \text{(C)}$$
$$ax^2 + by^2 + c = 0 \quad \text{(D)}$$

The *unknowns* (or *variables*) are x, y and z, while a, b, c and d are *coefficients* usually given by theory. Typical numerical versions of these equations are

$$4x - 2y + 4z - 14 = 0 \quad \text{(A)}$$
$$2x^2 - 5x - 3 = 0 \quad \text{(B)}$$
$$2x^3 - 22x^2 + 72x - 72 = 0 \quad \text{(C)}$$
$$11x^2 + 36y^2 - 396 = 0 \quad \text{(D)}$$

Some solutions to these equations are

(A) $x = 1$ $y = -1$ $z = 2$
(B) $x = 3$ and $x = -0.5$
(C) $x = 2$ and $x = 3$ and $x = 6$
(D) $x = 4.80$ $y = 1.99$

If we substitute the values in each case, the equations are satisfied: for example, testing (A) gives

$$4(1) - 2(-1) + 4(2) - 14 = 4 + 2 + 8 - 14 = 0$$

An equation of type (A) is called a *linear equation* because the three variables appear only directly, unlike the other three *non-linear* types in which *higher powers* of the unknowns than the first are present. Again equations (B) and (C) only contain one variable x. There are usually two solutions for type (B) which is called a *quadratic equation,* and three solutions for type (C) which is called a *cubic equation.* Equation (D) is called a *second-order* equation in two variables. In this chapter we discuss some common methods of solving such equations.

12.2 Linear equations

The treatment of linear equations requires an understanding of matrix algebra (see Chapter 7). Linear equations are of the form

$$ax + by + cz + d = 0$$

They contain no powers of the variables x, y and z higher than the first. If there are only two variables x and y, equations of the form

$$ax + by = c$$

can be represented as straight lines on paper, so are of interest to cartographers. An equation in three variables

$$ax + by + cz + d = 0$$

can represent a plane in three-dimensional space. If there are more than three variables, no simple *geometrical* interpretation can be made.

12.3 Simultaneous linear equations

Sets of equations which have to be satisfied by the same values of the variables are called *simultaneous equations*. Consider for example the linear equations

$$ax + by + c = 0$$
$$dx + ey + f = 0 \qquad (12.1)$$

We can represent these as straight lines on a graph as in Figure 12.1. Since the point P lies on both lines, the coordinates of $P(x, y)$ satisfy these equations.

Exercise 1 Plot the lines representing the equations

$$x + 2y = 5 \qquad \text{(E)}$$
$$3x + y = 5 \qquad \text{(F)}$$

and show that they intersect at the point (1,2).

Exercise 2 Solve the equations (E) and (F) by simple elimination. If we multiply each term of (E) by 3 we have

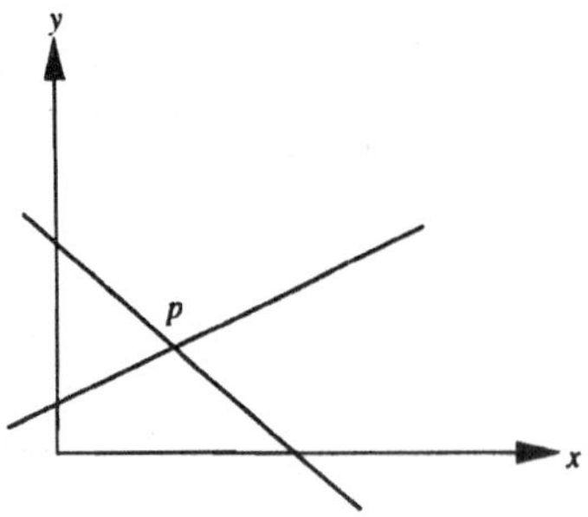

Figure 12.1

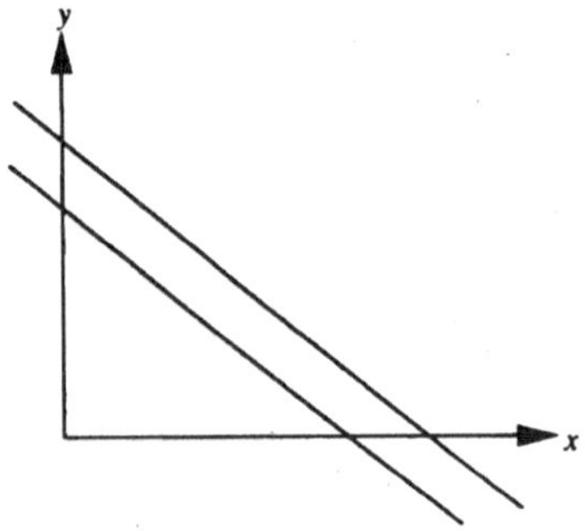

Figure 12.2

$$3x + 6y = 15$$

We have used the multiplier 3 so that we can eliminate x from the equations by subtracting (F) from this new equation. That is

$$3(E) - (F) = 3x + 6y - (3x + y) = 15 - 5 = 10$$

or

$$5y = 10$$

therefore

$$y = 2$$

Substituting for $y = 2$ in (F), we have

$$3x + 2 = 5$$

$$x = 1$$

It may appear that two equations in two variables, x and y say, are sufficient to obtain a solution. Whilst two equations are *necessary*, they may not be *sufficient*. Consider the two equations

$$ax + by + c = 0$$

and

$$nax + nby + d = 0$$

If we plot these on a graph we obtain the parallel lines shown in Figure 12.2 and because they do not intersect there is no point P and therefore no solution of the two equations.

Exercise 3 Show that there is no solution to the equations

$$x + 2y = 3$$

$$2x + 4y = 7$$

Exercise 4 Show that the determinant (see Section 6.6) formed by the coefficients of the variables is zero. We have the determinant Δ given by

$$\Delta = \begin{vmatrix} 1 & 2 \\ 2 & 4 \end{vmatrix}$$

$$\Delta = 1 \times 4 - (2 \times 2) = 0$$

When the terms in x and y in one equation are simple multiples of those in another, we say their coefficients are *linearly dependent*. For a solution we need that the rows and columns of the coefficients of the variables are *linearly independent*.

Whilst it is clear that two independent equations are necessary to obtain a solution, this does not rule out the possibility that several equations have a common solution. For example, the point P of Figure 12.3 may lie on three lines intersecting at one point. Any two of the lines can give a solution for P: thus one equation is *redundant*. We say that the problem is *overdetermined*.

In surveying, this extra information is used to check the result of measurements. For example, Figure 12.4 might represent the three rays observed from

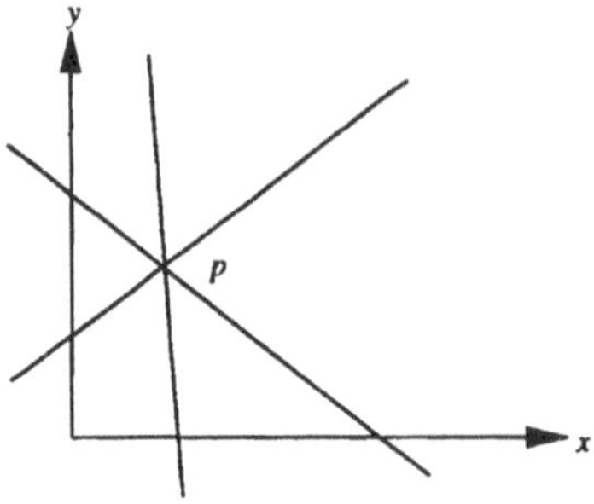

Figure 12.3

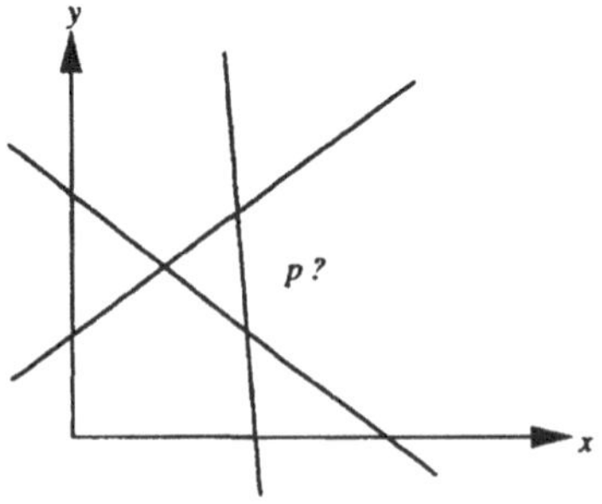

Figure 12.4

three survey stations. In the unlikely event of three rays intersecting exactly, we might say that the observations are free of error and that the equations are *consistent*. However since small observational errors are likely, a usual outcome is for three lines not to meet at one point as shown in Figure 12.4. The mismatch of the lines shows that the equations are *inconsistent*.

In the measurement sciences, the mismatch is used to give a statistical estimate of the precision of the measurements. To obtain a unique solution, the original equations are reduced in number to the minimum by replacing them with new equations, which are solved. The statistical process used for this replacement is called the *Principle of Least Squares* (see Chapter 13). See Allan (1997) for more information.

Sets of linear equations can be *underdetermined*, if there are fewer equations than variables, or *overdetermined* if there are more equations than variables, or *normal* if there is the same number of equations as variables. Ultimately the problem is to solve the normal case, with an equal number of variables as equations. These are, in matrix form (see Chapter 7)

$$\boldsymbol{N x} = \boldsymbol{b} \tag{12.2}$$

in which the vector of unknowns $\boldsymbol{x}$ is related to the square matrix $\boldsymbol{N}$ and the absolute vector $\boldsymbol{b}$. A typical set of such equations is

$$\begin{bmatrix} 1 & 2 \\ 3 & 1 \end{bmatrix} \begin{bmatrix} x \\ y \end{bmatrix} = \begin{bmatrix} 5 \\ 5 \end{bmatrix}$$

of which the solution is $x = 1,\ y = 2$, or in matrix notation

$$\begin{bmatrix} x \\ y \end{bmatrix} = \begin{bmatrix} 1 \\ 2 \end{bmatrix}$$

A formal statment of the solution is

$$\boldsymbol{x} = \boldsymbol{N}^{-1}\boldsymbol{b} \tag{12.3}$$

where the *inverse* of the matrix $\boldsymbol{N}$ is

$$\boldsymbol{N}^{-1}$$

Note: If the determinant of N is zero, its inverse is not defined, and the matrix is said to be singular. Refer to Exercise 4 above.

Although much use is now made of computer packages and spreadsheets to obtain solutions, it is none the less necessary to write solution algorithms when compiling software of one's own, or dealing with large problems. The matrix N is always square and often *symmetric*, that is

$$N^T = N \tag{12.4}$$

If N is *orthogonal* or *diagonal* or can be factorised into a product of these two types of matrix, the inverse can be written down directly (see 7.20 and 7.22). Usually problems are not so simple. There are two basic approaches to solution

(1) to obtain a solution without evaluating the inverse explicitly,
(2) to obtain a solution by first forming the inverse.

Generally, the first method is preferred because it gives a more accurate arithmetic solution. However, the discussion can be quite academic when a few *well-conditioned* equations, say up to thirty in number, have to be solved. Often a satisfactory method is to employ the matrix inversion routines available in spreadsheet software.

12.4 Cramer's rule for the solution of linear equations

The solution of the above two simple equations is quite trivial and may be effected by any simple method such as elimination by inspection. Consider the equations again

$$\begin{bmatrix} 1 & 2 \\ 3 & 1 \end{bmatrix} \begin{bmatrix} x \\ y \end{bmatrix} = \begin{bmatrix} 5 \\ 5 \end{bmatrix}$$

One approach is to use Cramer's rule to obtain the inverse, in this case as

$$\frac{1}{-5} \begin{bmatrix} 1 & -2 \\ -3 & 1 \end{bmatrix}$$

The way to obtain this inverse is as follows. First evaluate the determinant of coefficients of the original matrix. In the example

$$\Delta = (1 \times 1) - (3 \times 2) = -5$$

The reciprocal is multiplied by another matrix obtained by interchanging the diagonal terms in the original and changing the signs of the off-diagonal terms. The solution is then obtained from

$$\boldsymbol{x} = \boldsymbol{N}^{-1}\boldsymbol{b}$$

$$\begin{bmatrix} x \\ y \end{bmatrix} = \frac{1}{-5} \begin{bmatrix} 1 & -2 \\ -3 & 1 \end{bmatrix} \begin{bmatrix} 5 \\ 5 \end{bmatrix} = \begin{bmatrix} 1 \\ 2 \end{bmatrix}$$

This simple routine, called *Cramer's rule,* which is easy to remember, is probably worth using for this simple type of problem, which surprisingly does often arise. Although the value of Cramer's rule to invert a three-by-three matrix is debatable, its use is certainly out of the question for a larger problem, so it will be given no more attention here.

Exercise 1 Find the inverse of the following matrix by Cramer's Rule

$$\begin{bmatrix} 3 & 2 \\ 3 & -1 \end{bmatrix}$$

$$\Delta = 3(-1) - 3(2) = -9$$

thus the inverse is

$$\frac{1}{-9}\begin{bmatrix} -1 & -2 \\ -3 & 3 \end{bmatrix}$$

Check by multiplying these two matrices together to obtain the unit matrix ***I***.

12.5 Derivation of Cramer's Rule for a (2 × 2) matrix

Cramer's rule is derived as follows. Consider the two linear equations

$$a_1x + b_1y + c_1 = 0 \qquad \text{(G)}$$

$$a_2x + b_2y + c_2 = 0 \qquad \text{(H)}$$

To eliminate y we multiply (G) by b_2 and (H) by b_1 to give

$$b_2a_1x + b_2b_1y + b_2c_1 = 0$$

$$b_1a_2x + b_1b_2y + b_1c_2 = 0$$

and subtracting gives

$$(b_2a_1 - b_1a_2)x + b_2c_1 - b_1c_2 = 0$$

This can be written in determinants as

$$Dx + D_1 = 0 \qquad \text{(I)}$$

$$\text{where } D = \begin{vmatrix} a_1 & b_1 \\ a_2 & b_2 \end{vmatrix} \text{ and } D_1 = -\begin{vmatrix} b_1 & c_1 \\ b_2 & c_2 \end{vmatrix}$$

To eliminate x we multiply (G) by a_2 and (H) by a_1 to give

$$Dy + D_2 = 0 \qquad \text{(J)}$$

$$\text{where } D = \begin{vmatrix} a_1 & b_1 \\ a_2 & b_2 \end{vmatrix} \text{ and } D_2 = \begin{vmatrix} a_1 & c_1 \\ a_2 & c_2 \end{vmatrix}$$

Hence we have the solution

$$x = -D_1/D \quad \text{and} \quad y = -D_2/D$$

or in matrix form combining (I) and (J)

$$\begin{bmatrix} x \\ y \end{bmatrix} = \frac{-1}{D}\begin{bmatrix} b_2 & -b_1 \\ -a_2 & a_1 \end{bmatrix}\begin{bmatrix} c_1 \\ c_2 \end{bmatrix}$$

which is Cramer's rule for a two by two matrix. (Note in this derivation we have placed c_1 and c_2 on the left hand side of the equations (G) and (H).)

12.6 Note on consistency of linear equations

Consider equations (G) and (H) again. If a third linear equation

$$a_3x + b_3y + c_3 = 0 \qquad \text{(K)}$$

has the same solution as (G) and (H), the three equations are said to be consistent. For this to be possible

$$-a_3D_1/D - b_3D_2/D + c_3 = 0$$

or

$$-a_3D_1 - b_3D_2 + c_3D = 0 \qquad \text{(L)}$$

Writing ths is full we have

$$+a_3\begin{vmatrix} b_1 & c_1 \\ b_2 & c_2 \end{vmatrix} - b_3\begin{vmatrix} a_1 & c_1 \\ a_2 & c_2 \end{vmatrix} + c_3\begin{vmatrix} a_1 & b_1 \\ a_2 & b_2 \end{vmatrix} = 0$$

or more neatly

$$\begin{vmatrix} a_1 & b_1 & c_1 \\ a_2 & b_2 & c_2 \\ a_3 & b_3 & c_3 \end{vmatrix} = 0$$

Thus for the three equations

$$a_1x + b_1y + c_1 = 0 \qquad \text{(G)}$$
$$a_2x + b_2y + c_2 = 0 \qquad \text{(H)}$$
$$a_3x + b_3y + c_3 = 0 \qquad \text{(K)}$$

to be consistent, the determinant formed from their coefficients is zero. The result can be extended to four equations in three variables and so on to $(n + 1)$ equations in n variables.

12.7 Direct solution by *LU* decomposition

The two most common methods available for a direct solution of linear equations are due to the mathematicians Gauss and Cholesky. We treat only the former in this book. Both methods are versions of a process in which the original matrix $\boldsymbol{N}$ is *decomposed* into the product of two *triangular matrices*: a *lower triangular* matrix $\boldsymbol{L}$, and an *upper triangular* matrix $\boldsymbol{U}$ both of the same dimensions as the original. These terms will become clear as we go along. Let

$$\boldsymbol{N} = \boldsymbol{L}\,\boldsymbol{U} \qquad (12.5)$$

The original equations become

$$\boldsymbol{L}\,\boldsymbol{U}\boldsymbol{x} = \boldsymbol{b} \tag{12.6}$$

and when the further substitution is made

$$\boldsymbol{U}\boldsymbol{x} = \boldsymbol{f} \tag{12.7}$$

We have finally

$$\boldsymbol{L}\boldsymbol{f} = \boldsymbol{b} \tag{12.8}$$

The intermediate vector $\boldsymbol{f}$ is derived first by the *forward solution* of (12.8), then $\boldsymbol{x}$ by the *back solution* of (12.7). The meaning of these terms will become obvious when the following example is worked through. We describe these techniques in turn with reference to the same symmetric normal equations. Although it is unlikely that a hand solution will be carried out today, the explanation should be of value in coding an algorithm for a computer. The set of equations to be solved is

$$\begin{aligned} 4x - 2y + 4z &= 14 \\ -2x + 5y - 4z &= -15 \\ 4x - 4y + 14z &= 36 \end{aligned} \tag{12.9}$$

The solution is $x = 1$, $y = -1$, $z = 2$. This can be verified by substitution in equations (12.9).

12.8 Gaussian elimination

This method is capable of solving equations with a non-symmetric matrix of coefficients whilst Cholesky's is not. For information on Cholesky's method see Allan (1997). It must be pointed out that the only process involved in these methods is the *multiplication of matrices*. The multiplications are made in unconventional order as soon as numerical values become available. In these methods there is no inefficient duplication of effort such as is found using Cramer's rule.

The first stage is to decompose the normal matrix $\boldsymbol{N}$ into the two triangular matrices $\boldsymbol{L}$ and $\boldsymbol{U}$ as follows

$$\begin{bmatrix} N_{11} & N_{12} & N_{13} \\ N_{21} & N_{22} & N_{23} \\ N_{31} & N_{32} & N_{33} \end{bmatrix} = \begin{bmatrix} L_{11} & 0 & 0 \\ L_{21} & L_{22} & 0 \\ L_{31} & L_{32} & L_{33} \end{bmatrix} \begin{bmatrix} U_{11} & U_{12} & U_{13} \\ 0 & U_{22} & U_{23} \\ 0 & 0 & U_{33} \end{bmatrix}$$

In the above matrices, the twelve elements of the $\boldsymbol{L}$ and $\boldsymbol{U}$ matrices have to be evaluated from the nine known elements of the original $\boldsymbol{N}$ matrix. Thus we can choose reasonable arbitrary numbers for any three of these twelve. Gauss made the convenient choice of unity along the diagonal of the $\boldsymbol{U}$ matrix. The scheme is then to use

$$\begin{bmatrix} N_{11} & N_{12} & N_{13} \\ N_{21} & N_{22} & N_{23} \\ N_{31} & N_{32} & N_{33} \end{bmatrix} = \begin{bmatrix} L_{11} & 0 & 0 \\ L_{21} & L_{22} & 0 \\ L_{31} & L_{32} & L_{33} \end{bmatrix} \begin{bmatrix} 1 & U_{12} & U_{13} \\ 0 & 1 & U_{23} \\ 0 & 0 & 1 \end{bmatrix}$$

It is equally possible to make this choice for the diagonal of the $\boldsymbol{L}$ matrix instead, but not both.

12.9 Decomposition

The starting values for the decomposition process of equations (12.9) are

$$\begin{bmatrix} 4 & -2 & 4 \\ -2 & 5 & -4 \\ 4 & -4 & 14 \end{bmatrix} = \begin{bmatrix} L_{11} & 0 & 0 \\ L_{21} & L_{22} & 0 \\ L_{31} & L_{32} & L_{33} \end{bmatrix} \begin{bmatrix} 1 & U_{12} & U_{13} \\ 0 & 1 & U_{23} \\ 0 & 0 & 1 \end{bmatrix}$$

We multiply the first row of $\boldsymbol{L}$ with the three columns of $\boldsymbol{U}$ in turn, using each value as it is obtained, and equate the answers to the first row of $\boldsymbol{N}$. Thus we obtain from the first column of $\boldsymbol{U}$

$$L_{11}(1) + 0(0) + 0(0) = 4$$

Hence

$$L_{11} = 4$$

From the second column of $\boldsymbol{U}$ we have

$$L_{11}U_{12} + 0(1) + 0(0) = -2$$

therefore

$$4U_{12} = -2$$
$$U_{12} = -0.5$$

From the third column of $\boldsymbol{U}$ we have

$$L_{11}U_{13} + 0(U_{23}) + 0(1) = 4$$
$$4U_{13} = 4$$
$$U_{13} = 1$$

We have now carried out the following decomposition

$$\begin{bmatrix} 4 & -2 & 4 \\ -2 & 5 & -4 \\ 4 & -4 & 14 \end{bmatrix} = \begin{bmatrix} 4 & 0 & 0 \\ L_{21} & L_{22} & 0 \\ L_{31} & L_{32} & L_{33} \end{bmatrix} \begin{bmatrix} 1 & -\frac{1}{2} & 1 \\ 0 & 1 & U_{23} \\ 0 & 0 & 1 \end{bmatrix}$$

Notice that the first row of $\boldsymbol{U}$ equals the first row of $\boldsymbol{N}$ divided by its leading term 4. Next, multiplying the second row of $\boldsymbol{L}$ with the three columns of $\boldsymbol{U}$ and equating to the second row of $\boldsymbol{N}$ gives

$$\begin{bmatrix} 4 & -2 & 4 \\ -2 & 5 & -4 \\ 4 & -4 & 14 \end{bmatrix} = \begin{bmatrix} 4 & 0 & 0 \\ -2 & 4 & 0 \\ L_{31} & L_{32} & L_{33} \end{bmatrix} \begin{bmatrix} 1 & -\frac{1}{2} & 1 \\ 0 & 1 & -\frac{1}{2} \\ 0 & 0 & 1 \end{bmatrix}$$

And continuing the same process we complete the decomposition as follows

$$\begin{bmatrix} 4 & -2 & 4 \\ -2 & 5 & -4 \\ 4 & -4 & 14 \end{bmatrix} = \begin{bmatrix} 4 & 0 & 0 \\ -2 & 4 & 0 \\ 4 & -2 & 9 \end{bmatrix} \begin{bmatrix} 1 & -\frac{1}{2} & 1 \\ 0 & 1 & -\frac{1}{2} \\ 0 & 0 & 1 \end{bmatrix}$$

Thus we have decomposed the original matrix into the product of two triangular matrices using only the processes of matrix multiplication which can be programmed in a computer algorithm.

Exercise 1 Verify by multiplying $\boldsymbol{LU}$ together that the above decomposition is correct.

12.10 Forward solution

If the calculation is being carried out by hand, the final result should be checked by multiplying the $\boldsymbol{L}\ \boldsymbol{U}$ matrices together. The next stage is to solve for the values of $\boldsymbol{f}$ from

$$\boldsymbol{Lf} = \boldsymbol{b}$$

We have the numerical values

$$\begin{bmatrix} 4 & 0 & 0 \\ -2 & 4 & 0 \\ 4 & -2 & 9 \end{bmatrix} \begin{bmatrix} f_1 = \\ f_2 = \\ f_3 = \end{bmatrix} = \begin{bmatrix} b_1 = & 14 \\ b_2 = & -15 \\ b_3 = & 36 \end{bmatrix}$$

Multiplying out the first row of $\boldsymbol{L}$ and the $\boldsymbol{f}$ vector and equating to $b_1 = 14$ gives

$$\begin{bmatrix} 4 & 0 & 0 \\ -2 & 4 & 0 \\ 4 & -2 & 9 \end{bmatrix} \begin{bmatrix} f_1 = & 7/2 \\ f_2 = & \\ f_3 = & \end{bmatrix} = \begin{bmatrix} b_1 = & 14 \\ b_2 = & -15 \\ b_3 = & 36 \end{bmatrix}$$

And so on to the end when we have

$$\begin{bmatrix} 4 & 0 & 0 \\ -2 & 4 & 0 \\ 4 & -2 & 9 \end{bmatrix} \begin{bmatrix} f_1 = & 7/2 \\ f_2 = & -2 \\ f_3 = & 2 \end{bmatrix} = \begin{bmatrix} b_1 = & 14 \\ b_2 = & -15 \\ b_3 = & 36 \end{bmatrix}$$

This completes the *forward solution*.

12.11 Back solution

The back solution is carried out in a similar manner to the forward solution except that the matrix multiplication is carried out backwards, or from the bottom up. We start with the following

$$\begin{bmatrix} 1 & -\frac{1}{2} & 1 \\ 0 & 1 & -\frac{1}{2} \\ 0 & 0 & 1 \end{bmatrix} \begin{bmatrix} x_1 = & \\ x_2 = & \\ x_3 = & \end{bmatrix} = \begin{bmatrix} f_1 = & 7/2 \\ f_2 = & -2 \\ f_3 = & 2 \end{bmatrix}$$

Multiplying row 3 of $\boldsymbol{U}$ with $\boldsymbol{x}$ and equating to $\boldsymbol{f}$ we have

$$\begin{bmatrix} 1 & -\frac{1}{2} & 1 \\ 0 & 1 & -\frac{1}{2} \\ 0 & 0 & 1 \end{bmatrix} \begin{bmatrix} x_1 = & \\ x_2 = & \\ x_3 = & 2 \end{bmatrix} = \begin{bmatrix} f_1 = & 7/2 \\ f_2 = & -2 \\ f_3 = & 2 \end{bmatrix}$$

And continuing upwards we obtain finally the vector x.

$$\begin{bmatrix} 1 & -\frac{1}{2} & 1 \\ 0 & 1 & -\frac{1}{2} \\ 0 & 0 & 1 \end{bmatrix} \begin{bmatrix} x_1 = & 1 \\ x_2 = & -1 \\ x_3 = & 2 \end{bmatrix} = \begin{bmatrix} f_1 = & 7/2 \\ f_2 = & -2 \\ f_3 = & 2 \end{bmatrix}$$

Therefore we have the values of the variables $x = 1, y = -1$ and $z = 2$.

This process may seem a little complicated, but once programmed into a computer, there is almost no limit to the number of equations that can be solved, subject to storage problems in the computer.

12.12 Inverse of N

Although we do not require to find the inverse to obtain a solution, its approximate value is often needed to provide statistical information. In this case the inverse will be computed separately from the decomposed matrix $\boldsymbol{LU}$. We make use of the fact that the inverse of an upper triangular matrix is also an upper triangular matrix, whose *diagonal elements* are the *reciprocals* of the diagonal elements of the original triangular matrix. This can be verified quite easily by multiplication. Consider the product

$$\boldsymbol{UU}^{-1}$$

or for the **3 × 3** case

$$\begin{bmatrix} U_{11} & U_{12} & U_{13} \\ 0 & U_{22} & U_{23} \\ 0 & 0 & U_{33} \end{bmatrix} \begin{bmatrix} 1/U_{11} & ? & ? \\ 0 & 1/U_{22} & ? \\ 0 & 0 & 1/U_{33} \end{bmatrix} \tag{12.10}$$

The off-diagonal non-zero elements of the inverse of $\boldsymbol{U}$ are marked with a question mark, because they are never required if the original matrix $\boldsymbol{N}$ is symmetrical. In statistical least squares estimation problems this is always the case, and it is only in such problems that the inverse is required in mapping. For these reasons we restrict the treatment from now to the special case of a symmetric matrix N. Remembering that

$$UU^{-1} = \mathrm{I}$$

we see that

$$U_{11}(1/U_{11}) + U_{12}(0) + U_{13}(0) = 1$$
$$0(?) + U_{22}(1/U_{22}) + U_{23}(0) = 1$$
$$0(?) + 0(?) + U_{33}(1/U_{33}) = 1$$

Therefore (12.10) is valid.

Consider the inverse of the symmetric matrix N. By definition we have

$$N^{-1}N = I$$

or

$$N^{-1}L\,U = I$$

Thus

$$N^{-1}L\,U\,U^{-1} = U^{-1}$$
$$N^{-1}L = U^{-1}$$

Putting

$$M = N^{-1}$$

or in full

$$\begin{bmatrix} M_{11} & M_{12} & M_{13} \\ M_{21} & M_{22} & M_{23} \\ M_{31} & M_{32} & M_{33} \end{bmatrix} \begin{bmatrix} L_{11} & 0 & 0 \\ L_{21} & L_{22} & 0 \\ L_{31} & L_{32} & L_{33} \end{bmatrix} = \begin{bmatrix} 1/U_{11} & ? & ? \\ 0 & 1/U_{22} & ? \\ 0 & 0 & 1/U_{33} \end{bmatrix}$$

Those off-diagonal non-zero elements of U^{-1} marked by a question mark are not required if N and M are symmetrical. Since this is always the case in least squares problems the discussion is now confined to symmetric matrices. The problem therefore reduces to finding the six unique elements of M. This is expressed by

$$\begin{bmatrix} M_{11} & & \\ M_{21} & M_{22} & \\ M_{31} & M_{32} & M_{33} \end{bmatrix} \begin{bmatrix} L_{11} & 0 & 0 \\ L_{21} & L_{22} & 0 \\ L_{31} & L_{32} & L_{33} \end{bmatrix} = \begin{bmatrix} 1/U_{11} & ? & ? \\ 0 & 1/U_{22} & ? \\ 0 & 0 & 1/U_{33} \end{bmatrix}$$

Note that we have removed the upper triangle of coefficients from M because they are identical to those of its lower triangle and in the Gaussian method the diagonal values of U are each 1.Therefore the numerical version is

$$\begin{bmatrix} M_{11} & & \\ M_{21} & M_{22} & \\ M_{31} & M_{32} & M_{33} \end{bmatrix} \begin{bmatrix} 4 & 0 & 0 \\ -2 & 4 & 0 \\ 4 & -2 & 9 \end{bmatrix} = \begin{bmatrix} 1 & ? & ? \\ 0 & 1 & ? \\ 0 & 0 & 1 \end{bmatrix}$$

Carrying out a back multiplication of ML and equating to U^{-1} gives the required elements of the inverse as follows

$$\begin{bmatrix} M_{11} & & -1/12 \\ M_{21} & M_{22} & 1/18 \\ -1/12 & 1/18 & 1/9 \end{bmatrix} \begin{bmatrix} 4 & 0 & 0 \\ -2 & 4 & 0 \\ 4 & -2 & 9 \end{bmatrix} = \begin{bmatrix} 1 & ? & ? \\ 0 & 1 & ? \\ 0 & 0 & 1 \end{bmatrix}$$

Remember that $\boldsymbol{M}$ is symmetric about the diagonal. And the final solution

$$\begin{bmatrix} 3/8 & 1/12 & -1/12 \\ 1/12 & 5/18 & 1/18 \\ -1/12 & 1/18 & 1/9 \end{bmatrix} \begin{bmatrix} 4 & 0 & 0 \\ -2 & 4 & 0 \\ 4 & -2 & 9 \end{bmatrix} = \begin{bmatrix} 1 & ? & ? \\ 0 & 1 & ? \\ 0 & 0 & 1 \end{bmatrix}$$

Thus we have the inverse of $\boldsymbol{N}$ as

$$\begin{bmatrix} 3/8 & 1/12 & -1/12 \\ 1/12 & 5/18 & 1/18 \\ -1/12 & 1/18 & 1/9 \end{bmatrix}$$

Exercise 1 Verify the above inverse by premultiplication and postmultiplication of $\boldsymbol{M}$ by $\boldsymbol{N}$ to obtain the unit matrix. That is show that

$$\boldsymbol{MN} = \boldsymbol{NM} = \boldsymbol{I}$$

There are other methods of solving normal equations, by direct and iterative methods. Special methods are employed to suit sets of linear equations with the special structures which arise in photogrammetric and geodetic problems. It is not appropriate to discuss these here but see Allan (1997) for more information.

12.13 The solution of a quadratic equation

Surprisingly often in applied science, there is a need to solve an equation of the following kind

$$ax^2 + bx + c = 0 \tag{12.11}$$

In many cases there are two solutions to this *quadratic equation* given by

$$x = \frac{-b \pm \sqrt{(b^2 - 4ac)}}{2a} \tag{12.12}$$

The derivation is as follows. Divide equation (12.11) throughout by a to give

$$x^2 + \frac{b}{a}x + \frac{c}{a} = 0 \tag{12.13}$$

$$x^2 + \frac{b}{a}x = -\frac{c}{a} \tag{12.14}$$

To both sides of the equation we add a term to make a perfect square on the left side, thus

$$x^2 + \frac{b}{a}x + \left(\frac{b}{2a}\right)^2 = -\frac{c}{a} + \left(\frac{b}{2a}\right)^2 \qquad (12.15)$$

Then we have

$$\left(x + \frac{b}{2a}\right)^2 = -\frac{c}{a} + \left(\frac{b}{2a}\right)^2$$

$$= -\frac{c}{a} + \frac{b^2}{4a^2}$$

$$= \frac{-4ac + b^2}{4a^2}$$

therefore

$$x + \frac{b}{2a} = \pm\frac{\sqrt{(b^2 - 4ac)}}{2a}$$

$$x = -\frac{b}{2a} \pm \frac{\sqrt{(b^2 - 4ac)}}{2a}$$

$$= \frac{-b \pm \sqrt{(b^2 - 4ac)}}{2a}$$

If $(b^2 - 4ac)$ is negative, the square root is not defined in ordinary numbers, and there are no real solutions. If $b^2 = 4ac$ the two solutions are equal.

Exercise 1 Solve the quadratic equation

$$2x^2 - 5x - 3 = 0 \qquad (12.16)$$

Here $a = 2$, $b = -5$, and $c = -3$ therefore the solutions are

$$x = \frac{+5 \pm \sqrt{(25 + 24)}}{4}$$

$$x = \frac{+5 \pm 7}{4}$$

$$x = 3 \quad \text{or} \quad -0.5$$

12.14 Graphic solution of equations

Sometimes the quickest way to obtain an approximate solution to a high-order equation is to draw a graph of the function and find where it cuts the x axis. The plotting process is known as curve tracing. To illustrate the method we shall solve the quadratic equation (12.16) by the graphical method.

We draw up a table of values of the function

$$y = 2x^2 - 5x - 3$$

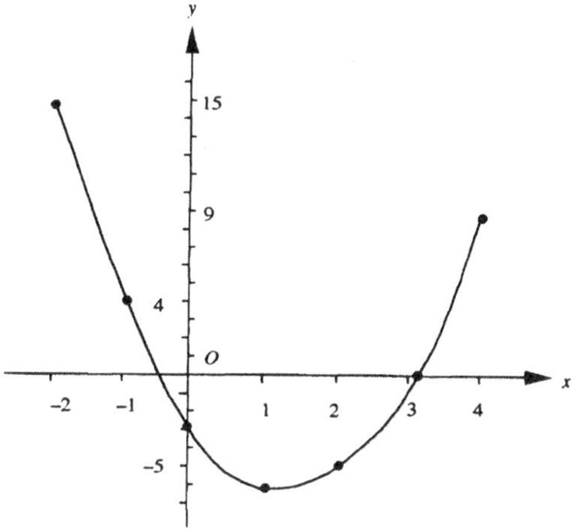

Figure 12.5

x	−2	− 1	0	1	2	3	4
y	15	4	− 3	− 6	− 5	0	9

To find the point on the curve where its gradient is zero we examine the differential of y with respect to x as follows

$$y = 2x^2 - 5x - 3$$

therefore

$$\frac{dy}{dx} = 4x - 5$$

When the gradient is zero, the curve runs parallel to the x axis, that is when

$$4x - 5 = 0$$

$$x = 1.25$$

This value of x gives us a point at which the gradient is zero. Such a point is called a *turning point* on the curve and y has a *turning value,* in this case a minimum. Using the values of x and y and the turning point as coordinates we can sketch the graph of the curve shown in Figure 12.5. At once we see that the curve crosses the x axis when

$$x = 3 \text{ and } -0.5$$

These are the solutions to the quadratic equation by the graphical method.

Exercise 2 Solve the equations

$$2x^2 - 5x - 3 = y \qquad (12.17)$$

$$x + 2y - 5 = 0 \qquad (12.18)$$

If we plot the graphs of both equations, as in Figure 12.6, their intersection yields the solutions

$$x = 3.2 \;\; y = 1 \quad \text{and} \quad x = -0.9 \;\; y = 2.75$$

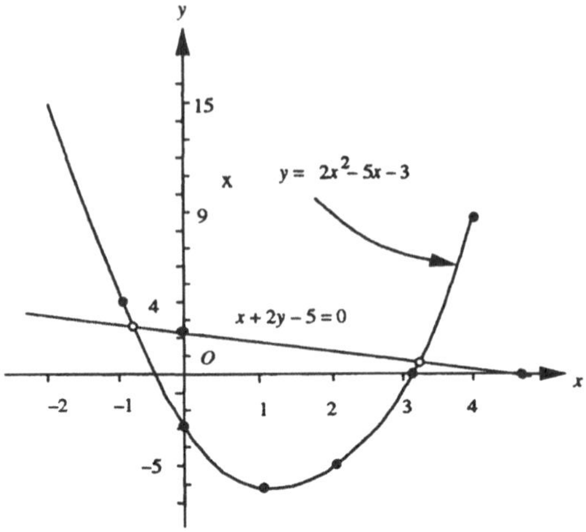

Figure 12.6

To obtain an exact analytical solution we proceed as follows
From (12.18)

$$y = 0.5\,(5\ - x)$$

Substituting in (12.17) for y we have a new quadratic equation in x

$$2x^2 - 5x - 3 - 2.5 + 0.5x = 0$$

$$4x^2 - 10x\ - 6 - 5 + x = 0$$

$$4x^2 - 9x - 11 = 0$$

which gives the solutions by (12.12)

$$x = \frac{9 \pm \sqrt{(81 + 176)}}{8}$$

$$= \frac{9 \pm \sqrt{(257)}}{8}$$

$$= \frac{9 \pm 16.03}{8}$$

$$= 3.13 \text{ or } -0.88$$

From (12.18) we have

$$x = 3.13,\ y = 0.94, \quad \text{or} \quad x = -0.88,\ y = 2.94$$

12.15 Solution of a cubic equation

The ability to solve a cubic equation is needed in error ellipsoid theory. Consider the equation

$$ax^3 + bx^2 + cx + d = 0 \tag{12.19}$$

Division throughout by a gives

$$x^3 + \frac{b}{a}x^2 + \frac{c}{a}x + \frac{d}{a} = 0$$

or

$$x^3 + b_1x^2 + c_1x + d_1 = 0 \tag{12.20}$$

If we complete the cube by making the substitution

$$x = y - \frac{1}{3}b_1$$

we obtain the equation in y

$$y^3 + \left(c_1 - \frac{1}{3}b_1^2\right)y + \left(\frac{2}{27}b_1^3 - \frac{1}{3}b_1c_1 + d_1\right) = 0$$

or

$$y^3 + c_2y + d_2 = 0 \tag{12.21}$$

Putting

$$y = k \cos q$$

$$k^3\cos^3 q + c_2k\cos q + d_2 = 0$$

and

$$\cos^3 q + \frac{c_2}{k^2}\cos q + \frac{d_2}{k^3} = 0 \tag{12.22}$$

we have

$$4\cos^3 q + 4\frac{c_2}{k^2}\cos q = -4\frac{d_2}{k^3} \tag{12.23}$$

and if we now put

$$c_2 = -\frac{3}{4}k^2$$

equation (12.23) becomes

$$4\cos^3 q - 3\cos q = -4\frac{d_2}{k^3}$$

But since

$$\cos 3q = 4\cos^3 q - 3\cos q$$

$$\cos 3q = -4\frac{d_2}{k^3}$$

and we obtain finally

$$q = \frac{1}{3}\text{arc cos}\left(-4\frac{d_2}{k^3}\right) \tag{12.24}$$

From which we obtain three values of q given by

$$q_1 = q \quad q_2 = q + \frac{2}{3}\pi \quad q_3 = q - \frac{2}{3}\pi$$

which yield the three values of x by back substitution first for

$$y = k \cos q$$

then

$$x = y - \frac{1}{3}b_1$$

Exercise 1 Find the values of x which satisfy the following cubic equation

$$2x^3 - 22x^2 + 72x - 72 = 0 \tag{12.25}$$

(a) by the graphical method
(b) by the analytical method.

Method (a) We draw up a table of the function

$$y = 2x^3 - 22x^2 + 72x - 72$$

x	0	1	2	3	4	5	6	7
y	–72	–20	0	0	-8	–12	0	40

The turning values are obtained from putting

$$\frac{dy}{dx} = 0$$

$$\frac{dy}{dx} = 6x^2 - 44x + 72 = 0$$

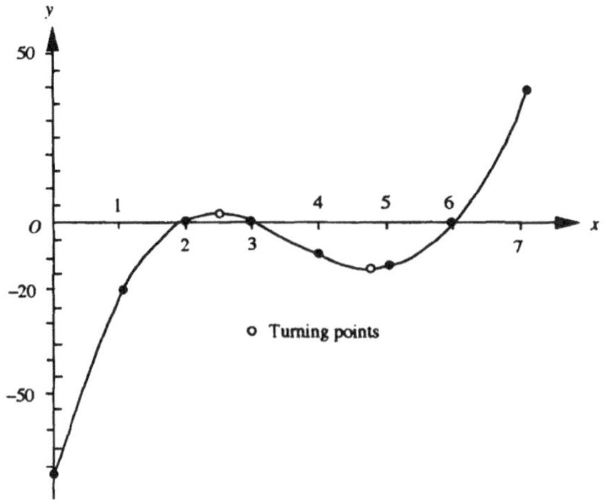

Figure 12.7

giving

$$x = \frac{44 \pm \sqrt{(1936 - 1728)}}{12}$$

$$x = \frac{44 \pm 14.42}{12}$$

$$x = 2.46 \quad x = 4.87$$

Corresponding values of y are

$$y = 1.76 \quad \text{and} \quad y = -12.13$$

Differentiating again, we have

$$\frac{d^2y}{dx^2} = 12x - 44$$

When $x = 4.87$ $$\frac{d^2y}{dx^2} = 14.44$$

indicating that y has a minimum turning value at this point.

When $x = 2.46$ $$\frac{d^2y}{dx^2} = -14.48$$

indicating that y has a maximum turning value at this point.

Figure 12.7 shows a rough sketch of the curve which crosses the x axis when

$$x = 2, 3 \text{ and } 6$$

These are the *approximate solutions* to the cubic. We say 'approximate' because they have been scaled from the graph. Substitution in the function gives $y = 0$ in each case, showing they are also the exact solutions. It should be noted that the graphical method of solution is unsuitable for computer use.

Method (b) To derive the solutions by computer we use the analytical method.

$$2x^3 - 22x^2 + 72x - 72 = 0 \tag{12.26}$$

Division by 2 gives

$$x^3 - 11x^2 + 36x - 36 = 0$$

$$b_1 = -11 \quad c_1 = 36 \quad d_1 = -36$$

therefore

$$y^3 + \left(c_1 - \frac{1}{3}b_1^2\right)y + \left(\frac{2}{27}b_1^3 - \frac{1}{3}b_1c_1 + d_1\right) = 0$$

becomes

$$y^3 + \left(36 - \frac{1}{3} \times 121\right)y + \left(-\frac{2}{27} \times 1331 + \frac{1}{3} \times 396 - 36\right) = 0$$

$$y^3 + (36 - 40.333)y + (-98.593 + 132 - 36) = 0$$

$$y^3 - 4.333y - 2.593 = 0$$

Now

$$c_2 = -\frac{3}{4}k^2$$

$$k^2 = -\frac{4}{3}c_2 = 5.778$$

therefore

$$k = 2.4037 \quad \text{and} \quad c_2 = -4.333$$

$$k^3 = 13.889$$

$$\cos 3q = -4\frac{d_2}{k^3} = \frac{4 \times 2.593}{13.889} = 0.7467$$

$$3q = 0.7277 \text{ radians} = 41.69° \quad q = 13.897°$$

We also have

$$q_1 = q \quad q_2 = q + \frac{2}{3}\pi \quad q_3 = q - \frac{2}{3}\pi$$

$$q_1 = 13.897°$$

$$q_2 = 13.897° + 120° = 133.897°$$

$$q_3 = 13.897° - 120° = -106.103°$$

$$k = 2.4037$$

These yield the three values of x by back substitution first for

$$y = k\cos q = 2.4037\cos q$$

then

$$x = y - \frac{1}{3}b_1 = y - \frac{1}{3}(-11)$$

$$\cos q_1 = 0.9707 \quad y = k\cos q_1 = 2.333 \quad x = 2.333 + 3.667 = 6.00$$

$$\cos q_2 = -0.6934 \quad y = k\cos q_2 = -1.667 \quad x = -1.667 + 3.667 = 2.00$$

$$\cos q_3 = -0.2774 \quad y = k\cos q_3 = -0.667 \quad x = -0.667 + 3.667 = 3.00$$

12.16 Newton's method of solution of non-linear equations

A most useful way to solve *non linear equations* is by *Newton's method*. The method is iterative, producing better solutions from previous ones. It is adopted for most problems in mapping involving least squares estimation.

Before giving a formal statement of the method we will work through an example. Consider the equations of the position fixing ellipse and hyperbola of Exercise 2 of Section 10.14. They are

$$11x^2 + 36y^2 = 396 \tag{12.27}$$

$$9x^2 - 16y^2 = 144 \tag{12.28}$$

At most, there are four solutions to these equations. By sketching the curves, as in Figure 10.16, we obtain an approximate solution

$$x = 4.75 \quad y = 2.05$$

Substituting these values in the left sides of (12.27) and (12.28) we get

$$399.4775 \quad \text{and} \quad 135.8225$$

Clearly the equations are not satisfied by these values. But they are close. Suppose the changes required to these values of x and y are δx and δy respectively, then the better solutions will be

$$x = 4.75 + \delta x \quad \text{and} \quad y = 2.05 + \delta y$$

Partially differentiating the equations (12.27) and (12.28) with respect to x and y we obtain the linear equations in δx and δy

$$22x\,\delta x + 72y\,\delta y = 396 -\!\!-399.4775 = -3.4775$$

$$18x\,\delta x - 32y\,\delta y = 144 - 135.8225 = 8.1775$$

The values of x and y to be used here are the approximate ones (4.75, 2.05), so the equations to be solved are

$$104.5\,\delta x + 147.6\,\delta y = -3.4775$$

$$85.5\,\delta x - 65.6\,\delta y = 8.1775$$

The solution is

$$\delta x = 0.050 \quad \delta y = -0.059$$

These are corrections to be made to the initial values x and y giving

$$x = 4.75 + 0.050 = 4.80 \quad \text{and} \quad y = 2.05 - 0.06 = 1.99$$

Checking by re-substitution in the left sides we obtain respectively

396.00 and 144.00

Sometimes, if a closer solution is needed, the approximation process is repeated. This type of calculation is ideally performed by computer.

12.17 Formal statement of Newton's method

The process usually involves the solution of non-linear equations. Suppose the equations to be solved can be written

$$F(x, y) = 0$$

Now let

$$x = x' + \delta x, \quad y = y' + \delta y$$

thus

$$F(x' + \delta x, y' + \delta y) = 0$$

$F(x, y)$ can be expanded by partial differentiation as

$$F(x,y) \approx F(x',y') + \frac{\partial F}{\partial x}\delta x + \frac{\partial F}{\partial y}\delta y \approx 0 \qquad (12.29)$$

This equation is approximate, but becomes more nearly true the closer x' comes to x, and y' to y; and correspondingly δx and δy become smaller. The procedure is to obtain better and better values for x' and y' by repeated solutions, i.e. by iteration.

Note: In virtually all practical cases come across in surveying and mapping, Newton's method provides an extremely efficient and rapid method for solving non-linear equations. Occasionally if the initial values are poor, the process will not converge to a solution. A consideration of the conditions for convergence is beyond the scope of this book.

To illustrate the need for iteration, we will select poor initial values for the intersection of the ellipse and hyperbola at the point

$$x = 4 \quad y = 2$$

In the case of the equation of the ellipse we have

$$F(x, y) = 11x^2 + 36y^2 - 396 = 0$$

$$F(x', y') = F(4, 2) = 11(4)^2 + 36(2)^2 - 396$$

$$= 320 - 396 = -76$$

Also

$$\frac{\partial}{\partial x}\left(11x^2 + 36y^2 - 396\right) = 22x = 22(4) = 88$$

$$\frac{\partial}{\partial y}\left(11x^2 + 36y^2 - 396\right) = 72y = 72(2) = 144$$

The linear equation to be solved is

$$\frac{\partial F}{\partial x}\delta x + \frac{\partial F}{\partial y}\delta y = -F(x', y') \qquad (12.30)$$

$$88\ \delta x + 144\ \delta y = 76$$

The second linear equation is obtained similarly from the equation of the hyperbola to be

$$72\ \delta x - 64\ \delta y = 64$$

Division by 88 and 72 respectively gives

$$\delta x + 1.6364\ \delta y = 0.8636$$

$$\delta x - 0.8889\ \delta y = 0.8889$$

$$2.5253\ \delta y = -0.0253$$

$$\delta y = -0.01 \text{ and } \delta x = 0.88$$

As before, these are corrections to be made to the initial values x' and y' giving

$$x = 4 + 0.88 = 4.88 \quad \text{and} \quad y = 2 - 0.01 = 1.99$$

If these new values are substituted into the equation of the ellipse, we find that

$$F(x, y) = 11 \times (4.88)^2 + 36 \times (1.99)^2 - 396 = 8.522$$

and if these new values are substituted into the equation of the hyperbola, we find that

$$F(x, y) = 9 \times (4.88)^2 - 16 \times (1.99)^2 - 144 = 6.968$$

Because these errors of 8.522 and 6.968 are unacceptable, a further iteration is necessary.

Two new equations can be formed using the best values for $x = 4.88$ and $y = 1.99$ and these small misclosures, to give

$$107.36\ \delta x + 143.28\ \delta y = -8.522$$

$$87.84\ \delta x - 63.68\ \delta y = -6.968$$

And a new solution

$$\delta y = -0.000 \text{ and } \delta x = -0.079$$

and acceptable values of

$$x = 4.801 \quad \text{and} \quad y = 1.990$$

We know they are acceptable because yet another iteration gives the equations

$$105.6\ \delta x + 143.28\ \delta y = -0.0036$$

$$86.4\ \delta x - 63.68\ \delta y = 0.0016$$

and the solution to four decimal places

$$\delta y = 0 \text{ and } \delta x = 0$$

Because computer algorithms to carry out such iterations are easy to write, Newton's method is widely used.

SUMMARY OF KEY WORDS

unknowns (or variables), coefficients, linear equation, quadratic equation, cubic equation, second order, simultaneous equations, necessary and sufficient, consistent, singular, linearly dependent, linearly independent, underdetermined, overdetermined, normal, symmetric, orthogonal, diagonal, well conditioned, Cramer's rule, decomposed, triangular matrices, lower triangular matrix, upper triangular matrix, forward solution, back solution, approximate solutions, non-linear equations, Newton's method

SUMMARY OF FORMULAE

Linear equations	$\boldsymbol{N x} = \boldsymbol{b}$	(12.2)
	$\boldsymbol{x} = \boldsymbol{N}^{-1}\boldsymbol{b}$	(12.3)
$\boldsymbol{L\,U}$ decomposition	$\boldsymbol{N} = \boldsymbol{L\,U}$	(12.5)
	$\boldsymbol{L\,U\,x} = \boldsymbol{b}$	(12.6)
Intermediate vector	$\boldsymbol{U x} = \boldsymbol{f}$	(12.7)
	$\boldsymbol{L f} = \boldsymbol{b}$	(12.8)
Quadratic equation	$ax^2 + bx + c = 0$	(12.11)
	$x = \dfrac{-b \pm \sqrt{\left(b^2 - 4ac\right)}}{2a}$	(12.12)
Cubic equation	$ax^3 + bx^2 + cx + d = 0$	(12.19)
Newton's Method	$F(x,y) \approx F(x',y') + \dfrac{\partial F}{\partial x}\delta x + \dfrac{\partial F}{\partial y}\delta y \approx 0$	(12.29)

Chapter 13
Least Squares Estimation

13.1 Introduction

In this book, we have been dealing so far with *ideal* mathematical entities such as lines, planes and circles. We now enter a different field, in which parameters are imperfect and have to be modelled from measurements. In this chapter we consider how to obtain the *best estimates* of *parameters*. These parameters may be *directly observed* (or measured) or may be *derived* from measured components. The technique, called Least Squares estimation, was developed by Legendre, Gauss and others in the nineteenth century. Until the advent of electronic computers, the tedium of calculation meant that the technique was applied only to specially important work. Today its application is quite general. The application of Least Squares techniques has two main purposes: to make the best use of all measured data in deriving results; and to obtain estimates of the quality of results and of quantities derived from them. In this book we can only introduce some basic essentials of a very large and important subject. There are two fundamental ways of dealing with Least Squares problems:

(1) by concentrating on the *observations* themselves or
(2) by concentrating on *constraints* among these observations.

We focus on (1), with only a short reference to (2) at the end of the chapter, in Section 13.33.

In Appendix A2, we develop some further matrix concepts and methods required for an understanding of the Least Squares estimation process described in this chapter. The discussion here is restricted to *classical matrices of full rank*. Although generalised inverses are now applied to mapping problems, they are not considered. Appendix A3 lists the notation employed by a few other authors and references to their work. Appendix A4 is concerned with the *error ellipse* and its *pedal curve*.

13.2 Calculation of examples

The Least Squares process involves calculations which are often exceedingly complex and extensive, and require the use of a computer. Spreadsheet systems, such as Excel, are of inestimable value in treating comparatively small problems. Although some of the examples of this chapter can be worked by pocket calculator, the reader is advised to use an Excel or other spreadsheet.

13.3 Measurement science

In most of this book we have been dealing with ideal mathematical functional models. For example, we have considered ideal versions of such entities as a straight line or a plane. In reality we are often never sure that we are dealing with these ideal entities, but only approximations to them. The entities under consideration will be described by *parameters* which may or may not be measured.

An example of a functional model is an equation which relates the distance between two points to their coordinate differences. Unlike matrices which are written in bold type, these models are written in ordinary type such as

$$s^2 - \Delta x^2 - \Delta y^2 = 0 \tag{13.1}$$

where s is the distance between the two points, and x and y are the usual coordinates. This equation is an *explicit statement* of the relationship between the distance and the coordinate differences. However it might be expressed in the *general form* as

$$F(s, \Delta x, \Delta y) = 0 \tag{13.2}$$

Equation (13.2) just tells us that the parameters are functionally related. The *observed parameter* may be the distance s, with Δx and Δy being the *unobserved parameters*. In this case it is helpful to rearrange the order of the parameters in equation (13.2) to be

$$F(\Delta x, \Delta y\text{: } s) = 0 \tag{13.3}$$

Here the unobserved parameters are placed first and separated from the observed parameter by a colon. This convention is helpful though not essential nor generally adopted by all authors.

Alternatively, the *observed parameters* might be the coordinate differences Δx and Δy which were measured from a map or aerial photograph, with s then being the distance derived from them, or the *unobserved parameter*. In this case we rearrange the functional model as

$$F(s\text{: } \Delta x, \Delta y) = 0 \tag{13.4}$$

Ultimately we reduce both cases to a standard mathematical form to be treated in matrix terms expressed in heavy (bold) type. Then the unobserved parameters will be *always* be expressed as the vector $\boldsymbol{x}$ and the observed parameter as the vector $\boldsymbol{s}$, writing the functional statement as

$$F(\boldsymbol{x}\text{: } \boldsymbol{s}) = 0 \tag{13.5}$$

This covers both cases. The change of notation from ordinary algebra to matrix algebra may confuse the beginner. To reduce the chances of possible confusion, we prefer to develop the full equations in the ordinary algebra of the customary notation for the problem, before converting to matrix notation. An example should help to make this clear.

Exercise 1 Express the bearing U between two points G and A as an explicit function of their coordinates. The formula linking two points A and G and their bearing U is

$$\tan U_{AG} = \frac{E_G - E_A}{N_G - N_A} \tag{13.6}$$

$$(N_G - N_A) \tan U_{AG} - E_G + E_A = 0 \tag{13.7}$$

This is the required explicit functional model.

Exercise 2 Express equation (13.7) in the general form, if the bearing U is the observed parameter and the coordinates the unobserved parameters. We have the general form in the notation of coordinate geometry

$$F(E_A, N_A, E_G, N_G: U) = 0 \tag{13.8}$$

and in matrix notation

$$F(\boldsymbol{x}: \boldsymbol{s}) = 0 \tag{13.5}$$

13.4 Statistical notation

It is a great pity that there is no generally agreed notation to be employed in Least Squares estimation, as can be seen from Appendix A3, which shows the notation adopted by several authors in this field. The topic is difficult enough for the beginner without this added confusion. We adopt the notation that has some general use in the UK and elsewhere. Once the topic is understood it is comparatively easy to adopt another notation, provided it is consistent within itself.

The reader is assumed to be familiar with the mathematical topics of *matrix algebra* and *partial differentiation*, which can be found in Chapters 7 and 9 respectively, and the concepts expressed in the Appendix A2.

In discussing the statistical models used in Least Squares analysis, a rather complex notation is required to distinguish between the *observed, provisional, estimated* and *population* versions of a parameter. Once the mathematical model has been cast into a standard form, the notation can be simplified without confusion. Figure 13.1 illustrates the notation used in this book to describe various statistical values, taking the measurement of a line as an example. If one end of a tape-measure is held fixed and the other end is free to rotate in a plane, it will describe a circle whose radius is s. Figure 13.1 shows short arcs (tangents) in the vicinity of the free end representing each of four versions of the length of s. These are, firstly, two *known* values

(1) an *observed value* $\overset{\circ}{s}$

(2) a selected *provisional value* $\overset{*}{s}$

and secondly, two values *to be estimated* from the observations

(3) a *least squares value* (*best estimate*) $\hat{s}$

(4) a *population value* $\bar{s}$

(**Note:** The population value is sometimes referred to as the *true value.*)
Differences between these fundamental parameters are

Errors in the best estimates $\quad ds = \bar{s} - \hat{s}$

Sample residuals $\quad v = \hat{s} - \overset{\circ}{s}$

Population residuals $\quad V = \bar{s} - \overset{\circ}{s}$

Differential changes $\quad \delta s = \hat{s} - \overset{*}{s} = \overset{\circ}{s} - \overset{*}{s} + v = L + v$

where $L = \overset{\circ}{s} - \overset{*}{s}$

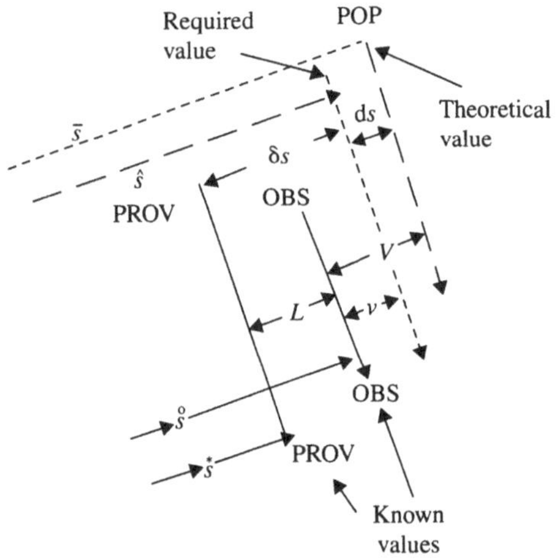

Figure 13.1

The objective in a Least Squares analysis is to find the *best estimates* of all parameters. The best estimates are invariably the best linear unbiased estimates (BLUE) obtained by applying the least squares principle to sampled known values of the observations.
In addition to the observed parameters we often have to estimate *additional parameters* for which no observations are available or required. The notation

for these is

Best estimates of the unobserved parameters	$\hat{x}$
Population values of the unobserved parameters	$\bar{x}$
Selected provisional values of these parameters	$\overset{*}{x}$

Derived from these fundamental parameters are

Errors in best estimates	$dx = \bar{x} - \hat{x}$
Differential changes	$\delta x = \hat{x} - \overset{*}{x}$

Note the absence of residuals in these expressions.

13.5 Newton's method – general theory

Because equations such as (13.1) and (13.6) are not linear, direct solution is complicated. (See Chapter 12, Sections 12.13 to 12.15.) Therefore we adopt Newton's method of solution (see Section 12.16). This involves first obtaining an approximate solution, which is then improved by iteration, using linear equations found by partially differentiating the relevant functional model. For example, consider equation (13.6) which is non-linear. It is written in general form as equation (13.8). Newton's method then takes the form

$$F(\hat{E}_A, \hat{N}_A, \hat{E}_G, \hat{N}_G : \hat{U}) = F(\overset{*}{E}_A, \overset{*}{N}_A, \overset{*}{E}_G, \overset{*}{N}_G : \overset{*}{U})$$
$$+ \frac{\partial F}{\partial E_A}\delta E_A + \frac{\partial F}{\partial N_A}\delta N_A + \frac{\partial F}{\partial E_G}\delta E_G + \frac{\partial F}{\partial N_G}\delta N_G + \frac{\partial F}{\partial U}\delta U \qquad (13.9)$$

The values with the 'hat', such as $\hat{E}_A$ are the final ones that we *choose to accept*, and those with the asterisks, such as $\overset{*}{E}_A$ are selected provisional values. These two sets of parameters are related by the fact that

$$\hat{E}_A = \overset{*}{E}_A + \delta E_A \quad \text{etc.} \qquad (13.10)$$

We say '*choose to accept*', because the iteration may need to be repeated. Sometimes, as in photogrammetry, many iterations are required before an acceptable solution is obtained. In this book generally one iteration will be acceptable.

To illustrate the method we now give a practical example from surveying.

13.6 Data for practical examples

Figure 13.2 shows a typical surveying network involving seven points A to G, which are located on a map grid. Following surveying conventions, the coordinates assigned are *Eastings E* and *Northings N*, quoted in that order, and the bearings U are reckoned clockwise from north. Table 13.1 lists the values in metres of the coordinates of points A to F, the bearings U (in sexagesimal degrees), and their tangents.

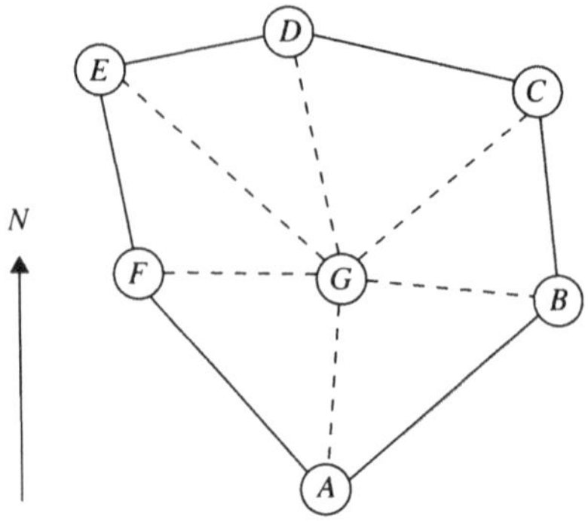

Figure 13.2

Table 13.1

Point	Easting E	Northing N	Ray	U = OBS BG to G degrees	tan U
A	512.402	386.280	AG	3.5588	0.0621928
B	567.895	443.275	BG	276.3248	-9.0220816
C	564.439	487.776	CG	231.1238	1.2403675
D	500.458	507.498	DG	164.8525	-0.2707103
E	457.825	503.912	EG	133.2231	-1.0640318
F	474.929	454.756	FG	98.0069	-7.1091575

Exercise 1 Assuming that the coordinates of A and B are known, calculate the coordinates of point G, using the observed bearings from these points. This is the standard surveying problem of *intersection*. Clearly we need two rays that intersect to give the position of G, and we could not obtain a solution if these rays were parallel.

Substituting the coordinates of A and tangent of the bearing of AG in equation (13.7) we have

$$(N_G - 386.28)\ 0.0621928 - E_G + 512.402 = 0$$

$$0.0621928\ N_G - 24.024 - E_G + 512.402 = 0$$

$$0.0621928\ N_G - E_G + 488.378 = 0 \qquad (13.11)$$

The equation of the ray BG is

$$(N_G - N_B) \tan U_{BG} - E_G + E_B = 0$$

and substituting values again we obtain the equation

$$(N_G - 443.275)(-9.0220816) - E_G + 567.895 = 0$$

$$-9.0220816\ N_G + 3999.263 - E_G + 567.895 = 0$$

$$-9.0220816\ N_G - E_G + 4567.158 = 0 \qquad (13.12)$$

Subtracting equation (13.11) from equation (13.12) gives

$$-9.0842744\, N_G + 4078.780 = 0$$

$$N_G = 448.993$$

Finally substituting in (13.11) gives

$$E_G = 488.378 + (0.0621928)\, 448.993 = 516.302$$

Thus the coordinates of point G derived from the two rays AG and BG are

$$(516.302,\ 448.993)$$

Exercise 2 Derive and solve the equations (13.11) and (13.12) by matrix methods.

In matrix form these equations are

$$\boldsymbol{N}\boldsymbol{x} = \boldsymbol{b} \qquad (13.13)$$

or specifically

$$\begin{bmatrix} -1 & 0.0621928 \\ -1 & -9.0220816 \end{bmatrix} \begin{bmatrix} E_G \\ N_G \end{bmatrix} = \begin{bmatrix} -488.378 \\ -4567.158 \end{bmatrix}$$

and the solution is

$$\boldsymbol{x} = \boldsymbol{N}^{-1}\boldsymbol{b}$$

or specifically

$$\begin{bmatrix} E_G \\ N_G \end{bmatrix} = \begin{bmatrix} -1 & 0.0621928 \\ -1 & -9.0220816 \end{bmatrix}^{-1} \begin{bmatrix} -488.378 \\ -4567.158 \end{bmatrix}$$

$$= \begin{bmatrix} -0.9931538 & -0.0068462 \\ 0.11008034 & -0.1100803 \end{bmatrix} \begin{bmatrix} -488.378 \\ -4567.158 \end{bmatrix}$$

$$= \begin{bmatrix} 516.302 \\ 448.993 \end{bmatrix}$$

Exercise 3 Show that the coordinates of G derived from the rays BG and CG are (516.330, 448.990).

13.7 Newton's method – application

In Section 13.6, we showed how to derive the provisional coordinates of point G. It will be seen that the right-hand sides of equations (13.13) are quite large numbers. In fact the coordinates used in mapping are often nine-digit numbers. An additional advantage of Newton's method is that it avoids the use of such large numbers, once the provisional values have been obtained. We remember that Newton's method involves the equation (13.9).

Because all final values must exactly satisfy the functional equation

$$F(\hat{E}_A, \hat{N}_A, \hat{E}_G, \hat{N}_G : \hat{U}) = 0$$

There are two ways of dealing with the expression linking the provisional values:

(1) we can calculate the provisional value of the 'observable' so that

$$F(\overset{*}{E}_A, \overset{*}{N}_A, \overset{*}{E}_G, \overset{*}{N}_G : \overset{*}{U}) = 0$$

(2) we can use the inconsistent provisional values along with the observed bearing to find

$$F(\overset{*}{E}_A, \overset{*}{N}_A, \overset{*}{E}_G, \overset{*}{N}_G : \overset{o}{U}) = K$$

We shall use the second method for the moment because it is slightly easier to calculate. *Normally* in this calculation we use the given coordinates of A, the provisional coordinates of G and the observed value of U. However, *in this case*, because we used the observed value of U to obtain the provisional coordinates of G, *the result* K will be zero. (The reader should verify this.) So, to illustrate the general method, we will select other provisional values close to the computed ones. We select

$\overset{*}{E}_G = 516.302 - 0.102 = 516.200$ and $\overset{*}{N}_G = 448.993 - 0.093 = 448.900$

Then

$$F(\overset{*}{E}_A, \overset{*}{N}_A, \overset{*}{E}_G, \overset{*}{N}_G : \overset{o}{U}_{AG}) = (448.900 - 386.28)\,0.0621928 - 516.200 + 512.402$$
$$= 0.09651 = K_{AG}$$

In a similar way for the ray BG

$$F(\overset{*}{E}_B, \overset{*}{N}_B, \overset{*}{E}_G, \overset{*}{N}_G : \overset{o}{U}_{BG}) = (\overset{*}{N}_G - \overset{*}{N}_B)\tan\overset{o}{U} - \overset{*}{E}_G + \overset{*}{E}_B = 0.94579 = K_{BG}$$

Exercise 1 Show that the K values for the other four rays are
$K_{CG} = 0.01847307$, $K_{DG} = 0.12108216$, $K_{EG} = 0.15951738$,
$K_{FG} = 0.36022632$

All that remains is to evaluate the partial differentials. We proceed as follows by partially differentiating the equation for the ray AG, i.e. equation

$$(N_G - N_A)\tan\overset{o}{U}_{AG} - E_G + E_A = 0 \qquad (13.7)$$

Thus we obtain

$$\frac{\partial F}{\partial E_A} = 1; \quad \frac{\partial F}{\partial N_A} = -\tan\overset{o}{U}_{AG}; \quad \frac{\partial F}{\partial E_G} = -1$$

$$\frac{\partial F}{\partial N_G} = \tan\overset{o}{U}_{AG}; \quad \frac{\partial F}{\partial U} = (N_G - N_A)\sec^2\overset{o}{U}_{AG}$$

Substitution in (13.9) gives

$$\delta E_A - \delta E_G - \delta N_A \tan \overset{o}{U}_{AG} + \delta N_G \tan \overset{o}{U}_{AG} + (N_G - N_A)\sec^2 \overset{o}{U}_{AG} \delta U_{AG} + K_{AG} = 0 \quad (13.14)$$

The similar equation for the ray BG is

$$\delta E_B - \delta E_G - \delta N_B \tan \overset{o}{U}_{BG} + \delta N_G \tan \overset{o}{U}_{BG} + (N_G - N_B)\sec^2 \overset{o}{U}_{BG} \delta U_{BG} + K_{BG} = 0 \quad (13.15)$$

These equations (13.14) and (13.15) are the general expressions for the rays AG and BG in terms of all the parameters. In the case of simple intersection, the points A and B are kept fixed so

$$\delta E_A = 0, \quad \delta N_A = 0, \quad \delta E_B = 0, \quad \delta N_B = 0$$

and we have

$$-\delta E_G + \delta N_G \tan U_{AG} + (N_G - N_A) \sec^2 U_{AG}\, \delta U_{AG} + K_{AG} = 0 \quad (13.16)$$

$$-\delta E_G + \delta N_G \tan U_{BG} + (N_G - N_B) \sec^2 U_{BG}\, \delta U_{BG} + K_{BG} = 0 \quad (13.17)$$

Where we have only two rays necessary to fix the point G the bearings U will not be changed so

$$\delta U_{AG} = 0 \quad \text{and} \quad \delta U_{BG} = 0$$

Hence equations (13.16) and (13.17) greatly simplify to

$$-\delta E_G + \delta N_G \tan U_{AG} + K_{AG} = 0 \quad (13.18)$$

$$-\delta E_G + \delta N_G \tan U_{BG} + K_{BG} = 0 \quad (13.19)$$

The numerical versions of these are

$$-\delta E_G + 0.0621928\, \delta N_G + 0.09651 = 0$$

$$-\delta E_G - 9.0220816\, \delta N_G + 0.94579 = 0$$

The solution is

$$\delta E_G = 0.102 \quad \text{and} \quad \delta N_G = 0.093$$

And the final coordinate values are

$$\hat{E}_G = \overset{*}{E}_G + \delta E_G = 516.200 + 0.102 = 516.302$$

$$\hat{N}_G = \overset{*}{N}_G + \delta N_G = 448.900 + 0.093 = 448.993$$

Which, as expected, are the same as those obtained by the direct solution in Section 13.6.

13.8 Summary and revision

In the above paragraphs we have introduced the reader to much of the notation and some concepts involved in the Least Squares estimation process, but have not actually begun to explain the method itself. To summarise, the essential mechanisms required to reach an understanding of what follows are:

(1) the solution of equations by Newton's method
(2) the need for partial differentiation in this method
(3) the great convenience of treating equations in matrix form
(4) the practical benefits of calculating results by a spreadsheet system.

13.9 Redundant measurements: degrees of freedom

To calculate the coordinates of point G any two rays that intersect will suffice. We can select two rays from six in any of 15 ways which will give 15 different versions of the position of G. Using the data from Table 13.1, Figure 13.3 shows a scale diagram of how the various rays intersect in the vicinity of G. (The intersection of BG with FG cannot be shown at this scale.) Clearly we can say that there is no unique solution to the problem as it stands, or that the position of G is *overdetermined.* This redundancy of information can be put another way. We say that, because there are six equations in all, only two of which are necessary and sufficient, there are $6 - 2 = 4$ *degrees of freedom* in the problem. In general if we have m equations in n variables, $m > n$, the number of degrees of freedom is $m - n$.

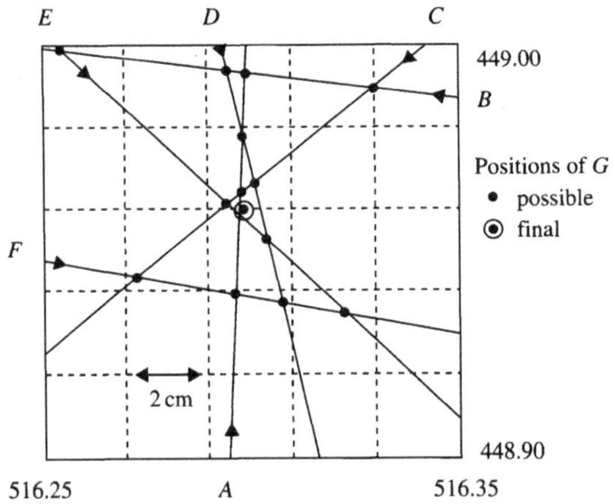

Figure 13.3

On this Figure 13.3, the reader can verify that rays AG and BG meet at the point whose coordinates are (516.302, 448.993), and rays BG and CG meet at (516.330, 448.990). The final objective of our analysis is to select one unique position for G which uses all the rays in some regular systematic way. Suppose the finally accepted position of G, resulting from the Least Squares process, is

$$E_G = 516.300 \quad \text{and} \quad N_G = 448.960$$

Table 13.2

	Best bearing $\hat{U}$ (degrees)	Observed bearing $\overset{o}{U}$ (degrees)	Residual $v=\hat{U}-\overset{o}{U}$ (degrees)	Residual (seconds of arc)
AG	3.558302	3.558756	–0.0004535	–1.6
BG	276.287579	276.324829	–0.0372503	–134.1
CG	231.119795	231.123833	–0.0040380	–14.5
DG	164.857584	164.852482	0.0051026	18.3
EG	133.221335	133.223144	–0.0018087	–6.5
FG	97.974722	98.006931	–0.0322095	–115.9

Clearly after this position has been found, the bearings of the rays from G to the points A to F will not generally equal the observed values. Table 13.2 shows the new 'best bearings' calculated from A to G etc. and the differences between the two sets of bearings. These differences are called 'Least Squares residuals', or just 'residuals' for short. Hence we define a residual as

Residual = *Best estimate of a parameter* – *Observed value of the parameter*

$$v=\hat{U}-\overset{o}{U} \tag{13.20}$$

(Note that some authorities define the residual in the opposite sense. This does not affect the treatment, provided the convention is consistent throughout the whole analysis.)

So how was the position of G found? It was found by making the sum of the squares of these residuals a minimum. For brevity, we usually let

the sum of the squares of residuals = Ω

Thus, in our example,

$$v_{AG}^2+v_{BG}^2+v_{CG}^2+v_{DG}^2+v_{EG}^2+v_{FG}^2=\Omega$$

and we have obtained the coordinates of G by making Ω a minimum.
As was shown in Chapter 7 Section 7.15, if $\boldsymbol{v}$ is a column matrix of residuals, Ω is the square of the length of the vector given by

$$\Omega = \boldsymbol{v}^{\mathrm{T}}\boldsymbol{v} \tag{13.21}$$

13.10 Observation equations

Returning to the practical problem, we form observation equations for all the observed bearings. Consider the ray AG for which we derived the equation (13.16).

$$-\delta E_G + \delta N_G \tan U_{AG} + (N_G - N_A) \sec^2 U_{AG}\delta U_{AG} + K_{AG} = 0 \tag{13.16}$$

We now see that the change to the bearing, after the Least Squares solution has been carried out, will be its residual, i.e. $\delta U_{AG} = v_{AG}$ hence (13.16) becomes

$$-\delta E_G + \delta N_G \tan U_{AG} + (N_G - N_A) \sec^2 U_{AG}\, v_{AG} + K_{AG} = 0 \tag{13.22}$$

Division of (13.22) by the coefficient of the residual v_{AG}, and rearranging th order of terms, gives

$$\frac{\delta E_G}{(N_G - N_A)\sec^2 U_{AG}} - \frac{\delta N_G \tan U_{AG}}{(N_G - N_A)\sec^2 U_{AG}} - \frac{K_{AG}}{(N_G - N_A)\sec^2 U_{AG}} = v_{AG} \quad (13.23)$$

Now, although these coefficients can be recast in simpler forms fo computation, and we will do so later, they can be expressed in algebrai form as

$$a_{11}\,\delta E_G + a_{12}\,\delta N_G + L_1 = v_1 \qquad (13.24)$$

where $a_{11} = +\dfrac{1}{(N_G - N_A)\sec^2 U_{AG}}, \quad a_{12} = -\dfrac{\tan U_{AG}}{(N_G - N_A)\sec^2 U_{AG}}$

and $L_1 = -\dfrac{K_{AG}}{(N_G - N_A)\sec^2 U_{AG}}$

In the same way we can form equations for the other rays and obtain the fu array of six equations as follows

$$\begin{aligned}
a_{11}\delta E_G + a_{12}\delta N_G + L_1 &= v_1\\
a_{21}\delta E_G + a_{22}\delta N_G + L_2 &= v_2\\
a_{31}\delta E_G + a_{32}\delta N_G + L_3 &= v_3\\
a_{41}\delta E_G + a_{42}\delta N_G + L_4 &= v_4\\
a_{51}\delta E_G + a_{52}\delta N_G + L_5 &= v_5\\
a_{61}\delta E_G + a_{62}\delta N_G + L_6 &= v_6
\end{aligned} \qquad (13.25)$$

Which are neatly expressed in matrix form as

$$\boldsymbol{Ax} + \boldsymbol{L} = \boldsymbol{v} \qquad (13.26)$$

where $\boldsymbol{A}$ is the (6 × 2) matrix of coefficients,
$\boldsymbol{x}$ is the (2 × 1) column vector of variables $(\delta E_G \;\; \delta N_G)^{\mathrm{T}}$ and
$\boldsymbol{v}$ is the (6 × 1) column vector of residuals $(v_1 \;\; v_2 \;\; v_3 \;\; v_4 \;\; v_5 \;\; v_6)^{\mathrm{T}}$.
(We write a column vector as the transpose of a row to save printing space.

Exercise 1 Express equations (13.25) in numerical form from the data i Table 13.1.
Before doing so we can recast the coefficients of $\boldsymbol{A}$ in neater forms.
Since

$$\sec U_{AG} = \frac{S_{AG}}{(N_G - N_A)}$$

where S_{AG} is the length of AG

and
$$\tan U_{AG} = \frac{E_G - E_A}{N_G - N_A}$$

$$a_{11} = +\frac{1}{(N_G - N_A)\sec^2 U_{AG}} = \frac{N_G - N_A}{S_{AG}^2}$$

$$a_{12} = -\frac{\tan U_{AG}}{(N_G - N_A)\sec^2 U_{AG}} = \frac{E_G - E_A}{S_{AG}^2}$$

$$L_1 = -\frac{K_{AG}}{(N_G - N_A)\sec^2 U_{AG}} = \frac{N_G - N_A}{S_{AG}^2} K_{AG}$$

Table 13.3 shows the calculations of the coefficients for all six equations.

Table 13.3

	Ray	Diff Easting	Diff Northing	Distance	K_i	a_{i1}	a_{i2}	L_i
1	*AG*	3.798	62.620	62.735	0.09651314	0.015911	–0.000965	0.001535
2	*BG*	–51.695	5.625	51.600	0.94579100	0.002080	0.019118	0.002005
3	*CG*	–48.239	–38.876	61.954	0.01847307	–0.010128	0.012567	–0.000187
4	*DG*	15.742	–58.598	60.675	0.12108216	–0.015917	–0.004276	–0.001933
5	*EG*	58.375	–55.012	80.212	0.15951738	–0.008550	–0.009073	–0.001364
6	*FG*	41.271	–5.856	41.685	0.36022632	–0.003370	–0.023752	–0.001193

13.11 Normal equations

The observation equations (13.25) are six equations in two variables of the form

$$\boldsymbol{Ax} + \boldsymbol{L} = \boldsymbol{v}$$

or in full they are

$$\begin{bmatrix} a_{11} & a_{12} \\ a_{21} & a_{22} \\ a_{31} & a_{32} \\ a_{41} & a_{42} \\ a_{51} & a_{52} \\ a_{61} & a_{62} \end{bmatrix} \begin{bmatrix} x_1 \\ x_2 \end{bmatrix} + \begin{bmatrix} L_1 \\ L_2 \\ L_3 \\ L_4 \\ L_5 \\ L_6 \end{bmatrix} = \begin{bmatrix} v_1 \\ v_2 \\ v_3 \\ v_4 \\ v_5 \\ v_6 \end{bmatrix}$$

where $m = 6$ and $n = 2$ and the variables are $x_1 = \delta E_G$ and $x_2 = \delta N_G$. It is shown in Appendix A2 that by minimising $\Omega = \boldsymbol{v}^{\mathrm{T}}\boldsymbol{v}$ we reduce the number of equations to be solved down to two in this case, that is to the usual *normal* case in which there is the same number of equations as variables. The solution of these *normal equations* gives the required parameters.

As shown in Appendix A2, these normal equations are of the form

$$\boldsymbol{A}^{\mathrm{T}}\boldsymbol{A}\boldsymbol{x} + \boldsymbol{A}^{\mathrm{T}}\boldsymbol{L} = \mathbf{0} \tag{13.27}$$

The dimensions of these matrices are as follows

$$\boldsymbol{A}^{\mathrm{T}}\boldsymbol{A} = (2 \times 6)(6 \times 2) = (2 \times 2) \quad \text{and} \quad \boldsymbol{A}^{\mathrm{T}}\boldsymbol{L} = (2 \times 6)\,(6 \times 1) = (2 \times 1)$$

The equations in full are

$$0.00069787\ \delta E_G + 0.0001228\ \delta N_G = 7.6929\text{E–}05 \tag{13.28}$$

$$0.0001228\ \delta E_G + 0.00118912\ \delta N_G = 8.3487\text{E–}05 \tag{13.29}$$

Their solution is

$$\delta E_G = 0.100 \quad \text{and} \quad \delta N_G = 0.060$$

and as before

$$\hat{E}_G = \overset{*}{E}_G + \delta E_G = 516.2 + 0.100 = 516.300 \tag{13.30}$$

$$\hat{N}_G = \overset{*}{N}_G + \delta N_G = 448.9 + 0.060 = 448.960 \tag{13.31}$$

13.12 Calculation of the residuals

After the variables $\delta E_G = 0.100$ and $\delta N_G = 0.060$ have been found we substitute them in the observation equations (13.25) to obtain the residuals. As a numerical check we show that $\boldsymbol{A}^{\mathrm{T}}\boldsymbol{v} = \mathbf{0}$.

Exercise 1 Evaluate the residuals in the example and verify that $\boldsymbol{A}^{\mathrm{T}}\boldsymbol{v} = \mathbf{0}$. The residuals are shown in Table 13.4 to be

Table 13.4

Residuals (radians)	Residuals (seconds of arc)
–7.915E–06	–1.6
–0.0006501	–134.1
–7.048E–05	–14.5
8.9057E–05	18.3
–3.157E–05	–6.5
–0.0005622	–115.9

13.13 Summary to date

(1) We form equations (functional models) relating the various observed and unknown parameters. Since the equations are generally non-linear, Newton's method of solution by partial differentiation is used. This has the effect of making us look at small changes to the variables, and has the added convenience that sizes of numbers are reduced.

(2) These observation equations are based on imperfect observations and so different combinations will give different answers. Each observation in turn differs from the 'best' answer by a small amount or residual.
(3) For ease of calculation we can express these equations, usually known as 'observation equations', in matrix form. We have an $\boldsymbol{A}$ matrix involving the coefficients of our unknowns – in this case the Easting and Northing of the intersected point. We also have a known $\boldsymbol{L}$ matrix which contains the observed data, here related to the observed bearings to the intersected point.
(4) The principle of Least Squares allows us to condense all the many (m) observation equations into just the right number (n) of 'Normal' equations to solve for the (n) unknowns. Finally we can calculate the (m) residuals – the differences between the observed values and best estimates – which reflect the quality of the original observations.

13.14 Calculation of L: an alternative approach

We suggest that the beginner should ignore this section and return to it some time later, as it does not add anything vital to the development of concepts.

We shall now describe an alternative way how to derive the absolute term $\boldsymbol{L}$ in observation equations. It is a method used widely in surveying literature, and has merit in its own right in special cases. In Section 13.7 we showed how to obtain K from

$$F(\overset{*}{E}_A, \overset{*}{N}_A, \overset{*}{E}_G, \overset{*}{N}_G : \overset{o}{U}) = K_{AG}$$

Now, instead, if $\overset{*}{U}$ *is chosen* so that

$$F(\overset{*}{E}_A, \overset{*}{N}_A, \overset{*}{E}_G, \overset{*}{N}_G : \overset{*}{U}) = 0$$

and we put $\overset{o}{U} = \overset{*}{U} + \delta U$, then

$$K_{AG} = F(\overset{*}{E}_A, \overset{*}{N}_A, \overset{*}{E}_G, \overset{*}{N}_G : \overset{o}{U}) = F(\overset{*}{E}_A, \overset{*}{N}_A, \overset{*}{E}_G, \overset{*}{N}_G : \overset{*}{U} + \delta U)$$

We expand this by Taylor's theorem (see Section 9.14) as

$$F(\overset{*}{E}_A, \overset{*}{N}_A, \overset{*}{E}_G, \overset{*}{N}_G : \overset{o}{U}) = F(\overset{*}{E}_A, \overset{*}{N}_A, \overset{*}{E}_G, \overset{*}{N}_G : \overset{*}{U}) + \boldsymbol{F}^1(\overset{*}{E}_A, \overset{*}{N}_A, \overset{*}{E}_G, \overset{*}{N}_G : \overset{*}{U})\,\delta U$$

where $\boldsymbol{F}^1$ is the differential of F with respect to U only.

Therefore $$K_{AG} = 0 + (N_G - N_A)\sec^2 U_{AG}\,\delta U$$

But from (13.24) $$L_1 = \frac{K_{AG}}{(N_G - N_A)\sec^2 U_{AG}}$$

therefore $$L_1 = \delta U = \overset{o}{U} - \overset{*}{U}$$

Exercise 1 Derive L_1 = for the ray AG by this alternative method. The computed bearing is obtained from

$$\tan\overset{*}{U} = \frac{\overset{*}{E}_G - \overset{*}{E}_A}{\overset{*}{N}_G - \overset{*}{N}_A}$$

$$\tan\overset{*}{U} = \frac{516.2 - 512.402}{448.9 - 386.280} = \frac{3.789}{62.62} = 0.060652$$

$$\overset{*}{U} = 0.060578 \text{ radians}$$

The observed bearing $\overset{o}{U} = 3.5588° = 0.062113$ radians, so

$$\overset{o}{U} - \overset{*}{U} = 0.062113 - 0.060578 = 0.001535 = L_1 \quad \text{as before}$$

Notice that this method of approach fits well into the general treatment as illustrated in Figure 13.1.

13.15 More functional models

Before considering more aspects of the Least Squares process, we shall consolidate the theory developed so far, by considering two other functional models:

(1) when distances have been measured
(2) when coordinates have been measured directly.

Exercise 1 Derive the equation to be solved if the length of AG has been measured to fix the position of G. The functional model is

$$S_{AG}^2 - (E_G - E_A)^2 - (N_G - N_A)^2 = 0 \qquad (13.32)$$

In the statement of Newton's method of solution we obtain the general functional equation for a distance measurement as

$$F(\overset{*}{E}_A, \overset{*}{N}_A, \overset{*}{E}_G, \overset{*}{N}_G : \overset{o}{S}) + \frac{\partial F}{\partial E_A}\delta E_A + \frac{\partial F}{\partial N_A}\delta N_A$$
$$+ \frac{\partial F}{\partial E_G}\delta E_G + \frac{\partial F}{\partial N_G}\delta N_G + \frac{\partial F}{\partial S}\delta S = 0 \qquad (13.33)$$

Then

$$\frac{\partial F}{\partial E_A} = 2(E_G - E_A), \quad \frac{\partial F}{\partial N_A} = 2(N_G - N_A)$$

$$\frac{\partial F}{\partial E_G} = -2(E_G - E_A), \quad \frac{\partial F}{\partial N_G} = -2(N_G - N_A)$$

$$\frac{\partial F}{\partial S} = 2S_{AG}, \quad F(\overset{*}{E}_A, \overset{*}{N}_A, \overset{*}{E}_G, \overset{*}{N}_G : \overset{o}{S}) = K_{AG}$$

Thus the explicit observation equation is

$$2(E_G - E_A)\delta E_A + 2(N_G - N_A)\delta N_A - 2(E_G - E_A)\delta E_G$$

$$-2(N_G - N_A)\delta N_G + 2S_{AG}v_{AG} + K_{AG} = 0 \qquad (13.34)$$

Division throughout by $2S_{AG}$ gives the final form as

$$\frac{(E_G - E_A)}{(S_{AG})}\delta E_A + \frac{(N_G - N_A)}{S_{AG}}\delta N_A - \frac{(E_G - E_A)}{S_{AG}}\delta E_G$$

$$-\frac{(N_G - N_A)}{S_{AG}}\delta N_G + v_{AG} + L_{AG} = 0 \qquad (13.35)$$

which again is of the form

$$\boldsymbol{Ax} + \boldsymbol{L} = \boldsymbol{v} \qquad (13.26)$$

Exercise 1 Derive the observation equations representing the two measured distances

$$AG = 62.820\,\text{m} \quad DG = 60.606\,\text{m}$$

$$F(\overset{*}{E}_A, \overset{*}{N}_A, \overset{*}{E}_G, \overset{*}{N}_G : \overset{o}{S}) = K_{AG}$$

$$= (\overset{o}{S}_{AG})^2 - (\overset{*}{E}_G - \overset{*}{E}_A)^2 - (\overset{*}{N}_G - \overset{*}{N}_A)^2$$

Using values from Table 13.3

$$K_{AG} = 62.820^2 - 3.798^2 - 62.620^2 = -10.663$$

$$L_{AG} = K_{AG}/2S_{AG} = -0.085$$

Evaluating the coefficients of equations (13.35) the distance observation equations become

$$0.06054030\delta E_G + 0.99816575\delta N_G - 0.085 = v_{AG}$$

$$0.259445040\delta E_G - 0.96575787\delta N_G + 0.070 = v_{DG} \qquad (13.36)$$

Exercise 2 Derive the observation equations representing the position of G measured by a Global Positioning System (GPS) satellite receiver. In this simple case, there is no functional model, but merely an observational model. The residuals are by definition

$$v = \hat{E} - \overset{o}{E}$$

We let

$$\hat{E} = \overset{*}{E} + \delta E$$

therefore

$$\delta E + \overset{*}{E} - \overset{o}{E} = v$$

Thus the observation equations are of the form

$$\begin{aligned} \delta E_G + L_{E_G} &= v_{E_G} \\ \delta N_G + L_{N_G} &= v_{N_G} \end{aligned} \qquad (13.37)$$

Suppose the measured GPS values are

$$\overset{o}{E}_G = 516.285 \qquad \overset{o}{N}_G = 448.988$$

these equations become

$$\begin{aligned} \delta E_G &= L_{E_G} + v_{E_G} = 0.085 + v_{E_G} \\ \delta N_G &= L_{N_G} + v_{N_G} = 0.088 + v_{N_G} \end{aligned} \qquad (13.38)$$

13.16 Combined measurements

We have now seen how equations representing observed bearings, measured distances and positions can be formed, so it might be thought that they can be combined into an $\boldsymbol{A}$ matrix as they stand, and proceed with the Least Squares solution. This cannot be done for three reasons:

(1) the measurements are not all of the same kind (angles and lengths)
(2) they will not necessarily be of the same quality
(3) the measurements may not be independent of each other (This is especially true of the GPS coordinates).

To allow for these three factors, we have to introduce the concept of *weighted observations* and *correlated observations*. These two factors are treated by the introduction of a *dispersion matrix*, often known as a *variance–covariance matrix.*

13.17 The arithmetic mean

Before embarking upon a discussion of *variance*, we shall consider the simple example of a single parameter measured several (m) times, and where each measure is of the same quality. Table 13.5 shows nine such measures of an angle A in column 1. Ignoring the rest of the table for now, we wish to estimate the best value of the angle by the method of Least Squares. Let the nine observed values be A_i, where i = 1,…, 9, the best estimate be $\hat{A}$ and the provisional value be $\overset{*}{A}$, then we have

$$\hat{A} = \overset{*}{A} + \delta A$$

and the residuals are

$$v_1 = \hat{A} - A_1 = \overset{*}{A} - A_1 + \delta A$$

Thus the observation equations are of the form

$$\delta A + L_1 = v_1 \qquad (13.39)$$

where $$L_1 = \overset{*}{A} - A_1$$

Again the nine equations are in the general form

$$\boldsymbol{Ax} + \boldsymbol{L} = \boldsymbol{v} \qquad (13.26)$$

In full these equations are

$$\begin{bmatrix} 1 \\ 1 \\ 1 \\ \vdots \\ 1 \\ 1 \end{bmatrix}^{\delta A} + \begin{bmatrix} L_1 \\ L_2 \\ L_3 \\ \vdots \\ L_8 \\ L_9 \end{bmatrix} = \begin{bmatrix} v_1 \\ v_2 \\ v_3 \\ \vdots \\ v_8 \\ v_9 \end{bmatrix}$$

Thus the $\boldsymbol{A}$ matrix is the (9×1) column vector $(1\ 1\ 1\ 1\ 1\ 1\ 1\ 1\ 1)^{\mathrm{T}}$. Forming the normal equations, we have as usual

$$\boldsymbol{A}^{\mathrm{T}}\boldsymbol{Ax} + \boldsymbol{A}^{\mathrm{T}}\boldsymbol{L} = \boldsymbol{0} \qquad (13.27)$$

But $\boldsymbol{A}^{\mathrm{T}}\boldsymbol{A} = m$ and $\boldsymbol{A}^{\mathrm{T}}\boldsymbol{L} = L_1 + L_2 + L_3 + \cdots + L_9 = \Sigma L$

therefore $$m\delta A = -\Sigma L \qquad (13.40)$$

$$\delta A = -\frac{\Sigma L}{m}$$

Remembering that $L_1 = \overset{*}{A} - A_1$ then $\Sigma L = m\overset{*}{A} - \Sigma A$

Then $$\hat{A} = \overset{*}{A} + \delta A = \overset{*}{A} - \frac{\Sigma L}{m} = \frac{\Sigma A}{m}$$

Thus the Least Squares estimate of the angle is its *arithmetic mean.*

Exercise 1 From the data of Table 13.5, calculate the arithmetic mean of the angle A, the nine residuals, and check that $\boldsymbol{A}^{\mathrm{T}}\boldsymbol{v} = 0$ *within expected numerical limits.*

(**Note:** Hand calculation has rounding errors at various stages which give rise to imperfect checks. If a spreadsheet is used, the rounding error is insignificant here. But all practical computation suffers from rounding error at the limit of precision.)

We will also note for the moment that $\Omega = \Sigma v^2 = 30.64$.

Exercise 2 The reader is encouraged to form 'residuals' from two values which differ from the arithmetic mean and verify that the sums of their squares are greater than 30.64, thus demonstrating that Ω is a minimum.

Table 13. 5

Obs angle A 109° 25′ secs	$+v$	$-v$	v^2
06.3	1.4		1.96
07.2	0.5		0.25
10.4		2.7	7.29
04.3	3.4		11.56
09.6		1.9	3.61
05.8	1.9		3.61
07.9		0.2	0.04
08.3		0.6	0.36
09.1		1.4	1.96
sum = 68.9	sum = +7.2	sum = –6.8	sum = 30.64
Mean = 07.7	Sum = 7.2 – 6.8 = 0.4		

13.18 More about weights

Now suppose that all records of the nine observations have been lost, but we have only two values, one of which we know is the mean of the first six observations and the other the mean of the remaining three. How do we still obtain the best estimate from this information? Consider the example. Ignoring the degrees and minutes, the two remaining known values now are

$$A_{10} = \frac{06.3 + 07.2 + 10.4 + 04.3 + 09.6 + 05.8}{6} = \frac{43.4}{6} = 7.2666$$

and
$$A_{11} = \frac{07.9 + 08.3 + 09.1}{3} = \frac{25.3}{3} = 8.4333$$

Their arithmetic mean is 07.85 which is not the correct value 07.7. But if we take a *weighted mean*, each value weighted by the number of observations used in determining it, we have the result

$$\frac{(6 \times 7.2666) + (3 \times 8.4333)}{9} = \frac{68.9}{9} = 07.7$$

We can write this whole calculation in matrix notation. If we define a diagonal weight matrix $\boldsymbol{W}$ whose diagonal terms are 6 and 3, the numbers of values from which the observed values were derived, the Least Squares solution is given by the equations

$$\boldsymbol{A}^{\mathrm{T}}\boldsymbol{WAx} + \boldsymbol{A}^{\mathrm{T}}\boldsymbol{WL} = \boldsymbol{0} \qquad (13.41)$$

Since in this case the $\boldsymbol{A}$ matrix is a column matrix $(1 \quad 1)^{\mathrm{T}}$ we have in full

$$\boldsymbol{A}^{\mathrm{T}}\boldsymbol{W}\boldsymbol{A} = \begin{bmatrix}1 & 1\end{bmatrix}\begin{bmatrix}6 & 0\\0 & 3\end{bmatrix}\begin{bmatrix}1\\1\end{bmatrix} = 9$$

and

$$\boldsymbol{A}^{\mathrm{T}}\boldsymbol{W} = \begin{bmatrix}1 & 1\end{bmatrix}\begin{bmatrix}-7.2666\\-8.4333\end{bmatrix} = 68.9$$

The solution is

$$\boldsymbol{x} = -(\boldsymbol{A}^{\mathrm{T}}\boldsymbol{W}\boldsymbol{A})^{-1}\boldsymbol{A}^{\mathrm{T}}\boldsymbol{W}\boldsymbol{L} = \frac{68.9}{9} = 7.7$$

This may seem like using a heavy hammer to crack a nut, but the structure is important and we anticipate, as we shall see later, that the general case follows that of (13.41).

13.19 Variance

The residuals show how much the observations are spread about the mean, and we say they give an indication of the *dispersion* of the observations. However, a better indicator of dispersion is the squares of the residuals, so their average is used as a dispersion indicator. This average of the sum of the squares of residuals is called the *variance*. For a limited number of observations it is usually denoted by s^2. Such a limited number is called a *sample* of the total possible number of all possible observations – *the population*. Thus for m observations

$$\text{Sample variance} = s^2 = \frac{\Sigma v^2}{m} \qquad (13.42)$$

Exercise 1 Calculate the sample variance of the observations in Table 13.5. The squares of the residuals are given in column. Their sum is 30.64. Thus the sample variance is

$$s^2 = \frac{\Sigma v^2}{m} = \frac{30.64}{9} = 3.40$$

The quantity s is called the *standard deviation* of the sample and in this case

$$s = 1.845$$

Exercise 2 Divide all residuals by s to give new residuals u and calculate $\Sigma u^2/m$

The calculations are shown in Table 13.6.

Table 13.6

v	v^2	$u = v/s$	u^2
1.4	1.96	0.76	0.58
0.5	0.25	0.27	0.07
–2.7	7.29	–1.46	2.14
3.4	11.56	1.84	3.40
–1.9	3.61	–1.03	1.06
1.9	3.61	1.03	1.06
–0.2	0.04	–0.11	0.01
–0.6	0.36	–0.33	0.11
–1.4	1.96	–0.76	0.58
	$\Sigma v^2 = 30.64$		$\Sigma u^2 = 9$
	$s^2 = 3.40$		
	$s = 1.845$		

Thus
$$\frac{\Sigma u^2}{m} = 1$$

Hence scaling the sample residuals by their standard deviation gives a new sample whose variance (and standard deviation) are one.

Exercise 3 Prove this result by working through the above procedure in terms of the algebra only.

$$\Sigma u^2 = u_1^2 + u_2^2 + \cdots = \frac{v_1^2}{s^2} + \frac{v_2^2}{s^2} + \cdots \text{ to } m \text{ terms}$$

$$\frac{\Sigma u^2}{m} = \frac{\Sigma v^2}{ms^2} = 1$$

13.20 Expectation

In Section 13.17 we calculated the mean value of a sample of m observations, and showed that this equals the best estimate obtained from an application of the principle of Least Squares. If we increased the sample size to infinite proportions, we would calculate the mean value of the whole population. In mathematical language this idea is expressed as *expectation*. The mathematical expectation E(y) of a function y is defined as the *average value* to which y will tend when an infinite number of values is taken.

Thus, for example,

if $y = A_i$ (a measured parameter such as an angle),
the expectation of y, written $E(y) = E(A_i)$,
is the limit to which the mean of the m values A_i ($i = 1$ to m)
will tend as m tends to infinity, or in other words it is the population mean.
The concept of expectation is written as $E(A_i) = \bar{A}$

Again, suppose we have taken the means $\hat{A}$ (i = 1 to n) of n different samples then it is reasonable to see that the mean of all samples will tend again to the population mean and we can write

$$E(\hat{A}_i) = \bar{A}$$

Again if we increase the number of observations in a sample to infinity, the mean of the observations will tend to the population mean, and each residual will become a population residual, denoted by V. Thus a population residual is defined as

$$V_i = \bar{A} - A_i$$

and a sample residual defined as before by

$$v_i = \hat{A}_i - A_i$$

In both cases $E(v_i) = 0$ and $E(V_i) = 0$ because $\Sigma v = 0$ and $\Sigma V = 0$.
Now, since we define variance as the average value of v^2, as we increase the sample size, each value of v_i tends to the population residual V_i and the expectation of V^2 is the population variance, usually denoted by σ^2. Thus we can say

$$E(V^2) = \sigma^2$$

In matrix terminology this is expressed as

$$E(\mathbf{V}^T\mathbf{V}) = \sigma^2$$

It might be thought that, just as the expectation of the sample mean is the population mean, the expectation of the sample variance will be the population variance. But this is not so, and

$$E(s^2) \neq \sigma^2$$

What $E(s^2)$ says is that we calculate all the sample variances for the whole population and then take their mean. Since the residuals of each sample are biased by their sample mean this is not the same thing as first obtaining the population residuals and then calculating the variance. We shall return to this topic in Section 13.26.

13.21 Covariance and correlation

Now consider two sets of observed parameters A and B. Each will have a population variance – call these

$$\sigma_A^2 = E(V_A^2) \quad \text{and} \quad \sigma_B^2 = E(V_B^2)$$

We wish to find out if these observed parameters are connected in any way. A way of finding if there is a link is to calculate a statistic by cross-multiplying all combinations of their residuals, i.e. to find the *covariance* defined by

$$\text{covar} = E(V_A V_B) = \sigma_{AB}$$

This notation for covariance is misleading because it is a second-orde quantity like variance itself. In older books it was written with a bette notation as a double sigma $\sigma\sigma_{AB}$. This is thought to be unnecessary today. I two parameters are quite unrelated then their covariance is zero. This resul depends on the fact that the sum of the residuals of each parameter is zero Calculating the covariance, each residual from one sample is multiplied by al residuals from the other for which the sum is zero. All such products are zer so, for *uncorrelated* parameters,

$$\text{covar} = E(V_A V_B) = \sigma_{AB} = 0$$

Exercise 1 Example of covariance Consider the case of three direction observed by theodolite to points A, B and C as shown in Figure 13.4. Assum that the three pointings are quite independent, i.e. if we ignore centerin errors and refraction. For each direction, let three typical residuals be

$$V_A,\ V_B \text{ and } V_C$$

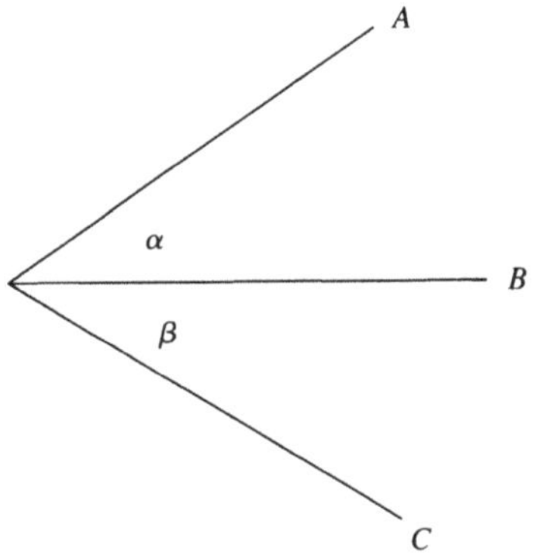

Figure 13.4

Then the residuals of the two included angles α and β will be given by

$$V_\alpha = V_B - V_A \quad \text{and} \quad V_\beta = V_C - V_B$$

then

$$V_\alpha V_\beta = (V_B - V_A)(V_C - V_B)$$

$$V_\alpha V_\beta = V_B V_C - V_A V_C + V_A V_B - V_B V_B$$

Taking expectations we have

$$E(V_\alpha V_\beta) = E(V_B V_C) - E(V_A V_C) + E(V_A V_B) - E(V_B V_B)$$

Assuming that the directions are independent then

$$E(V_B V_C) = 0, \quad E(V_A V_C) = 0, \quad \text{and} \quad E(V_A V_B) = 0$$

$$E(V_\alpha V_\beta) = -E(V_B V_B)$$

Therefore $$\sigma_{\alpha\beta} = -\sigma_B^2 \tag{13.43}$$

The result is partly to be anticipated from common sense because we would expect that the statistical connection between the two angles has something to do with their common direction. On the other hand, if entirely separate measures had been made for each angle, i.e. if the direction to B was observed separately for each angle, there would be no correlation and the covariance would be zero.

In the above example, we were able to predict the covariance from theory. In practice we often do not know whether or not there is any correlation between two measurements, nor do we know the value of the covariance. Sometimes in research work finding the correlation is the object of the analysis. Often we have to assume no correlation at all.

Exercise 2 Example of variance An angle is the difference between two directions read on a theodolite. Let such an angle be given by

$$\alpha = R_B - R_A$$

Then the angle residual will be related to the direction residuals by

$$V_\alpha = V_B - V_A$$

Now $$\sigma_\alpha^2 = \mathrm{E}(V_\alpha^2), \quad \sigma_A^2 = \mathrm{E}(V_A^2) \quad \text{and} \quad \sigma_B^2 = \mathrm{E}(V_B^2)$$

Taking expectations we have

$$\sigma_\alpha^2 = \mathrm{E}(V_\alpha^2) = \mathrm{E}(V_B - V_A)^2 = \mathrm{E}(V_A^2) - 2\mathrm{E}(V_A V_B) + \mathrm{E}(V_B^2)$$

$$\sigma_\alpha^2 = \sigma_A^2 - 2\sigma_{AB} + \sigma_B^2$$

But since the directions to A and B are independent

$$\sigma_{AB} = 0$$

so

$$\sigma_\alpha^2 = \sigma_A^2 + \sigma_B^2$$

To the beginner, this may appear to be a rather unexpected result. We might have anticipated a minus sign for the variance of the difference of two observed quantities.

In the same way, the variance of the second angle is given by

$$\sigma_\beta^2 = \sigma_C^2 + \sigma_B^2$$

Exercise 3 Suppose we use the sum of these two angles in a calculation. What is the correct variance of this sum to be used? Let their sum be

$$\gamma = \alpha + \beta$$

We can answer this question in two ways. From first principles we have

$$\gamma = R_C - R_A$$

hence, from the previous result,

$$\sigma_\gamma^2 = \sigma_C^2 + \sigma_A^2$$

Alternatively, it is instructive to obtain this result indirectly, working from the variances and covariance of the component angles, for

$$V_\gamma = V_\alpha + V_\beta$$

therefore

$$\sigma_\gamma^2 = \sigma_\alpha^2 + 2\sigma_{\alpha\beta} + \sigma_\beta^2$$

$$= \sigma_A^2 + \sigma_B^2 - 2\sigma_B^2 + \sigma_C^2 + \sigma_B^2$$

$$= \sigma_C^2 + \sigma_A^2$$

as before.

Exercise 4 Find the variance of the mean of an angle observed m times. Suppose A_i is a typical independent observation ($i = 1$ to m) and that each has the same variance. From an extension of the above result, the variance of the sum S of the m observations each with the same variance σ_A^2 is given by

$$\sigma_S^2 = \Sigma\sigma_A^2 = m\sigma_A^2$$

The mean is given by $\bar{A} = S/m$ therefore

$$V_{\bar{A}} = \frac{V_S}{m} \quad \text{and} \quad V_{\bar{A}}^2 = \frac{V_S^2}{m^2}$$

Now the variance of each independent observation is σ_A^2 so taking expectations we have

$$\sigma_{\bar{A}}^2 = \mathrm{E}\left(\frac{V_S^2}{m^2}\right) = \frac{m\sigma_A^2}{m^2} = \frac{\sigma_A^2}{m} \tag{13.44}$$

In Section 13.18 we showed that the weights can be assigned to observations according to the number of observations of which they are the mean. This result (13.44) shows that weights can also be assigned in proportion to the reciprocals of their variances, or

$$w_A = \frac{1}{\sigma_A^2} \tag{13.45}$$

Exercise 5 Obtain the coordinates of G by Least Squares if the standard errors of the bearings AG and BG are 1″ and 60″ for all the other bearings. The two variances are 1 and 3600 or in radians 1/206265 and 3600/206265. (There are 206265 seconds of arc in a radian.) Hence the weights are 206265 and 206265/3600 = 57.2958. If the computation is by hand calculator, the simplest way is to multiply each observation equation by the square root of its weight, then treat these weighted observation equations as before. This working is shown in Table 13.7.

The original observation equations are given in cells A1 to B6 and the values of L in column C. Cells A1 to C2 are multiplied by the square root of the weight $\sqrt{206265}$ to give the weighted observation equations in cells D1 to F2. The remaining four weighted observation equations in cells D3 to F6 are obtained by multiplying cells A3 to C6 by the square root of their weights $\sqrt{57.2958}$. These are the equations

$$\sqrt{\boldsymbol{W}}\boldsymbol{Ax} + \sqrt{\boldsymbol{W}}\boldsymbol{L} = 0 \qquad (13.46)$$

Table 13.7

	A	B	C	D	E	F
1	0.015911	−0.000965	0.001535	7.22620447	−0.4382683	0.69714184
2	0.002080	0.019118	0.002005	0.94466126	8.68270864	0.91059895
3	−0.010128	0.012567	−0.000187	−0.0766629	0.09512466	−0.0014155
4	−0.015917	−0.004276	−0.001933	−0.1204822	−0.0323668	−0.0146317
5	−0.008550	−0.009073	−0.001364	−0.0647184	−0.0686772	−0.0103247
6	−0.003370	−0.023752	−0.001193	−0.0255089	−0.1797884	−0.0090303

Refer to Table 13.8. The weighted normal equations formed in cells A1 to C2 are

$$\boldsymbol{A}^{\mathrm{T}}\boldsymbol{WAx} = -\boldsymbol{A}^{\mathrm{T}}\boldsymbol{WL} \qquad (13.47)$$

$$\boldsymbol{Nx} = \boldsymbol{b}$$

Table 13.8

	A	B	C
1	53.1356483	5.04083991	5.90066698
2	5.04083991	75.6286452	7.60360177
3	0.01893951	−0.0012624	0.10215724
4	−0.0012624	0.01330664	0.09372961

The inverse of $\boldsymbol{N}$ is located in cells A3 to B4 and the solution in C3 and C4. However, this method is only possible if the observations are uncorrelated (covariances are zero) so the more applicable alternative is to be preferred, although in this case the calculation is somewhat longer.

The standard method of calculation by spreadsheet or computer programme is to form the square weight matrix $\boldsymbol{W}$ and perform the calculations by matrix methods. Table 13.9 shows the weight matrix in cells A1 to F6.

Table 13.9

	A	B	C	D	E	F
1	206265	0	0	0	0	0
2	0	206265	0	0	0	0
3	0	0	57.2958	0	0	0
4	0	0	0	57.2958	0	0
5	0	0	0	0	57.2958	0
6	0	0	0	0	0	57.2958

Performing the matrix multiplications $\boldsymbol{A}^{\mathrm{T}}\boldsymbol{WA}$ and $\boldsymbol{A}^{\mathrm{T}}\boldsymbol{WL}$ yields the same weighted normal equations as before and of course the same solution. This is $\delta E = 0.102$ and $\delta N = 0.094$ and final coordinates of G as

$$E = 516.302 \quad \text{and} \quad N = 448.994$$

It will be noticed that these values are virtually the same as those using the rays AG and BG only. This is to be expected, because we gave such high weights to these directions compared with the others. It might be thought, therefore, that we should just have ignored the other rays. Had we used only two rays, there would be no check on the result, so that, if a mistake had been made in copying down a reading say, a gross error could result for the position of G and, what's more, we would have no indication of it. Such a result therefore is said to be *unreliable*. It may of course be perfectly satisfactory, but we are not sure of this. The extra rays and their residuals do give us a check against a mistake, and enable us to say we have a *reliable* result.

13.22 Summary

We discussed *mathematical expectation, variance* and *covariance* and we have shown that the *variance* is the expectation of the square of a residual or

$$\sigma_a^2 = \text{Expectation}(V_{a_i}V_{a_i}) \quad \text{or} \quad = \text{E}(V_{a_i}^2)$$

And in the same way we defined the *covariance* of two different parameters such as a_i and b_i to be

$$s_{ab} = \text{covariance of } a_i \text{ and } b_i = \text{E}(V_{a_i}V_{b_i})$$

We also showed how observations of different quality can be incorporated using a weight matrix. We now extend our ideas to consider how to treat *correlated* observations.

13.23 The dispersion matrix or variance – covariance matrix

Although the following discussion is limited to a vector of three population residuals, simply to contain the explanation to a manageable size and in a simple form, the theory is valid for any number of variables. Consider three population residuals $\boldsymbol{V}$ arranged as a column matrix. We can write

$$\boldsymbol{V} = \begin{bmatrix} V_1 \\ V_2 \\ V_3 \end{bmatrix}$$

Now, if we postmultiply this column vector by its transpose (a row vector) we obtain a square matrix as follows

$$\boldsymbol{VV}^{\mathrm{T}} = \begin{bmatrix} V_1 \\ V_2 \\ V_3 \end{bmatrix} \begin{bmatrix} V_1 & V_2 & V_3 \end{bmatrix}$$

Taking mathematical expectations we have

$$\mathrm{E}(\boldsymbol{VV}^{\mathrm{T}}) = \mathrm{E}\begin{bmatrix} V_1^2 & V_1V_2 & V_1V_3 \\ V_2V_1 & V_2^2 & V_2V_3 \\ V_3V_1 & V_3V_2 & V_3^2 \end{bmatrix}$$

$$= \begin{bmatrix} \sigma_1^2 & \sigma_{12} & \sigma_{13} \\ \sigma_{21} & \sigma_2^2 & \sigma_{23} \\ \sigma_{31} & \sigma_{32} & \sigma_3^2 \end{bmatrix}$$

This matrix is called the *variance–covariance matrix* or, better, the *dispersion matrix* of the observations, and is usually denoted by $\boldsymbol{D}_{\mathrm{o}}$.
This square matrix is symmetrical because

$$V_1V_2 = V_2V_1$$

and therefore

$$\mathrm{E}(V_1V_2) = \mathrm{E}(V_2V_1)$$

$$\sigma_{12} = \sigma_{21}$$

and so on. When the correlations are zero (if we consider the observations to be independent of each other), these matrices are diagonal and therefore easy to invert.
The dispersion matrix is then given by

$$\boldsymbol{D}_{\mathrm{o}} = \begin{bmatrix} \sigma_1^2 & 0 & 0 \\ 0 & \sigma_2^2 & 0 \\ 0 & 0 & \sigma_3^2 \end{bmatrix}$$

and

$$D_o^{-1} = \begin{bmatrix} 1/\sigma_1^2 & 0 & 0 \\ 0 & 1/\sigma_2^2 & 0 \\ 0 & 0 & 1/\sigma_3^2 \end{bmatrix}$$

In this case of independent observations

$$D_o^{-1} = W \qquad (13.48)$$

$$D_o = W^{-1} \qquad (13.49)$$

$$D_o D_o^{-1} = \begin{bmatrix} 1 & 0 & 0 \\ 0 & 1 & 0 \\ 0 & 0 & 1 \end{bmatrix} = I$$

Where I is a (3 × 3) unit matrix in this case. If there are m observations I would be a unit matrix of dimensions $(m \times m)$, written as I_m.
We remind readers that a square symmetric matrix is equal to its transpose: for example, the normal equation's matrix N and the weight matrix W are square symmetric, so

$$N = N^T \quad \text{and} \quad W = W^T$$

This is also true of their inverses:

$$N^{-1} = (N^{-1})^T \quad \text{and} \quad W^{-1} = (W^{-1})^T$$

13.24 The dispersion matrix of derived parameters

We are now going to show that the dispersion matrix of the derived parameters (in the example, the coordinates of G) is obtained from the inverse of the normal equations. If, as usual, the variables x are obtained from a solution of the normal equations

$$Nx = b$$

the dispersion matrix D_x of these unknowns is given by

$$D_x = \sigma_o^2 N^{-1}$$

where σ_o^2 is a parameter estimated from the residuals (see Section 13.26). We denote the dispersion matrix of the observed parameters by D_o and the dispersion matrix of the derived parameters by D_x. Because the dispersion matrix of the observed parameters can only be estimated from some data from previous experience, we make allowance for an error in estimation by introducing an unknown scaling factor, σ_o^2, called the *variance of an observation of unit weight*. We relate the inverse of the weight matrix W^{-1} to the *a priori* estimated dispersion matrix $\boldsymbol{D}_o$ by

$$\sigma_o^2 \boldsymbol{W}^{-1} = \boldsymbol{D}_o$$

The data of any Least Squares calculation will be used to *estimate* this scaling factor from the sample residuals, and thus gain a better value for the dispersion matrix of the observations. If the calculated value of σ_o^2, is equal to 1 then the original estimated weight matrix is correct and

$$\boldsymbol{W}^{-1} = \boldsymbol{D}_o$$

In practice we are happy with a value of σ_o^2 which is close to 1.

The next step in the discussion is to examine how the best estimates obtained from the sample of actual observations differ from their theoretical, or population, values.

The observations made are only a *sample* of all the possible observations that might be made in theory. We must now distinguish between the population and the sample.

If $\bar{s}$ is the population value of an observed parameter $\overset{o}{s}$, then we define the population residual V by

$$\bar{s} = \overset{o}{s} + V$$

Correspondingly, the vectors representing all the observations are written in bold notation

$$\bar{\boldsymbol{s}} = \overset{o}{\boldsymbol{s}} + \boldsymbol{v}$$

Similarly the vector of data for the sample is

$$\hat{\boldsymbol{s}} = \overset{o}{\boldsymbol{s}} + \boldsymbol{V}$$

We use capital $\boldsymbol{V}$ for population residuals and lower case $\boldsymbol{v}$ for the sample residuals. Then the differences **ds** between the Least Squares values and the population values will be given by

$$\mathbf{d}\boldsymbol{s} = \bar{\boldsymbol{s}} - \hat{\boldsymbol{s}} = \boldsymbol{V} - \boldsymbol{v}$$

Also, we can write the respective sample and population models as

$$\boldsymbol{Ax} + \boldsymbol{L} = \boldsymbol{v}$$

$$\boldsymbol{AX} + \boldsymbol{L} = \boldsymbol{V} \qquad (13.50)$$

where $\boldsymbol{X}$ are the population values of the derived parameters and $\boldsymbol{x}$ are the sample values of these parameters derived from a Least Squares treatment. Subtracting gives

$$\boldsymbol{A}(\boldsymbol{X} - \boldsymbol{x}) = \boldsymbol{V} - \boldsymbol{v}$$

or

$$\boldsymbol{A}\mathbf{d}\boldsymbol{x} = \boldsymbol{V} - \boldsymbol{v} \qquad (13.51)$$

where $\mathbf{d}\boldsymbol{x}$ is the vector of the differences between the population parameters and their corresponding values estimated by Least Squares.

Now, assume there are two parameters to be derived and therefore two elements in the vector of errors, so that

$$\mathbf{dxdx}^{\mathrm{T}} = \begin{bmatrix} \mathrm{d}x_1 \\ \mathrm{d}x_2 \end{bmatrix} \begin{bmatrix} \mathrm{d}x_1 & \mathrm{d}x_2 \end{bmatrix}$$

We have chosed two for convenience. Taking expectations we have

$$\mathrm{E}(\mathbf{dxdx}^{\mathrm{T}}) = \mathrm{E}\begin{bmatrix} \mathrm{d}x_1^2 & \mathrm{d}x_1\mathrm{d}x_2 \\ \mathrm{d}x_2\mathrm{d}x_1 & \mathrm{d}x_2^2 \end{bmatrix}$$

$$= \begin{bmatrix} \sigma_{x_1}^2 & \sigma_{x_1 x_2} \\ \sigma_{x_1 x_2} & \sigma_{x_1}^2 \end{bmatrix}$$

$$= \boldsymbol{D_x} \tag{13.52}$$

which is the dispersion matrix of the derived parameters.

The next stage is to link the two dispersion matrices together. First of all, we have to estimate the variances on the diagonal of the dispersion matrix of the *observations* using samples of previous data, such as repetitive measures with the instrument or similar instrument, or results from its previous use. Depending on the way the observation equations have been treated we may have some idea about the covariances between measures. Quite often we make the simple assumption that these are zero, which means that the observations are uncorrelated. But this is unlikely with rounds of angles, and especially in GPS measures. This topic is one that attracts a lot of thought and discussion. The *mechanism* for treating the dispersion matrix is straight forward, so that's what we will concentrate on here.

Given that we have some acceptable values for variances and covariances, we obtain a weight matrix $\boldsymbol{W}$ from

$$\boldsymbol{W} = \boldsymbol{D}_{\mathrm{o}}^{-1}$$

Now from (13.51)

$$\mathbf{A}\mathbf{dx} = \boldsymbol{V} - \boldsymbol{v}$$

Premultiplying both sides by $\mathbf{A}^{\mathrm{T}}\boldsymbol{W}$ we obtain

$$\mathbf{A}^{\mathrm{T}}\boldsymbol{W}\mathbf{A}\mathbf{dx} = \mathbf{A}^{\mathrm{T}}\boldsymbol{WV} - \mathbf{A}^{\mathrm{T}}\boldsymbol{Wv}$$

and remembering that

$$\mathbf{A}^{\mathrm{T}}\boldsymbol{Wv} = \mathbf{0} \tag{A2.8}$$

$$\mathbf{A}^{\mathrm{T}}\boldsymbol{W}\mathbf{A}\mathbf{dx} = \mathbf{A}^{\mathrm{T}}\boldsymbol{WV}$$

or say

$$\boldsymbol{N}\mathbf{dx} = \mathbf{A}^{\mathrm{T}}\boldsymbol{WV}$$

and

$$\mathbf{dx} = \boldsymbol{N}^{-1}\mathbf{A}^{\mathrm{T}}\boldsymbol{WV}$$

Then

$$\mathbf{dxdx}^{\mathrm{T}} = \boldsymbol{N}^{-1}\boldsymbol{A}^{\mathrm{T}}\boldsymbol{WV}\,(\boldsymbol{N}^{-1}\boldsymbol{A}^{\mathrm{T}}\boldsymbol{WV})^{\mathrm{T}}$$

$$= \boldsymbol{N}^{-1}\boldsymbol{A}^{\mathrm{T}}\boldsymbol{WVV}^{\mathrm{T}}\boldsymbol{WAN}^{-1}$$

Taking expectations we have

$$\boldsymbol{D_x} = \boldsymbol{N}^{-1}\boldsymbol{A}^{\mathrm{T}}\boldsymbol{WD}_{\mathrm{o}}\boldsymbol{WAN}^{-1}$$

$$= \boldsymbol{N}^{-1}\boldsymbol{A}^{\mathrm{T}}\boldsymbol{WW}^{-1}\boldsymbol{WAN}^{-1}$$

$$= \boldsymbol{N}^{-1}$$

It is quite amazing that this result eventually comes out as such a simple expression.

For completeness, we should modify the result by allowing for poor estimates of the variances and covariances and put

$$\boldsymbol{D_x} = \sigma_{\mathrm{o}}^2 \boldsymbol{N}^{-1} \qquad (13.53)$$

Exercise 1 What are the variances, the standard errors and the covariance of the coordinates of *G*?

From Table 13.8 the inverse of $\boldsymbol{N}$ is

$$\begin{bmatrix} 0.01893951 & -0.0012624 \\ -0.0012624 & 0.01330664 \end{bmatrix}$$

$$\boldsymbol{N}^{-1} = \begin{bmatrix} \sigma_{x_1}^2 & \sigma_{x_1 x_2} \\ \sigma_{x_1 x_2} & \sigma_{x_1}^2 \end{bmatrix}$$

Identifying the variables in terms of the coordinates of *G* we have

$$\boldsymbol{N}^{-1} = \begin{bmatrix} \sigma_{E_G}^2 & \sigma_{E_G N_G} \\ \sigma_{E_G N_G} & \sigma_{N_G}^2 \end{bmatrix}$$

Then

$$\sigma_{E_G}^2 = 0.01893951 \quad \text{and} \quad \sigma_{E_G} = 0.139$$

$$\sigma_{N_G}^2 = 0.01330664 \quad \text{and} \quad \sigma_{N_G}^2 = 0.115$$

$$\sigma_{E_G N_G} = -0.0012624$$

This means that, in general terms, the position is only good to about 10 cm, which is obvious from the Figure 13.3. It shows that we really do not need to draw such a figure at all, and rely on the calculated values of the variances and the covariance for quality assessment. (Appendix A4 shows how these results can also be depicted by an *error ellipse* and its *pedal curve*.)

The significance of the covariance is that it can be used in a subsequent Least Squares estimation involving the point G, and other measurements, such as a GPS fix.

13.25 Error propagation

In the exercises of Section 13.21 we demonstrated from first principles how error propagation can be determined in individual cases. We now show how a matrix treatment can be used to generalise the process. Consider again the directions A, B and C and their included angles α, β and γ. The relationships among residuals are

$$V_\alpha = V_B - V_A$$

$$V_\beta = V_C - V_B$$

$$V_\gamma = V_C - V_A$$

In matrix form these equations are

$$\begin{bmatrix} V_\alpha \\ V_\beta \\ V_\gamma \end{bmatrix} = \begin{bmatrix} -1 & 1 & 0 \\ 0 & -1 & 1 \\ -1 & 0 & 1 \end{bmatrix} \begin{bmatrix} V_A \\ V_B \\ V_C \end{bmatrix}$$

in short they can be written as

$$\boldsymbol{V} = \boldsymbol{Av}$$

Then

$$\boldsymbol{VV}^{\mathrm{T}} = \boldsymbol{Av}(\boldsymbol{Av})^{\mathrm{T}} = \boldsymbol{Avv}^{\mathrm{T}}\boldsymbol{A}^{\mathrm{T}} \tag{13.54}$$

By the same reasoning as in (13.26)

$$\mathrm{E}(\boldsymbol{vv}^{\mathrm{T}}) = \mathrm{E}\begin{bmatrix} V_A^2 & V_AV_B & V_BV_C \\ V_BV_A & V_B^2 & V_BV_C \\ V_CV_A & V_CV_B & V_C^2 \end{bmatrix}$$

$$= \begin{bmatrix} \sigma_A^2 & \sigma_{AB} & \sigma_{AC} \\ \sigma_{BA} & \sigma_B^2 & \sigma_{BC} \\ \sigma_{CA} & \sigma_{CB} & \sigma_C^2 \end{bmatrix} = \begin{bmatrix} \sigma_A^2 & 0 & 0 \\ 0 & \sigma_B^2 & 0 \\ 0 & 0 & \sigma_C^2 \end{bmatrix}$$

Because the directions A, B and C are independent, their respective covariances are zero

$$\sigma_{AB} = 0, \quad \sigma_{AC} = 0, \quad \sigma_{BC} = 0$$

Similarly

$$E(\boldsymbol{VV}^{T}) = \begin{bmatrix} \sigma_\alpha^2 & \sigma_{\alpha\beta} & \sigma_{\alpha\gamma} \\ \sigma_{\beta\alpha} & \sigma_\beta^2 & \sigma_{\beta\gamma} \\ \sigma_{\gamma\alpha} & \sigma_{\gamma\beta} & \sigma_\gamma^2 \end{bmatrix}$$

$$\begin{bmatrix} \sigma_\alpha^2 & \sigma_{\alpha\beta} & \sigma_{\alpha\gamma} \\ \sigma_{\beta\alpha} & \sigma_\beta^2 & \sigma_{\beta\gamma} \\ \sigma_{\gamma\alpha} & \sigma_{\gamma\beta} & \sigma_\gamma^2 \end{bmatrix} = \begin{bmatrix} -1 & 1 & 0 \\ 0 & -1 & 1 \\ -1 & 0 & 1 \end{bmatrix} \begin{bmatrix} \sigma_A^2 & 0 & 0 \\ 0 & \sigma_B^2 & 0 \\ 0 & 0 & \sigma_C^2 \end{bmatrix} \boldsymbol{A}^{T}$$

$$= \begin{bmatrix} -\sigma_A^2 & \sigma_B^2 & 0 \\ 0 & -\sigma_B^2 & \sigma_C^2 \\ -\sigma_A^2 & 0 & \sigma_C^2 \end{bmatrix} \begin{bmatrix} -1 & 0 & -1 \\ 1 & -1 & 0 \\ 0 & 1 & 1 \end{bmatrix} = \begin{bmatrix} \sigma_A^2+\sigma_B^2 & -\sigma_B^2 & \sigma_A^2 \\ -\sigma_B^2 & \sigma_B^2+\sigma_C^2 & \sigma_C^2 \\ \sigma_A^2 & \sigma_C^2 & \sigma_A^2+\sigma_C^2 \end{bmatrix}$$

Equating terms we obtain the results of the separate exercises together with expressions for $\sigma_{\alpha\gamma}$ and $\sigma_{\beta\gamma}$ not previously derived from first principles.

13.26 Estimation of σ_o^2

Because the weights might not be estimated correctly we introduce a scaling parameter called the variance of unit weight σ_o^2 given by

$$\sigma_o^2 = E\left(\frac{\boldsymbol{v}^{T}\boldsymbol{W}\boldsymbol{v}}{m-n}\right) \qquad \text{(A2.26)}$$

is an *unbiased estimator* of the variance of unit weight σ_o^2. This result is derived in Appendix A2, Section A2.6. If we let

$$s_o^2 = \frac{\boldsymbol{v}^{T}\boldsymbol{W}\boldsymbol{v}}{m-n}$$

then the expectation of s_o^2 is σ_o^2 and we say that s_o^2 is an *unbiased estimator* of σ_o^2.

The greater the sample we take, the nearer we get to an estimate of the population unit variance. Also the numerator $(m-n)$ shows the degree of freedom of the problem. The greater it is, the better the estimate will be. For example, in the case of m observations of one parameter, (A2.26) reduces to the familiar expression

$$s_o^2 = \frac{\boldsymbol{v}^{T}\boldsymbol{W}\boldsymbol{v}}{m-1} \qquad (13.55)$$

That's why there are two common formulae in use, one for the sample variance and another for an estimate of the population variance, which are respectively

$$s^2 = \frac{\boldsymbol{v}^{\mathrm{T}}\boldsymbol{W}\boldsymbol{v}}{m} \quad \text{and} \quad s_o^2 = \frac{\boldsymbol{v}^{\mathrm{T}}\boldsymbol{W}\boldsymbol{v}}{m-1}$$

The use of the scaling parameter σ_o^2 only allows for a general failure to estimate the weight matrix properly. With a complicated problem, involving different types of measurements, we can use different scaling factors for different types of measurement. A lot of research into GPS error sources uses this concept, but this topic is out with the scope of a basic textbook such as this.

13.27 Combined Least Squares estimation

We now have the theory to tackle the problem of combining observed parameters of different quality and of different kind, and of estimating the quality of the results. We shall use the original data in Table 13.1, also incorporating distance and GPS measurements into the original intersection problem. These extra equations are (13.36) and (13.38) above.

One might argue that the GPS positions are not 'observations', because we actually have results derived from a great many observations, which have been numerically treated in a sophisticated manner. The same thing can be said of distances, and come to think of it of angles too. But we must settle for the numerical output by the systems as 'observations'. Perhaps that's why many authors refer to them not as 'observations' but as 'observables'.

There are the six direction equations already dealt with, and two distance and the two position equations, giving ten equations in two variables, i.e. $m = 12$ and $n = 2$. These are shown in part of the Excel spreadsheet of Table 13.10.

Table 13.10

OBS	a_{i1}	a_{i2}	L	Residuals[a], v	Shifts, Sv
AG	0.015911	–0.000965	0.001535	–9.485E–05	–0.0059502
BG	0.00208	0.019118	0.002005	–0.0002247	–0.0116832
CG	–0.010128	0.012567	–0.000187	0.00025913	0.01605399
DG	–0.015917	–0.004276	–0.001933	5.8688E–05	0.00356095
EG	–0.00855	–0.009073	–0.001364	–0.0002034	–0.0163144
FG	–0.00337	–0.023752	–0.001193	–0.0010939	–0.0455997
Dist *AG*	0.0605403	0.99816575	0.085	0.00336109	
Dist *DG*	0.25944504	–0.9657579	–0.070	0.0148884	
Easting	1	0	0.085	0.01053062	
Northing	0	1	0.088	–0.0052706	

[a] The residuals are computed later from the solution.

We must now select suitable elements for the dispersion matrix. Let's say that the directions are uncorrelated and each has a standard error of 20 seconds of arc, that the distances are also uncorrelated with a standard error of 0.01 m and let's also say that the position of point G is obtained by differential GPS relative to point A with a standard error of 0.014 m in Easting and a more accurate 0.0014 in Northing.

We shall see what value we obtain for σ_o^2. However, we really must also include a value for the covariance of the GPS results. This would be output from the GPS software anyway. Let's say –0.000006. Also, since we worked the directions in radians, their standard error of 20 seconds has to be converted to radians thus

$$\frac{20\pi}{180 \times 60 \times 60} = 9.70\text{E}-05 = S \quad \text{say}$$

and the variance

$$P = S^2 = 9.4012\text{E}{-09}$$

This deals with $\boldsymbol{D}_o$, the dispersion matrix of the observations, which we show in Table 13.11 as part of the spreadsheet. The rest of the process is purely arithmetical. The various stages are shown in Tables 13.12 to 13.14 of the spreadsheet.

Table 13.11

Dispersion matrix $\boldsymbol{D}_o$

S = 9.70E–05, $P = S^2$ = 9.4012E–09

1	P	0	0	0	0	0	0	0	0	0
2	0	P	0	0	0	0	0	0	0	0
3	0	0	P	0	0	0	0	0	0	0
4	0	0	0	P	0	0	0	0	0	0
5	0	0	0	0	P	0	0	0	0	0
6	0	0	0	0	0	P	0	0	0	0
7	0	0	0	0	0	0	0.0001	0	0	0
8	0	0	0	0	0	0	0	0.0001	0	0
9	0	0	0	0	0	0	0	0	0.0002	–0.000006
10	0	0	0	0	0	0	0	0	–0.000006	0.000002
	1	2	3	4	5	6	7	8	9	10

The normal equations of Table 13.13 are

$$\boldsymbol{Nx} = \boldsymbol{b}$$

where

$$\boldsymbol{N} = \boldsymbol{A}^{\mathrm{T}}\boldsymbol{WA} \quad \text{and} \quad \boldsymbol{b} = \boldsymbol{A}^{\mathrm{T}}\boldsymbol{WL}$$

Table 13.12

Weight matrix $\boldsymbol{W} = \boldsymbol{D}_o^{-1}$
$1/P = 106368929$

	1	2	3	4	5	6	7	8	9	10
1	1/P	0	0	0	0	0	0	0	0	0
2	0	1/P	0	0	0	0	0	0	0	0
3	0	0	1/P	0	0	0	0	0	0	0
4	0	0	0	1/P	0	0	0	0	0	0
5	0	0	0	0	1/P	0	0	0	0	0
6	0	0	0	0	0	1/P	0	0	0	0
7	0	0	0	0	0	0	10000	0	0	0
8	0	0	0	0	0	0	0	10000	0	0
9	0	0	0	0	0	0	0	0	5494.50549	16483.5165
10	0	0	0	0	0	0	0	0	16483.5165	549450.549

Table 13.13

Normals		
80436.277	27645.6036	9971.23131
27645.6036	695226.164	60156.6311

And the solution is

$$\boldsymbol{x} = \boldsymbol{N}^{-1}\boldsymbol{b}$$

The final position of point G is

516.296   448.983

We calculate the residuals from

$$\boldsymbol{v} = \boldsymbol{Ax} + \boldsymbol{L}$$

which are placed in Table 13.10. And from these residuals we calculate s_o^2, an estimate of σ_o^2 from the formula

$$s_o^2 = \frac{\boldsymbol{v}^T\boldsymbol{Wv}}{m-n} \tag{13.55}$$

In this case $s_o^2 = 20.237100$

The fact that this is not close to unity indicates that the dispersion matrix has not been modelled very well, or that the mathematical model is imperfect, which we shall see is the case.

Table 13.14

Inverse		Solution
1.2604E–05	–5.012E–07	0.09553062
–5.012E–07	1.4583E–06	0.08272938

13.28 Statistical tests for outliers

A huge benefit arising from a Least Squares estimation of results is the statistical information available as a bi-product. Some of this will now be described. We start by assuming that residuals have a *normal distribution* (see Section 9.30). We must stress that for most tests to be valid we should be dealing with a sample of more than about 30 individual measurements. In our simple example we have only ten observations, in which case we should be discussing the Student's '*t*' distribution but, as most Least Squares problems involve many more than 30 variables, we will stick to the normal distribution. We cannot deal with theoretical matters now. We can only explain the principles involved and explain how to operate the tests.

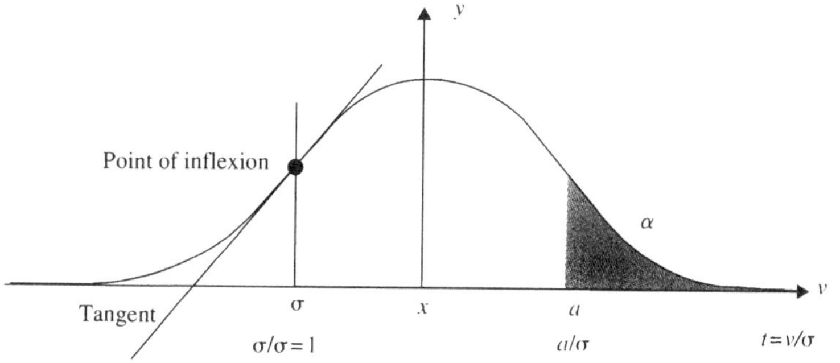

Figure 13.5

From the Least Squares process we obtain the residuals $\boldsymbol{v}$. We assume these are normally distributed about an origin x. A theoretical graph of these residuals about an origin at x is shown in Figure 13.5. This is a normal distribution curve showing the probability y of the occurrence of a residual $\boldsymbol{v}$. The area under the curve represents the probability of all errors occurring, or certainty, or $P = 1$. The unshaded area represents the probability of a residual being less than or equal to a, and conversely the shaded area (α) represents the probability of a residual being greater than or equal to a.

The points of inflexion occur when $\boldsymbol{v} = \sigma$ the standard error (see Section 9.30). Every graph will be the same shape but a different size, so we scale each graph by dividing each residual by σ, including, of course, σ itself. This gives what we call a standardised normal distribution with zero origin and unit standard error (and also unit variance) (see Section 13.26). The probability function is tabulated for this standardised normal distribution, written as N(0, 1). The original curve would be written as N($\boldsymbol{x}$, σ). We use the curve to illustrate our ideas. A normal distribution table gives the actual numbers.

The reasoning goes like this. Suppose we want to know the probability of there being a residual greater than the standard error σ. We look up the tables with $t = \boldsymbol{v}/\sigma = 1$. The tabular entry is $P = 0.8413$. Complete normal distribution tables are quite readily available in many textbooks on surveying or elsewhere. Table 13.15 shows selected values from such a table.

Table 13.15 Table of cumulative normal probability

t	0	1	2	3	4	5	6	7	8	9
0.0	.500	.504	.508	.512	.516	.520	.524	.528	.532	.536
0.5	.674									
0.6								.750		
1.0	.841	.844	.846	.848	.851	.853	.855	.858	.860	.862
1.6					.950					
1.7						.960				
1.9							.975			
2.0	.977	.978	.978	.979	.979	.980	.981	.981	.981	.982
3.0	.999	.999	.999	.999	.999	.999	.999	.999	.999	.999

The chance of a residual being *greater than* σ is therefore

$$P = 1 - 0.8413 = 0.16 \text{ or } 16\%.$$

And the probability of it being both greater than +σ or less than –σ is 32%. Or, put another way, the probability of a residual being *numerically* less than σ is 68%.

(Note that the value of t, when $P = 0.5$, is 0.6745, was the basis of the concept of *probable error* PE = 0.6745σ used in old Least Squares literature.)

Consideration of the normal distribution is used to detect *outliers* in the data. By an *outlier*, we mean a suspect observation or result.

If we look at Figure 13.3, some directions seem wrong. But we need a test which is not arbitrary, and which can be applied automatically, before we can reject an observation. Although we can use a simple test such as 'reject residuals greater or less than three times the standard error', such a test has limited application, as we shall see. It is almost as easy to test the probability of a residual exceeding a selected value, or the value outside a given probability. Suppose, for example, that we decide to reject all residuals whose probability of occurrence is greater than or equal to 5%. This means that a two-sided test for 2.5% is applied. The table value corresponding to 97.5% is 1.96. Then we apply the test:

$$\text{is } t = \boldsymbol{v}/\sigma \text{ greater than or equal to } 1.96?$$

If so, we reject the observation.

To apply this test to our example we have first to calculate the standard errors of each residual. We assume the formula for the moment so that the flow of the discussion is not interrupted. We give the derivation in Section A2.8 below.

The variance of the residuals are readily obtained as the diagonal terms of the matrix

$$\boldsymbol{D}_v = \boldsymbol{D}_o - \boldsymbol{D}_s \tag{13.56}$$

where

$$\boldsymbol{D}_s = \boldsymbol{A}\boldsymbol{D}_x\,\boldsymbol{A}\boldsymbol{T} \tag{13.57}$$

$\boldsymbol{D}_s$ is the dispersion matrix of the observables, and $\boldsymbol{D}_o$ the originally estimated dispersion matrix of the observations. The square roots of the diagonals of $\boldsymbol{D}_v$ are the required standard errors. We summarise all these matrices and parameters in Table 13.16.

Table 13.16

	Parameter		Matrix	Diagonal variance	Square root standard error
(1)	observed	$\boldsymbol{s}_o$	$\boldsymbol{D}_o$	$\sigma^2_{\boldsymbol{s}_o}$	$\sigma_{\boldsymbol{s}_o}$
(2)	best	$\boldsymbol{s}$	$\boldsymbol{D}_s$	$\sigma^2_{\boldsymbol{s}}$	$\sigma_{\boldsymbol{s}}$
(3)	unobserved	$\boldsymbol{x}$	$\boldsymbol{D}_x$	$\sigma^2_{\boldsymbol{x}}$	$\sigma_{\boldsymbol{x}}$
(4)	residuals	$\boldsymbol{v}$	$\boldsymbol{D}_v$	$\sigma^2_{\boldsymbol{v}}$	$\sigma_{\boldsymbol{v}}$

In calculating

$$\boldsymbol{D}_x = \sigma_o^2 \boldsymbol{N}^{-1}$$

we will use $\sigma_o^2 = 1$ instead of the clearly incorrect value of s_o^2.

Not all the figures have been shown but, as we are interested only in the diagonal terms, we will concentrate on them and display them as rows, alongside the residuals from the Least Squares solution. See Table 13.18.

In Table 13.18 we have the following

Row 1 shows the diagonal elements of $\boldsymbol{D}_o$

Row 2 shows the diagonal elements of $\boldsymbol{D}_s$

Row 3 shows the difference of diagonal elements of $\boldsymbol{D}_v = \boldsymbol{D}_o - \boldsymbol{D}_s$

Row 4 shows the square roots of the diagonal elements of $\boldsymbol{D}_v$, i.e. σ_v

Row 5 shows the residuals v

Row 6 shows the statistical test ratio $t = v/\sigma_v$

These ratios t will be used to detect for blunders in the observations. Again it is assumed that the observations are normally distributed. We select a suitable rejection level of significance, usually 0.01% and test to see if any observations lie outside the test bound. From a normal distribution table we see that this critical value is 2.57 which is exceeded by three ratios in Table 13.18. We do not reject them all.

Table 13.17 Dispersion matrix of observables

$$D_s = AD_xA^T$$

3.21E–09	2.38E–10	–2.15E–09	–3.16E–09	–1.63E–09	–4.54E–10	2.80E–09	6.12E–08	2.01E–07	–9.38E–09
2.38E–10	5.47E–10	1.68E–10	–3.79E–10	–3.85E–10	–6.93E–10	2.77E–08	–2.16E–08	1.66E–08	2.68E–08
–2.15E–09	1.68E–10	1.65E–09	2.03E–09	9.33E–10	–1.04E–10	1.52E–08	–5.73E–08	–1.34E–07	2.34E–08
–3.16E–09	–3.7E–10	2.03E–09	3.15E–09	1.68E–09	6.27E–10	–1.02E–08	–5.31E–08	–1.98E–07	1.74E–09
–1.63E–09	–3.85E–10	9.33E–10	1.68E–09	9.63E–10	5.60E–10	–1.51E–08	–1.81E–08	–1.03E–07	–8.94E–09
–4.54E–10	–6.93E–10	–1.04E–10	6.27E–10	5.60E–10	8.85E–10	–3.47E–08	2.38E–08	–3.05E–08	–3.29E–08
2.80E–09	2.77E–08	1.52E–08	–1.02E–08	–1.51E–08	–3.47E–08	1.43E–06	–1.30E–06	2.62E–07	1.42E–06
6.12E–08	–2.16E–08	–5.73E–08	–5.31E–08	–1.81E–08	2.38E–08	–1.30E–06	2.45E–06	3.75E–06	–1.53E–06
2.01E–07	1.66E–08	–1.34E–07	–1.98E–07	–1.03E–07	–3.05E–08	2.62E–07	3.75E–06	1.26E–05	–5.01E–07
–9.38E–09	2.68E–08	2.34E–08	1.74E–09	–8.94E–09	–3.29E–08	1.42E–06	–1.53E–06	–5.01E–07	1.45E–06

Table 13.18

1	9.40E–09	9.40E–09	9.40E–09	9.40E–09	9.40E–09	9.40E–09	1.00E–04	1.00E–04	2.00E–04	2.00E–06
2	3.21E–09	5.48E–10	1.65E–09	3.15E–09	9.64E–10	8.86E–10	1.44E–06	2.46E–06	–5.01E–07	1.46E–06
3	6.19E–09	8.85E–09	7.75E–09	6.25E–09	8.44E–09	8.52E–09	9.86E–05	9.75E–05	2.01E–04	5.42E–07
4	7.87E–05	9.41E–05	8.80E–05	7.91E–05	9.19E–05	9.23E–05	9.93E–03	9.88E–03	1.42E–02	7.36E–04
5	9.48E–05	2.25E–04	2.59E–04	5.87E–05	2.03E–04	1.09E–03	3.36E–03	1.49E–02	1.05E–02	5.27E–03
6	1.2	2.4	2.9	0.7	2.2	11.8	0.3	1.5	0.8	7.2

We reject the observation with the largest rejection ratio, namely the direction from the sixth point F, and re-run the whole analysis. This is not a colossal piece of extra work for, with a spreadsheet calculation, all we have to do is to set the coefficients of equation 6 to zero while retaining the original weight matrix.

On recomputation the value we get for the variance of unit weight is 2.67 which is much better. Table 13.19 shows the calculations for the test ratios t, for the new situation. Now we see two directions failing to meet the test.

We now have two options: either to go on with the process and end up by removing the three directions from F, C and E, or to enquire more into the circumstances of the observations to see if there may be some reason to explain these large residuals.

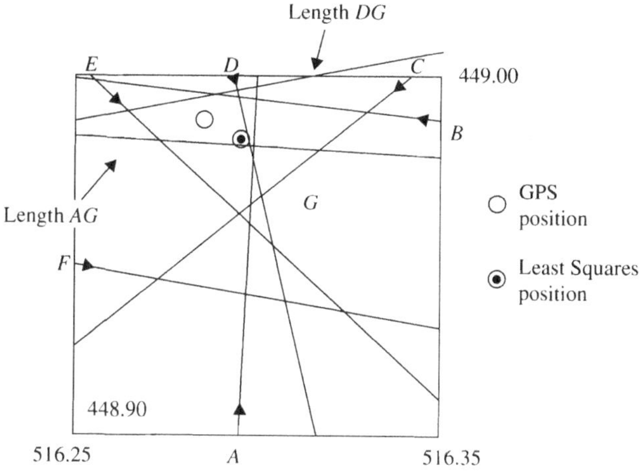

Figure 13.6

On doing so and referring to Figure 13.6, we found out that, at the time of the GPS fix, the Electromagnetic Distance Measurement (EDM) distances to A and D were measured, so there is no likelihood of miscentering at G. However, for use as a reference object later, a flag pole was erected at G and kept in place by guy wires. The inward angle observations were taken at different times, those at A and D in the same field visit as the EDM and GPS. The observations from B, C, E and F were taken when these stations were visited later when running the loop surround traverse. The centering of the pole was not checked. We should have looked more closely at the field books in the first place. But we could still go into the field and inspect the flag pole. In the case in point it was found that the pole could wobble by about 2 to 3 cm about the vertical position, so we should assign standard errors to the four poor observations in inverse proportion to the squares of the lengths of lines say by 0.025/distance^2. When we do that we obtain a different result.

Table 13.19 Network with direction from F removed

1	9.40E–09	9.40E–09	9.40E–09	9.40E–09	9.40E–09	9.40E–09	1.00E–04	1.00E–04	2.00E–04	2.00E–06
2	3.23E–09	6.04E–10	1.65E–09	3.20E–09	1.00E–09	0.00E+00	1.58E–06	2.53E–06	–3.83E–07	1.59E–06
3	6.17E–09	8.80E–09	7.75E–09	6.20E–09	8.40E–09	9.40E–09	9.84E–05	9.75E–05	2.00E–04	4.14E–07
4	7.85E–05	9.38E–05	8.80E–05	7.88E–05	9.17E–05	9.70E–05	9.92E–03	9.87E–03	1.42E–02	6.44E–04
5	3.64E–05	1.36E–04	2.73E–04	2.19E–05	2.75E–04	0.00E+00	7.82E–03	1.18E–02	1.45E–02	1.04E–03
6	0.46	1.45	3.10	0.28	3.00	0.00	0.79	1.20	1.02	1.61

Table 13.20

1	9.40E–09	2.31E–07	1.63E–07	9.40E–09	9.71E–08	3.60E–07	1.00E–04	1.00E–04	2.00E–04	2.00E–06
2	4.19E–09	6.67E–10	2.12E–09	4.12E–09	1.26E–09	1.08E–09	1.74E–06	3.04E–06	–5.89E–07	1.76E–06
3	5.22E–09	2.30E–07	1.61E–07	5.28E–09	9.59E–08	3.59E–07	9.83E–05	9.70E–05	2.01E–04	2.42E–07
4	7.22E–05	4.80E–04	4.01E–04	7.27E–05	3.10E–04	5.99E–04	9.91E–03	9.85E–03	1.42E–02	4.92E–04
5	6.28E–05	1.30E–04	2.95E–04	2.01E–06	2.66E–04	1.21E–03	8.20E–03	1.09E–02	1.28E–02	5.64E–04
6	0.87	0.27	0.74	0.03	0.86	2.03	0.83	1.11	0.91	1.15

The respective standard errors for directions *BG, CG, EG* and *FG* are

	BG	*CG*	*EG*	*FG*
Distances	52.0001312	61.9543743	80.211974	41.6843877
Variances	2.3114E–07	1.6283E–07	9.7141E–08	3.5969E–07

Re-running the computations gives a value for $s_o^2 = 1.07$ and the test results shown in row 6 of Table 13.20.

The value of s_o^2 seems fine now and for the ratio 1.07 is fine. The *t* tests are all acceptable. So, in practice, when the statistical ratio shows a problem in some of the residuals (and therefore in the original observations), we need to look at whether the problem is due to a blunder or an incorrect estimation of weights.

Common blunders are to mix up the order of the stations for an angle, or to mis-identify a station name. Correcting this often makes the network solution converge as if by magic. Over- or under-estimating the observation weights (in this case due to faulty equipment) can similarly cause major problems. The statistical test that we've outlined will point us towards the culprit, and we can make a better estimate of relevant weights.

13.29 Dispersion matrices of derived quantities

So far we have assumed or obtained measures of *precision*: variances found as the diagonal terms of the dispersion matrices, and covariances as their off-diagonal terms. It is also possible to extend the analysis to obtain measures of precision of any parameter derived from the final coordinates of a network. This parameter could well be an important dimension such as the location of bridge piers, or a bearing on which to base a tunnel drive.

Suppose we want to calculate the standard error in the length of the side *BG* which has not been directly measured for some reason. Its length is calculated from the coordinates in the usual way. We can also obtain its standard error by setting up the coefficients of an observation equation for this length *BG*. (See Section 13.15.) The elements of the $\boldsymbol{A}$ matrix are

$$\frac{(E_G - E_B)}{S} \quad \text{and} \quad \frac{(N_G - N_B)}{S}$$

where in this case S = length *BG*. We then apply the formulae

$$\boldsymbol{D_S} = \boldsymbol{A D_x A}^{\mathrm{T}} \tag{13.58}$$

Consider the following example.

Table 13.21

1	A	B	C	D	E
2	Point	*E*	*N*		
3	*B*	567.895	443.275	1.6454E–05	–5.887E–07
4	*G*	516.2	448.9	–5.887E–07	1.7579E–06
5	Delta	–51.695	5.625		
6	*S* = *BG*	52.000		Var	sd
7	Coef	–0.9941321	0.1081728	1.6408E–05	0.00405

Refer to Table 13.21. The inverse of the normal equations (the same normals of the previous example) is located in cells D3 to E4. The inverse is the dispersion matrix of the calculated coordinates derived by the Least Squares computation using the observed data.
The variance of the length of *BG* is then obtained from (13.58) or in full

$$\sigma_x^2 = \begin{bmatrix} -0.9941321 & 0.1081728 \end{bmatrix} \begin{bmatrix} 1.6454E{-}05 & 5.887E{-}07 \\ -5.887E{-}07 & 1.7579E{-}06 \end{bmatrix} \begin{bmatrix} -0.9941328 \\ 0.1081728 \end{bmatrix}$$

$$= 1.6408E{-}5$$

Giving the standard error of the length *BG*, $\sigma_x = 0.00405$.

In the same way we could estimate the standard error of a bearing such as *AG* using the formula for the coefficients in the **A** matrix

$$\frac{(N_G - N_A)}{AG^2} \quad \text{and} \quad \frac{(E_G - E_A)}{AG^2}$$

The rule is to derive the coefficients of the equation as if for an observed direction. In fact you can treat any function of the coordinates in this way, such as an area, or in the case of three dimensions, a calculated volume.

Many of the procedures we have been describing can be carried out without any observations at all. A pre-analysis can be made when designing a measurement programme, before ever going into the field. Such an analysis can usually save money and wasted effort.

13.30 Reliability testing

Reliability concerns the ability to check work. For example the *reliability* of a position fix is a measure of the ease with which gross errors may be detected. The greater the ease, the more reliable. Reliability also depends on the amount of redundancy in the problem. Imagine a point fixed by only two measured distances from two known points. There are no redundant measurements, so we have no residuals. Thus the test

$$t = \frac{v}{\sigma_v}$$

is indeterminate and no check for gross errors is possible. We consider such a fix as *unreliable.* It could in fact be very good, but there is also the thought that it could be very wrong! We need as much redundancy as is practicable. In our example we have eight degrees of freedom so the fix should be reliable. With this example we can make some other useful reliability statements to assess quality in comparative terms.

We need to recap a little. In Section 13.28 we described how we can detect outliers, based on a boundary statistic of 2.57σ, defined by the confidence limit of 99% (α = 1% significance level) obtained from a normal distribution table. We now extend this idea to consider the marginal error with respect to two hypotheses: (1) that there is no gross error in a measurement (the null hypothesis), and (2) that there is a gross error (the alternative hypothesis). This assumes that there is only one outlier in the data. Refer to Figure 13.7.

Here we illustrate two hypotheses about the normal distributions of residuals of a particular observation, call it the *i*th observation. The preferred hypotheses (the null hypothesis) shows the dispersion of residuals in this *i*th observation about a zero mean, and the other hypothesis (the alternative hypothesis) about a mean δ_i^u. This second distribution is believed to arise from the existence of a maximum gross error in the *i*th observation of Δ_i^u, where

$$\delta_i^u = \frac{\Delta_i^u}{\sigma_i} \tag{13.59}$$

and σ_i is the standard error of the *i*th residual. The 'u' indicates the upper limit. Considering the null hypothesis, if we reject residuals outside the bound (defined by α), we are rejecting some good data and we are said to have made a Type I error. According to statistical theory, very small or very large residuals are possible, but very unlikely. But this Type I error is a small price to pay for the detection of bad data.

However, sometimes outliers may be so small that the test is passed and data containing outliers is accepted. When this occurs we say a Type II error has occurred. Considering the alternative hypothesis, the probability of such an event is called β, or alternatively the probability of detecting the outlier is $(1 - \beta)$. This probability is called the *power of the test.* To quantify the extent to which outliers can be detected, the power of the test is selected (usually 80%) and we calculate the magnitude of the outlier that may be detected with this probability. This magnitude is called the marginal detectable error, MDE. The 80% probability is selected so that we can say the magnitude of the outlier can be found with reasonable certainty.

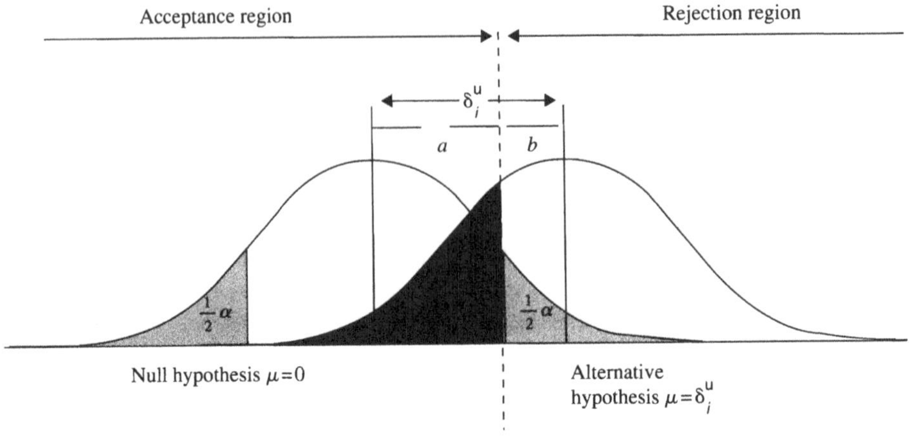

Figure 13.7

The choice of α determines the rejection limit and so affects our actions. The choice of β has no real effect on the rejection process; it merely affects what we say about the data quality. To make comparisons between different sets of data, it is usual to accept these probability levels for both distributions.

For example in Figure 13.7, we select both α and β, find a and b from normal tables and calculate $(a + b)$ and the MDE from

$$\text{MDE} = \Delta_i^u = \delta_i^u \sigma_i = (a+b)\sigma_i \qquad (13.60)$$

where σ_i is the standard error of the residual v_i.

Internal reliability

The *internal reliability* is expressed in terms of MDEs. It is common to test the assumption (hypothesis) that only one observation has a gross error and we calculate the MDE for each observation.

Table 13.22

σ	v	t
7.22E–05	6.28E–05	0.87
4.80E–04	1.30E–04	0.27
4.01E–04	2.95E–04	0.74
7.27E–05	2.01E–06	0.03
3.10E–04	2.66E–04	0.86
5.99E–04	1.21E–03	2.03
9.91E–03	8.20E–03	0.83
9.85E–03	1.09E–02	1.11
1.42E–02	1.28E–02	0.91
4.92E–04	5.64E–04	1.15

Table 13.22 is a summary of previous data. We have already shown that, since no value of t exceeds the test statistic 2.57, we reject no further data. From normal distribution tables $\alpha = 0.01$ (two tailed), the entry of 99.5% yields $a = 2.57$. Again from the tables we get $b = 0.84$, entry with 80% (one tailed). Hence we obtain the MDEs for each observation. The first MDE is given by

$$(2.57+0.84)\ 7.22\text{E}{-}05 = 2.47\text{E}{-}04$$

All the MDEs are shown in Table 13.23 together with some other data used later.

Table 13.23

	MDE	L	DE	DN
AG	2.47E–04	2.47E–04	6.89E–03	-2.90E–04
BG	1.64E–03	0.00E+00	1.63E–04	2.30E–04
CG	1.37E–03	0.00E+00	–1.46E–03	2.36E–04
DG	2.48E–04	0.00E+00	-6.85E–03	4.89E–05
EG	1.06E–03	0.00E+00	-1.47E–03	-1.19E–04
FG	2.05E–03	0.00E+00	-2.36E–04	-2.26E–04
Dist *AG*	3.39E–02	0.00E+00	1.38E–04	5.82E–04
Dist *DG*	3.36E–02	0.00E+00	1.63E–03	-6.22E–04
Easting	4.84E–02	0.00E+00	3.90E–03	1.25E–03
Northing	1.68E–03	0.00E+00	-8.78E–05	1.61E–03

The quality is an assessment of the size and nature of undetected errors that remain in the solution. In this case the largest MDE is 4.84E–02 or approximately 5 mm in the GPS Easting. Such a low figure indicates a high internal reliability. We can say of the quality that when outlier detection is carried out with a level of significance of 1% there is an 80% chance an outlier of 5 mm will be detected.

External reliability

External reliability is actually a more useful concept because a large undetected outlier may have little effect on the solution and conversely. External reliability is assessed by the largest effect of an observational MDE on the solution, in this case on the coordinates of G. We compute the separate effect of each MDE on the solution, and quote the largest to describe the quality of the fix. We make a series of computations to find the contribution of each MDE in turn using the formula

$$\mathbf{d}\boldsymbol{x} = \boldsymbol{N}^{-1}\boldsymbol{A}^{\mathrm{T}}\mathbf{W}\mathbf{d}\boldsymbol{L} \qquad (13.61)$$

where the column $\mathbf{d}\boldsymbol{L}$ has all zeros except for the observation under scrutiny, which its MDE in place. This is shown in column 2 of Table 13.23. The inverse and calculation of $\mathbf{d}\boldsymbol{x}$ is shown in Table 13.24.

Table 13.24

Inverse		**dx**
1.6454E–05	–5.887E–07	6.89E–03
–5.887E–07	1.7579E–06	–2.90E–04

Thus the effects of the MDE in the first observation on the coordinates of G are about 7 mm in Easting and 0.3 mm in Northing. Columns 3 and 4 of Table 13.23 show the MDEs in position caused by the MDE of each observation in turn. So we see that the External reliability is also good, because the largest MDEs in position are 7 mm in Easting and 2 mm in Northing.

13.31 Test on the variance, Fisher F test

Although this test has wide application, it is used mainly in Least Squares problems to see if the value calculated for the unit variance is 'reasonable', which in this context means close to the expected value of 'one'. The test generally is to see if we can accept that two sample variances are typical of one population, or alternatively that the evidence suggests they belong to populations with different variances.

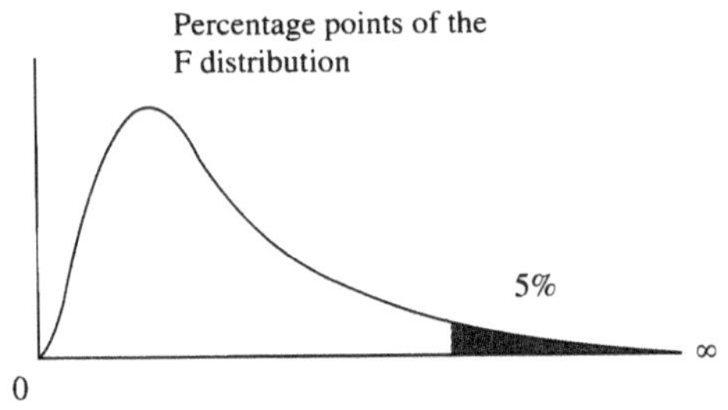

Figure 13.8

The test statistic t *is the ratio of the two variances always taking the ratio greater than unity.* Figure 13.8 shows the approximate shape of the F statistic for degrees of freedom greater than three. The actual shape varies for each degree of freedom. It is also tabulated in Table 13.25.

Example Before considering the unit variance in a Least Squares problem, let's consider a simpler case. Suppose we calculated a sample variance s_1^2 from nine observations to be 3.4, and suppose we have similar sample with a variance s_2^2 of 5.6 from seven observations. To test whether these samples come from populations with equal variances, we test the ratio of the larger variance to the smaller, i.e. the test statistic is:

$$t = \frac{s_1^2}{s_2^2} = \frac{5.6}{3.4} = 1.64$$

The respective degrees of freedom are $r_1 = 6$ and $r_2 = 8$, giving a rejection limit of 3.58 from the Fisher Table 13.25, which has been tabulated for the 5% confidence level. Since 1.64 is well within the rejection limit of 3.58, we can accept that these samples come from populations with equal variances.

Table 13.25 5% points of Fisher table

r_1		6	7	8
r_2	7	3.87	3.79	3.73
	8	3.58	3.50	3.44
	9	3.37	3.29	3.23
	10	3.22	3.14	3.07
	.	.	.	.
	.	.	.	.
	.	.	.	.
	∞	2.10	2.01	1.94

In a Least Squares problem, the sample statistic is the unit variance calculated from the weighted residuals, i.e. from

$$\frac{\boldsymbol{v}^{\mathrm{T}}\boldsymbol{W}\boldsymbol{v}}{m-n}$$

This is compared with the theoretical value for the population variance. Thus the test statistic is given by

$$t = \frac{s_o^2}{\sigma_o^2}$$

Since $\sigma_o^2 = 1$ in theory the test is

$$t = \frac{s_o^2}{1}$$

In our example we want to see if the value for the unit variance of 1.07 is acceptable. We compare this value from $r_1 = 8$ degrees of freedom with the theoretical value of 1.0 from an infinite number of values, $r_2 = \infty$. From Table 13.25 we have the bound of 1.94. Thus the tested value is acceptable.

13.32 General Least Squares, the *C* matrix

In Appendix A2, Section A2.3 we derive the general equations for the Least Squares process. Here we repeat some of the concepts with reference to our explanatory problem of the surveying network. In the derivation of the observation equations we manipulated them to obtain the form which eliminated the coefficients of the unobserved parameters to give equations of the form

$$\boldsymbol{Ax} + \boldsymbol{L} = \boldsymbol{v} \tag{13.26}$$

Partial differentiation of a non-linear mathematical model such as the length equation

$$(E_B - E_A)^2 + (N_B - N_A)^2 = S^2$$

gives

$$\frac{\partial F}{\partial E_A}\delta E_A + \frac{\partial F}{\partial N_A}\delta N_A + \frac{\partial F}{\partial E_G}\delta E_G + \frac{\partial F}{\partial N_G}\delta N_G + \frac{\partial F}{\partial S}\delta S = 0 \tag{13.62}$$

and we said further that these equations are usually further simplified by dividing throughout by the coefficient of the observed parameter, S in this case, to give

$$\frac{\partial F}{\partial E_A}\frac{\partial S}{\partial F}\delta E_A + \frac{\partial F}{\partial N_A}\frac{\partial S}{\partial F}\delta N_A + \frac{\partial F}{\partial E_G}\frac{\partial S}{\partial F}\delta E_G + \frac{\partial F}{\partial N_G}\frac{\partial S}{\partial F}\delta N_G + \delta S = 0 \tag{13.63}$$

This simplification is only possible when there is only one measurement in each equation, as is the case of a distance or an angle.

To handle the more general case, the method is as follows. Equation (13.63) was derived from equation (13.62) by dividing the equation by the coefficient of the measured parameter S. To explain the more general approach we have to write the equation (13.62) in a more general form as

$$\left(\frac{\partial F}{\partial E_A}\delta E_A + \frac{\partial F}{\partial N_A}\delta N_A + \frac{\partial F}{\partial E_G}\delta E_G + \frac{\partial F}{\partial N_G}\delta N_G\right) + \left(\frac{\partial F}{\partial S}\right)\delta S = 0 \tag{13.64}$$

i.e. we separate the coefficients of the observed parameters δS from those of the required unobserved parameters $(\delta E_A \;\; \delta N_A \;\; \delta E_G \;\; \delta N_G) = \boldsymbol{x}^{\mathrm{T}}$ and rewrite equation (13.64) in the form

$$\boldsymbol{Bx} + \boldsymbol{C}\delta\boldsymbol{S} = 0 \tag{13.65}$$

To reduce this to the required form we premultiply by the inverse of the matrix $\boldsymbol{C}$, so we have

$$\boldsymbol{C}^{-1}\boldsymbol{Bx} + \boldsymbol{C}^{-1}\boldsymbol{C}\delta\boldsymbol{S} = 0$$

$$\boldsymbol{C}^{-1}\boldsymbol{Bx} + \delta\boldsymbol{S} = 0 \tag{13.66}$$

It will be seen that equation (13.66) is the same as equation (13.63).

To invert $\boldsymbol{C}$ is not always simple. A case of this arises in coordinate transformation and line-fitting problems, and stage (13.65) has to be included as a matrix operation to produce equations of the standard form

$$\boldsymbol{A}^{\mathrm{T}}\boldsymbol{WAx} = -\boldsymbol{A}^{\mathrm{T}}\boldsymbol{WL}$$

The following example will demonstrate the method.

13.33 Mathematical models with more than one observed parameter

In the two simple problems considered above, there was only one observed parameter in each equation, thus giving equations of the form

$$\boldsymbol{Ax} + \boldsymbol{L} = \boldsymbol{v}$$

To deal with problems with more than one observed parameter per equation we introduce equations of the form

$$\boldsymbol{Ax} + \boldsymbol{Cv} + \boldsymbol{CL} = \boldsymbol{0} \qquad \text{(A2.11)}$$

These equations ultimately lead to the solution of a set of symmetric equations of the same form as before, i.e.

$$\boldsymbol{N}_1\boldsymbol{x} = \boldsymbol{b}_1 \qquad \text{(A2.18)}$$

where

$$\boldsymbol{N}_1 = \boldsymbol{A}^{\mathrm{T}}\boldsymbol{N}^{-1}\boldsymbol{A}$$

$$\boldsymbol{b}_1 = \boldsymbol{A}^{\mathrm{T}}\boldsymbol{N}^{-1}\boldsymbol{b}$$

$$\boldsymbol{N} = \boldsymbol{CW}^{-1}\boldsymbol{C}^{\mathrm{T}}$$

$$\boldsymbol{b} = -\boldsymbol{CL}$$

$\boldsymbol{W}$ is the weight matrix of the observations.

We will derive these equations in the course of fitting a straight line to observed data points.

Example of fitting a straight line by Least Squares

In this section we use the example of fitting a straight line to data. Not only does this particular problem occur in several fields, such as surveying, cartography, engineering surveying and photogrammetry, but the method of treatment is the same for all manner of curve-fitting problems. Thus the following treatment of the straight line problem must be seen as typical of many.
Only two points are needed to define a straight line. For simplicity we confine our attention to the two-dimensional problem of fitting a straight line in a plane. Three cases of fitting the corresponding values of x and y, listed in the first two columns of Table 13.26, will be treated.

Table 13.26

x	y	Comp y	L	v	$\hat{y}$
1.000	2.812	2.5	0.312	–0.023	2.789
2.000	3.066	3.0	0.066	–0.035	3.030
3.000	3.218	3.5	–0.282	0.053	3.271
4.000	3.482	4.0	–0.518	0.030	3.513
5.000	3.713	4.5	–0.787	0.041	3.754
6.000	4.033	5.0	–0.967	–0.037	3.996
7.000	4.278	5.5	–1.222	–0.040	4.237
8.000	4.445	6.0	–1.555	0.034	4.478
9.000	4.783	6.5	–1.717	–0.063	4.720
10.000	4.920	7.0	–2.080	0.041	4.961

In two dimensions the equation of a straight line is

$$y = mx + c \qquad (13.66)$$

where m is the gradient and the intercept on the y axis is c.
(**Note**: This form of the equation of a line can cause problems if the line runs parallel to or nearly parallel to one of the axes. In such a case, the axes are rotated through some suitable angle such as 45°, before carrying out the analysis, and then rotated back to the original orientation. See Section 7.16.)
The functional equation is

$$F(m, c\colon x, y) = 0$$

Generally the problem is to find values for the coefficients m and c, from given values of x and y. There are two main cases of the line-fitting problem:

(a) when only one of the parameters x or y is observed,
(b) when both parameters x and y are observed.

In the first case, the Least Squares model follows the simple method already considered above. We shall deal with this briefly to set the scene for the more complex treatment which follows.

If a large-scale graph is drawn of the values of y and x it will be seen that a straight line cannot be found to fit the data exactly. This graph is illustrated in Figure 13.9 showing the eye-balled line 1–10, from which the parameters can be measured as a check to give

$$c = 2.5 \quad \text{and} \quad m = \text{arc tan}\,(13°)$$

(The figures of the example have been selected from an initial line whose equation is $y - 0.5x - 2.0 = 0$ with randomly generated errors added.) The Least Squares process is an analytical method of drawing this graph.

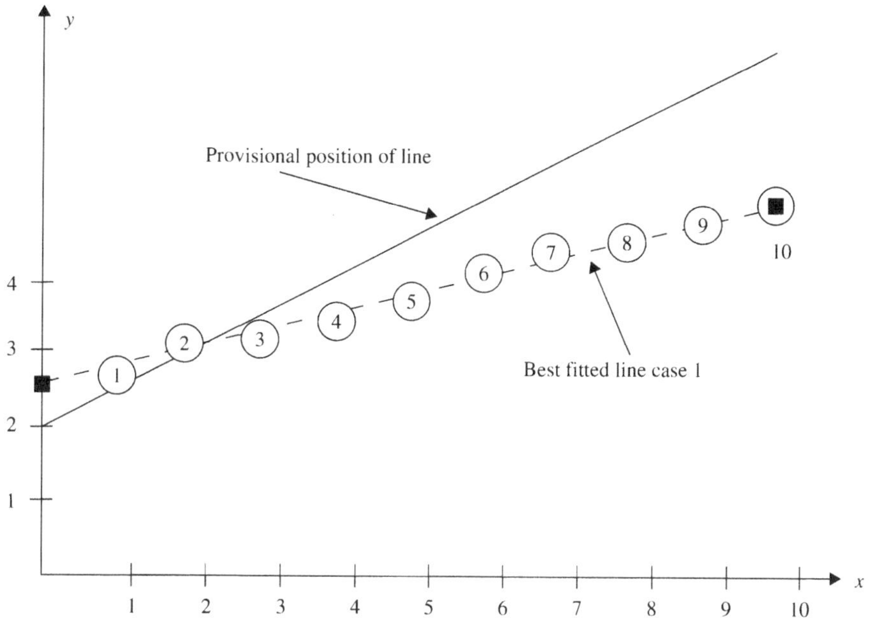

Figure 13.9

It produces unique results and will not depend on the skill of the person drawing it.

Exercise 1 The beginner is encouraged to draw this graph before continuing with the analytical method.

The first case is similar to a regression analysis of two variables, in which the small residuals are considered to be only in the direction of the y axis. The second requires that the residuals lie perpendicular to the line and thus needs to be treated by the general Least Squares method. In a later section we will deal with another case in which the solution is constrained in some way, for example to force the line to pass through a given fixed point. These examples illustrate the basic mechanisms for tackling a wide range of such problems.

Case 1: Simple case with one observable per equation

For each observed value, $\overset{o}{y}$, there is an observation equation. We select provisional values of the parameters to be

$$\overset{*}{m} = 0.5, \quad \overset{*}{c} = 2.0$$

Ten values of $\overset{*}{y}$, in column 3 of Table 13.26, are calculated using the equation

$$\overset{*}{y} = 0.5\overset{*}{x} + 2.0$$

of the respective values of x = 1 to 10.

The objective is to find the best estimates of the unobserved parameters $\hat{m}$, $\hat{c}$, and of the observed parameters $\hat{y}$. As usual the various quantities are related by the equations:

$$\hat{m} = \overset{*}{m} + \delta m$$

$$\hat{c} = \overset{*}{c} + \delta c$$

$$\hat{y} = \overset{*}{y} + \delta y$$

Since there is only one observed parameter, $\overset{\circ}{y}$, in each equation the residuals v are written

$$v = \hat{y} - \overset{\circ}{y} = \overset{*}{y} - \overset{\circ}{y} + \delta y$$

therefore

$$\delta y = \overset{\circ}{y} - \overset{*}{y} + v$$

$$= L + v$$

Because the best estimates and the provisional values satisfy equation we may write:

$$F(\hat{m}, \hat{c} : x, \hat{y}) = 0$$

$$F(\overset{*}{m}, \overset{*}{c} : x, \overset{*}{y}) = 0$$

Hence

$$\frac{\partial F}{\partial \overset{*}{m}} \delta m + \frac{\partial F}{\partial \overset{*}{c}} \delta c + \frac{\partial F}{\partial \overset{*}{y}} \delta y = 0 \qquad (13.67)$$

which gives

$$(x)\ \delta m + (1)\ \delta c = L + v \qquad (13.68)$$

It will be remembered that, in this simple case, x is considered error free because it is not an observed parameter. Again we have equations of the standard form

$$\mathbf{Ax} + \mathbf{L} = \mathbf{v}$$

$$\begin{bmatrix} 1 & 1 \\ 2 & 1 \\ 3 & 1 \\ 4 & 1 \\ 5 & 1 \\ 6 & 1 \\ 7 & 1 \\ 8 & 1 \\ 9 & 1 \\ 10 & 1 \end{bmatrix} \begin{bmatrix} \delta m \\ \delta c \end{bmatrix} + \begin{bmatrix} -0.312 \\ -0.066 \\ 0.282 \\ 0.518 \\ 0.787 \\ 0.967 \\ 1.222 \\ 1.555 \\ 1.717 \\ 2.080 \end{bmatrix} = \boldsymbol{v}$$

The normal equations, on the assumption that observations are all of equal weight, are:

$$\boldsymbol{Nx} = \boldsymbol{b}$$

$$\begin{bmatrix} 385 & 55 \\ 55 & 10 \end{bmatrix} \begin{bmatrix} \delta m \\ \delta c \end{bmatrix} = \begin{bmatrix} -69.458 \\ -8.75 \end{bmatrix}$$

Giving the solution $\delta m = -0.25858$

$$\delta c = 0.5472$$

Back substitution in the observation equations gives the residuals of the fifth column in Table 13.26. Thus we can calculate the variance factor s_o^2 from

$$s_o^2 = \frac{\boldsymbol{v}^{\mathrm{T}}\boldsymbol{v}}{m-n}$$

$$= 0.017228/8$$

$$= 0.0021535$$

and $s_o = 0.044$

Finally the best estimates of the derived parameters are obtained from

$$\hat{m} = \overset{*}{m} + \delta m = 0.2414 \quad \hat{a} = 13^{\circ}.5716$$

$$\hat{c} = \overset{*}{c} + \delta c = 2.5472$$

and the best estimates of the observed parameters from

$$\hat{y} = \overset{o}{y} + v_y$$

These are listed in the last column of Table 13.26.

Case 2: Both *x* and *y* are observed parameters

We now extend the treatment to the more general case in which both x and y are observed. The principles involved are also common to problems in astronomy and coordinate transformation. The treatment is the same as before with the addition of the extra fact that we have to estimate

$$\hat{x} = \overset{o}{x} + v_x$$

The functional models to be satisfied are

$$F(\hat{m}, \hat{c} : \hat{x}, \hat{y} = 0 \quad \text{and} \quad F(\overset{*}{m}, \overset{*}{c} : \overset{*}{x}, \overset{*}{y}) = 0$$

$$\frac{\partial F}{\partial \overset{*}{m}} \delta m + \frac{\partial F}{\partial \overset{*}{c}} \delta c + \frac{\partial F}{\partial \overset{*}{x}} \delta x + \frac{\partial F}{\partial \overset{*}{y}} \delta y = 0$$

$$(\overset{*}{x})\,\delta m + (1)\,\delta c + (\overset{*}{m})\,\delta x - \delta y = 0$$

The notation indicates that the coefficients of the variables are evaluated using the provisional values adopted for the problem. More casually the equation may be written without the asterisks as:

$$(x)\delta m + (1)\delta c + (m)\delta x - \delta y = 0 \qquad (13.69)$$

Separating the unobserved from the observed parameters we have:

$$\begin{bmatrix} x & 1 \end{bmatrix} \begin{bmatrix} \delta m \\ \delta c \end{bmatrix} + \begin{bmatrix} m & -1 \end{bmatrix} \begin{bmatrix} \delta x \\ \delta y \end{bmatrix} = 0$$

These equations can be written

$$\boldsymbol{Ax} + \boldsymbol{Cs} = \boldsymbol{0}$$

Notice the change in the notation from the ordinary algebra of the problem, in this case m, c, x and y, to that of the matrix notation in which we always write the derived vector as $\boldsymbol{x}$ and the observed vector as $\boldsymbol{s}$. The dimensions of the matrices and vectors are

$$\boldsymbol{A} \text{ matrix} = 10 \times 2, \; \boldsymbol{C} \text{ matrix} = 10 \times 20$$

$$\boldsymbol{x} \text{ vector} = 2 \times 1, \; \boldsymbol{s} \text{ vector} = 20 \times 1$$

Forming ten equations for every one of the ten observed points gives

$$\begin{bmatrix} 1 & 1 \\ 2 & 1 \\ 3 & 1 \\ \vdots & \vdots \\ 10 & 1 \end{bmatrix} \begin{bmatrix} \delta m \\ \delta c \end{bmatrix} + \begin{bmatrix} 0.5 & -1 & 0 & 0 & \cdots & & 0 \\ 0 & 0 & 0.5 & -1 & & & 0 \\ 0 & 0 & 0 & 0 & & & 0 \\ \vdots & \vdots & \vdots & \vdots & & & \vdots \\ 0 & 0 & \cdots & \cdots & & 0.5 & -1 \end{bmatrix} \begin{bmatrix} \delta x_1 \\ \delta y_1 \\ \delta x_2 \\ \delta y_2 \\ \vdots \\ \delta x_{10} \\ \delta y_{10} \end{bmatrix} = 0$$

$$(r \times n) \quad (n \times 1) \qquad\qquad (r \times m) \qquad\qquad (m \times 1)$$

In general there are r equations in m observed parameters to estimate n derived parameters. As usual the small changes to the observed parameters are split into two parts: the known part $\boldsymbol{L}$, and the unknown part $\boldsymbol{v}$.
Thus for example

$$\delta x_1 + Lx_1 = vx_1$$

The final equations are of the form

$$\boldsymbol{Ax} + \boldsymbol{Cv} = -\boldsymbol{CL}$$
$$\boldsymbol{Ax} + \boldsymbol{Cv} = \boldsymbol{b}$$

We obtain the normal equations

$$\boldsymbol{N}_1\boldsymbol{x} = \boldsymbol{b}_1$$

where

$$\boldsymbol{N}_1 = \boldsymbol{A}^{\mathrm{T}}\boldsymbol{N}^{-1}\boldsymbol{A} \quad \text{and} \quad \boldsymbol{b}_1 = \boldsymbol{A}^{\mathrm{T}}\boldsymbol{N}^{-1}\boldsymbol{b}$$
$$\boldsymbol{N} = \boldsymbol{CW}^{-1}\boldsymbol{C}^{\mathrm{T}} \quad \text{and} \quad \boldsymbol{b} = -\boldsymbol{CL}$$

The weight matrix is $\boldsymbol{W}$. It is not practicable to give all the working of the above example here. Instead we give the key information only.

Approximate solution with weights Suppose we assign weights of 1 to x and 2 to y giving a (20×20) weight matrix with these values alternating down the diagonal. In this example, because the numbers have been chosen to simplify the arithmetic, the matrix $\boldsymbol{N}$ (10×10) is diagonal with one value (0.75) for all of its diagonal terms. Its inverse is therefore the simple:

$$\boldsymbol{N}^{-1} = (0.75)\boldsymbol{I}$$

where $\boldsymbol{I}$ is the (10×10) unit matrix.

All values of the $\boldsymbol{L}$ vector are zero for each value of x because the provisional values were chosen to be the same as the observed values. The $\boldsymbol{L}$ values for each value of y are the same as before.

The problem finally reduces to a solution of the normal equations

$$\boldsymbol{N}_1\boldsymbol{x} = \boldsymbol{b}_1 \qquad \text{(A2.18)}$$

which are

$$\begin{bmatrix} 513.33 & 73.33 \\ 73.33 & 13.33 \end{bmatrix} \begin{bmatrix} \delta m \\ \delta c \end{bmatrix} = \begin{bmatrix} -77.138 \\ -9.086 \end{bmatrix}$$

The solution is

$$\delta m = -0.246\ 86$$

$$\delta c = 0.676$$

Although this solution is not very different from the previous method, in this case the residuals are the *perpendicular distances* from the observed points to the line.

Case 3: Solution with additional constraint

Sometimes new measurements have to be constrained to fit previous work upon which maps will have been plotted, or building construction has begun.

Suppose that the line just fitted to data has to pass through a fixed point (x', y'). This means that there is one equation which must be satisfied exactly: i.e.

$$y' - mx' - c = 0$$

This is a constraint equation which the finally estimated parameters must also satisfy exactly, i.e. we have

$$y' - \hat{m}x' - \hat{c} = 0$$

One practical way to treat the problem is to assign a very high weight to the fixed point coordinates and treat them as observations in the usual way. If we hold the tenth point nearly fixed by assigning it a weight of 100, we obtain the solution:

$$\delta m = -0.2901362$$

$$\delta c = 0.8348$$

Although this procedure is often acceptable in practice, it is incorrect theoretically because an infinite weight cannot be handled computationally. However, the solution is seen to be acceptable when compared with the exact solution which follows.

Additional constraints Since a theoretically correct treatment is not difficult, it should certainly be employed in scientific studies. Two theoretically correct methods are available.

A conceptually simple method is to eliminate one parameter from the problem by expressing it in terms of the other. Although this would be the simplest way to treat the particular problem of our example, which has only two parameters, it is not convenient in complex cases. For this reason we present a general alternative method of treatment.

Constraints by Lagrangian multipliers As is explained in Appendix A2 the use of Lagrange's method yields equations of the form:

$$\boldsymbol{N}_1\boldsymbol{x} + \boldsymbol{E}^{\mathrm{T}}\boldsymbol{k} = \boldsymbol{b}_1 \tag{13.70}$$

$$\boldsymbol{E}\boldsymbol{x} + \boldsymbol{d} = \boldsymbol{0} \tag{13.71}$$

These can be combined into one hyper-matrix as follows

$$\begin{bmatrix} \boldsymbol{N}_1 & \boldsymbol{E}^{\mathrm{T}} \\ \boldsymbol{E} & \boldsymbol{0} \end{bmatrix} \begin{bmatrix} \boldsymbol{x} \\ \boldsymbol{k} \end{bmatrix} = \begin{bmatrix} \boldsymbol{b}_1 \\ -\boldsymbol{d} \end{bmatrix}$$

If we hold point 10 fixed, its observation equation becomes the constraint equation, k is the Lagrangian multiplier, and the hyper-matrix is:

$$\begin{bmatrix} 513.33 & 73.33 & 1 \\ 73.33 & 13.33 & 10 \\ 1 & 10 & 0 \end{bmatrix} \begin{bmatrix} \delta m \\ \delta c \\ k \end{bmatrix} = \begin{bmatrix} -77.138 \\ -9.086 \\ -2.080 \end{bmatrix}$$

the solution of which is

$$\delta m = -0.2922$$

$$\delta c = 0.8426$$

This compares very well with the previous weighted solution.

13.34 Further reading

The literature dealing with Least Squares estimation is vast. In Appendix A3 we give the titles of some texts concerned with its surveying applications together with the notation used. Articles in journals, such as *Survey Review* or the *Photogrammetric Record*, should also be consulted.

SUMMARY OF KEY WORDS

Least Squares estimation, classical matrices of full rank, observed parameter, unobserved parameters, observed, provisional, estimated and population versions of a parameter, Least Squares value (best estimate), errors in the best estimates, sample residuals, population residuals, Newton's method, degrees of freedom, observation equations, normal equations, functional models, correlated observations, dispersion matrix or variance–covariance matrix, the arithmetic mean, weights, variance, standard deviation, expectation, covariance, unbiased estimator, weight matrix, outliers, normal distribution, dispersion matrix of the observables, reliability, null hypothesis, alternative hypothesis, Type I error, Type II error, power of the test, marginal detectable error, rejection limit, internal reliability, external reliability, Fisher F test, fitting a straight line, Lagrangian multipliers.

SUMMARY OF FORMULAE

Functional models	$F(\boldsymbol{x}: \boldsymbol{s}) = 0$	(13.5)
Normal equations	$\boldsymbol{Nx} = \boldsymbol{b}$	(13.13)
	$\Omega = \boldsymbol{v}^{\mathrm{T}}\boldsymbol{v}$	(13.21)
	$\boldsymbol{A}^{\mathrm{T}}\boldsymbol{v} = \boldsymbol{0}$	
Observation equations	$\boldsymbol{Ax} + \boldsymbol{L} = \boldsymbol{v}$	(13.26)
Normal equations	$\boldsymbol{A}^{\mathrm{T}}\boldsymbol{Ax} + \boldsymbol{A}^{\mathrm{T}}\boldsymbol{L} = \boldsymbol{0}$	(13.27)
Weighted normal equations	$\boldsymbol{A}^{\mathrm{T}}\boldsymbol{WAx} + \boldsymbol{A}^{\mathrm{T}}\boldsymbol{WL} = \boldsymbol{0}$	(13.41)
Sample variance	$s^2 = \dfrac{\Sigma v^2}{m}$	(13.42)
Expectation	$\mathrm{E}(\boldsymbol{V}^{\mathrm{T}}\boldsymbol{V}) = \sigma^2$	
Covariance	$\mathrm{covar} = \mathrm{E}(V_A V_B) = \sigma_{AB}$	
	$s_{\bar{A}}^2 = \mathrm{E}\left(\dfrac{V_S^2}{m^2}\right) = \dfrac{m s_A^2}{m^2} = \dfrac{s_A^2}{m}$	(13.44)
Weights	$w_A = \dfrac{1}{s_A^2}$	(13.45)
Dispersion matrix	$\boldsymbol{D}_{\mathrm{o}}^{-1} = \boldsymbol{W}$ and $\boldsymbol{D}_{\mathrm{o}} = \boldsymbol{W}^{-1}$	(13.49)
	$\boldsymbol{Adx} = \boldsymbol{V} - \boldsymbol{v}$	(13.51)
	$D_x = s_{\mathrm{o}}^2 N^{-1}$	(13.53)
	$\boldsymbol{VV}^{\mathrm{T}} = \boldsymbol{Av}(\boldsymbol{Av})^{\mathrm{T}} = \boldsymbol{Avv}^{\mathrm{T}}\boldsymbol{A}^{\mathrm{T}}$	(13.54)
Variance	$\sigma_{\mathrm{o}}^2 = \mathrm{E}\left(\dfrac{\boldsymbol{v}^{\mathrm{T}}\boldsymbol{Wv}}{m-n}\right)$	(A2.26)
Special case	$s_{\mathrm{o}}^2 = \dfrac{\boldsymbol{v}^{\mathrm{T}}\boldsymbol{Wv}}{m-1}$	(13.55)

Dispersion matrix of residuals	$\boldsymbol{D}_v = \boldsymbol{D}_o - \boldsymbol{D}_s$	(13.56)
	where $\boldsymbol{D}_s = \boldsymbol{A}\boldsymbol{D}_x\boldsymbol{A}^{\mathrm{T}}$	(13.57)
	$\boldsymbol{D}_s = \boldsymbol{A}\boldsymbol{D}_x\boldsymbol{A}^{\mathrm{T}}$	(13.58)
Test ratio	$t = \dfrac{v}{\sigma_v}$	
	$\delta_i^{\mathrm{u}} = \dfrac{\Delta_i^{\mathrm{u}}}{\sigma_i}$	(13.59)
	$\mathrm{MDE} = \Delta_i^{\mathrm{u}} = \delta_i^{\mathrm{u}}\,\sigma_i = (a+b)\sigma_i$	(13.60)
	$\mathbf{dx} = \boldsymbol{N}^{-1}\boldsymbol{A}^{\mathrm{T}}\mathbf{W}\mathbf{d}\boldsymbol{L}$	(13.61)
F test	$t = \dfrac{s_1^2}{s_2^2}$	
General Least Squares	$\boldsymbol{Bx} + \boldsymbol{C}\delta\boldsymbol{S} = 0$	(13.65)
	$\boldsymbol{C}^{-1}\boldsymbol{Bx} + \delta\boldsymbol{S} = 0$	(13.66)
	$\boldsymbol{Ax} + \boldsymbol{Cv} + \boldsymbol{CL} = 0$	(A2.11)
	$\boldsymbol{N}_1\boldsymbol{x} = \boldsymbol{b}_1$	(A2.18)
	where $\boldsymbol{N}_1 = \boldsymbol{A}^{\mathrm{T}}\boldsymbol{N}^{-1}\boldsymbol{A}$	
	$\boldsymbol{b}_1 = \boldsymbol{A}^{\mathrm{T}}\boldsymbol{N}^{-1}\boldsymbol{b}$	
	$\boldsymbol{N} = \boldsymbol{C}\boldsymbol{W}^{-1}\boldsymbol{C}^{\mathrm{T}}$	
	$\boldsymbol{b} = -\boldsymbol{CL}$	
Line fitting	$y = mx + c$	(13.66)
	$(x)\delta m + (1)\delta c = L + v$	(13.68)
	$(x)\delta m + (1)\delta c + (m)\delta x - \delta y = 0$	(13.69)
	$\boldsymbol{N}_1\boldsymbol{x} + \boldsymbol{E}^{\mathrm{T}}\boldsymbol{k} = \boldsymbol{b}_1$	(13.70)
	$\boldsymbol{Ex} + \boldsymbol{d} = \boldsymbol{0}$	(13.71)

References

Allan, A.L. (1997) *Practical Surveying and Computations*. Butterworth Heinemann.
Ayres, F. (1962) *Theory and Problems of Matrices*. Schaum Outline Series.
Cohn, P.M. (1961) *Solid Geometry*. Routledge and Kegan Paul.
Coulson, A.E. (1965) *An Introduction to Matrices*. Longman.
Formulae for Advanced Mathematics with Statistical Tables. (1989) 2nd edition. Cambridge University Press.
Geary, A., Lowry, H.V. and Hayden, H.A. (1961) *Advanced Mathematics for Technical Students*. Longman.
Heard, T.J. (1978) *Extending Mathematics* Oxford University Press.
Hirst, D.M. (1985) *Mathematics for Chemists*. Macmillan.
Kreysig,E. (1993) *Advanced Engineering Mathematics*. 7th edition. Wiley, New York.
Thompson, E.H. (1969) *An Introduction to the Algebra of Matrices*. Adam Hilger. Available from Department of Photogrammetry and Surveying, University College London.
Yule, G.U. and Kendall, M.G. (1950) *An Introduction to the Theory of Statistics*. xivth edition, Griffin, London.

(Note: These are the books used by the author, although the reader may find more recent editons of some titles.)

References for Least Squares Estimation

Allan, A.L. and Atkinson, N. (2003) 'Back to Basics' Series Nos 14 to 24 – Least Squares Statistics and all that, *Survey Review*, Vols 35 and 36, Nos 272 to 282. Appendix-3.qxd 18/08/03 8:21 PM Page 368.
American Society of Photogrammetry (1966) *Manual of Photogrammetry*, Library of Congress Catalog No 65–20813 Vol 1.
Cooper, M.A.R. (1987) *Control Surveys in Civil Engineering*, Collins (ISBN 0-00-383183-3, 381 pages).
Cooper, M.A.R. and Cross, P.A. (1988) Statistical Concepts and their Application in Photogrammetry and Surveying, *Photogrammetric Record*, Vol XIII, No 73, 645–678.
Koch, K.R. (1997) *Parameter Estimation and Hypothesis Testing in Linear Models*, Springer (ISBN 3–540–65257–4, 325 pages).
Mikhail, E.M. (1976) *Observations and Least Squares*, Dun-Donnelley, (ISBN 0-7002-2481-5, 497 pages).
Wolf, P.R. and Ghilani, C.D. (1997) *Adjustment Computations*, Wiley (ISBN 0-471-16833-5, 564 pages).

Appendix of Useful Data

Fractions and multiples of units

Note the symbol for units should *never* have the plural 's' attached. For example a distance is 30 km, **not** 30 kms.

power of 10	prefix	symbol
–1	deci	d
–2	centi	c
–3	milli	m
–6	micro	m
–9	nano	n
–12	pico	p
–15	femto	f
–18	atto	a
+ 1	deka	da
+ 2	hecto	h
+ 3	kilo	k
+ 6	mega	M
+ 9	giga	G
+12	tera	T

Units of length

1 metre (trad) = length of a quadrant of the Earth $\times 10^{-7} = 0.5\,\pi R\, 10^{-7}$
where R is the radius of the Earth

Radius of the Earth $R \approx 6.4 \times 10^{6}$ m

1 metre = distance travelled by light in vacuo during a period of 1/299 792 458 th of a second of time.

1 British foot = 0.3048 metre (m) exactly (one inch = 25.4 mm)

1 British statute mile = 5280 feet (ft) = 1760 yards (yd) = 1609.344 m $\approx \frac{8}{5}$ km

1 ft = 12 inches (in)
1 inch = 25.4 mm exactly
1 yard = 3 ft
1 chain = 22 yards = 66 feet

Units of Area

1 are (a) = 100 square metres (m^2)
1 hectare (ha) = 10^2 ares = 10^4 m^2 = the area of the average football pitch
1 acre = 10 square chains ≈ 0.405 ha = about half the area of a football pitch

Units of Angle

2 π radians =one cycle = 360 sexagesimal degrees (°) = 400 centesimal degrees (g)
60 sexagesimal minutes of arc (') = one sexagesimal degree
60 sexagesimal seconds of arc (") = one sexagesimal minute
100 centesimal minutes (c) = one centesimal degree (g)
100 centesimal seconds (cc) = one centesimal minute (c)
one sexagesimal minute ≈ 0.003 rad
one sexagesimal second ≈ 3 centesimal seconds (cc) ≈ 0.000 005 rad
1 radian (rad) = the angle subtended by the arc of a circle equal to its radius
1 radian = $\frac{180}{\pi}$ sexagesimal degrees (°) ≈ 57.3° ≈ 3438' ≈ 206 265"
= $\frac{200}{\pi}$ centesimal degrees (gon) (g) ≈ 63.66g.

Appendix A2: Further Matrix Algebra

In this Appendix we develop some further matrix concepts and methods required for an understanding of the Least Squares estimation process described in Chapter 13. The discussion here is restricted to classical matrices of full rank. Although generalised inverses are now applied to mapping problems, they are not considered.

A2.1 Trace of a matrix

The *trace* of a square matrix is the sum of its diagonal terms. The property we will use is

$$\boldsymbol{V}^{\mathrm{T}}\boldsymbol{W}\boldsymbol{V} = \mathrm{trace}(\boldsymbol{V}\boldsymbol{V}^{\mathrm{T}}\boldsymbol{W}) \tag{A2.1}$$

Consider first the simple case

$$\boldsymbol{V}\boldsymbol{V}^{\mathrm{T}} = \begin{bmatrix} V_1 \\ V_2 \\ V_3 \end{bmatrix} \begin{bmatrix} V_1 & V_2 & V_3 \end{bmatrix}$$

$$= \begin{bmatrix} V_1^2 & V_1V_2 & V_1V_3 \\ V_2V_1 & V_2^2 & V_2V_3 \\ V_3V_1 & V_3V_2 & V_3^2 \end{bmatrix}$$

Then by definition of the *trace* of a matrix

$$\mathrm{trace}(\boldsymbol{V}\boldsymbol{V}^{\mathrm{T}}) = \text{sum of the diagonal terms of}$$

$$\begin{bmatrix} V_1^2 & V_1V_2 & V_1V_3 \\ V_2V_1 & V_2^2 & V_2V_3 \\ V_3V_1 & V_3V_2 & V_3^2 \end{bmatrix}$$

$$= V_1^2 + V_2^2 + V_3^2 = \boldsymbol{V}^{\mathrm{T}}\boldsymbol{V}$$

Now consider the general case $\boldsymbol{V}^{\mathrm{T}}\boldsymbol{W}\boldsymbol{V} = \text{trace}(\boldsymbol{V}\boldsymbol{V}^{\mathrm{T}}\boldsymbol{W})$

$$\boldsymbol{V}\boldsymbol{V}^{\mathrm{T}}\boldsymbol{W} = \begin{bmatrix} V_1^2 & V_1V_2 & V_1V_3 \\ V_2V_1 & V_2^2 & V_2V_3 \\ V_3V_1 & V_3V_2 & V_3^2 \end{bmatrix}\begin{bmatrix} W_{11} & W_{12} & W_{13} \\ W_{21} & W_{22} & W_{23} \\ W_{31} & W_{32} & W_{33} \end{bmatrix}$$

The trace $\boldsymbol{T}$ of the resulting matrix is

$$\begin{aligned} \boldsymbol{T} = {} & V_1^2W_{11} + V_1V_2W_{21} + V_1V_3W_{31} \\ & + V_2V_1W_{12} + V_2^2W_{22} + V_2V_3W_{32} \\ & + V_3V_1W_{13} + V_3V_2W_{23} + V_3^2W_{33} \end{aligned} \qquad \text{(A2.2)}$$

Now

$$\boldsymbol{V}^{\mathrm{T}}\boldsymbol{W}\boldsymbol{V} = \begin{bmatrix} V_1 & V_2 & V_3 \end{bmatrix}\begin{bmatrix} W_{11} & W_{12} & W_{13} \\ W_{21} & W_{22} & W_{23} \\ W_{31} & W_{32} & W_{33} \end{bmatrix}\begin{bmatrix} V_1 \\ V_2 \\ V_3 \end{bmatrix}$$

$$\begin{aligned} = {} & (V_1W_{11} + V_2W_{21} + V_3W_{31})V_1 + (V_1W_{12} + V_2W_{22} + V_3W_{32})V_2 \\ & + (V_1W_{13} + V_2W_{23} + V_3W_{33})V_3 \end{aligned}$$

which on multiplying out is equal to $\boldsymbol{T}$.

A2.2 Differentiation of matrices: bilinear and quadratic forms

If $\boldsymbol{A}$ is a square matrix whose elements are functions of a variable x, then we take $\partial\boldsymbol{A}/\partial\boldsymbol{x}$ to be the matrix of the same dimensions as $\boldsymbol{A}$ containing elements which are the respective differentials of each element of $\boldsymbol{A}$ with respect to x. For example if

$$\boldsymbol{A} = \begin{bmatrix} A_{11} & A_{12} & A_{13} \\ A_{21} & A_{22} & A_{23} \\ A_{31} & A_{32} & A_{33} \end{bmatrix}$$

$$\frac{\partial\boldsymbol{A}}{\partial\boldsymbol{x}} = \begin{bmatrix} \dfrac{\partial A_{11}}{\partial x} & \dfrac{\partial A_{12}}{\partial x} & \dfrac{\partial A_{13}}{\partial x} \\ \dfrac{\partial A_{21}}{\partial x} & \dfrac{\partial A_{22}}{\partial x} & \dfrac{\partial A_{23}}{\partial x} \\ \dfrac{\partial A_{31}}{\partial x} & \dfrac{\partial A_{32}}{\partial x} & \dfrac{\partial A_{33}}{\partial x} \end{bmatrix}$$

If $\boldsymbol{x}$ and $\boldsymbol{y}$ are *column vectors* of variables of order $(m \times 1)$ and $(n \times 1)$ respectively, and $\boldsymbol{A}$ is a matrix of coefficients of order $(m \times n)$ the *bilinear form* U is defined as

$$U = \boldsymbol{x}^{\mathrm{T}}\boldsymbol{A}\boldsymbol{y}$$
$$= [x_1 \quad x_2 \quad x_3 \quad \cdots \quad x_m]\boldsymbol{A}\boldsymbol{y}$$

Then

$$\frac{\partial U}{\partial x_1} = [1 \quad 0 \quad 0 \quad \cdots \quad 0]\boldsymbol{A}\boldsymbol{y}$$
$$\frac{\partial U}{\partial x_2} = [0 \quad 1 \quad 0 \quad \cdots \quad 0]\boldsymbol{A}\boldsymbol{y}$$
$$\vdots$$
$$\frac{\partial U}{\partial x_m} = [0 \quad 0 \quad 0 \quad \cdots \quad 1]\boldsymbol{A}\boldsymbol{y}$$

Assembling the elements into a column vector (as is $\boldsymbol{x}$) we have

$$\left(\frac{\partial U}{\partial \boldsymbol{x}}\right)^{\mathrm{T}} = \begin{bmatrix} \frac{\partial U}{\partial x_1} \\ \frac{\partial U}{\partial x_2} \\ \vdots \\ \frac{\partial U}{\partial x_m} \end{bmatrix} = \boldsymbol{I}\boldsymbol{A}\boldsymbol{y} = \boldsymbol{A}\boldsymbol{y}$$

where $\boldsymbol{I}$ is the unit matrix of dimensions $(m \times m)$. Or writing $\partial U/\partial \boldsymbol{x}$ for the *row vector* of partial derivatives we have

$$\frac{\partial U}{\partial \boldsymbol{x}} = (\boldsymbol{A}\boldsymbol{y})^{\mathrm{T}} = \boldsymbol{y}^{\mathrm{T}}\boldsymbol{A}^{\mathrm{T}} \tag{A2.3}$$

In the same way we have

$$\frac{\partial U}{\partial y_1} = \boldsymbol{x}^{\mathrm{T}}\boldsymbol{A}\begin{bmatrix}1\\0\\0\\\vdots\\0\end{bmatrix} \qquad \frac{\partial U}{\partial y_2} = \boldsymbol{x}^{\mathrm{T}}\boldsymbol{A}\begin{bmatrix}0\\1\\0\\\vdots\\0\end{bmatrix}$$

and so on. These are the first and second columns of $\boldsymbol{x}^{\mathrm{T}}\boldsymbol{A}$.

Writing $\partial U/\partial \boldsymbol{y}$ as a row vector we have

$$\frac{\partial U}{\partial \boldsymbol{y}} = \boldsymbol{x}^{\mathrm{T}}\boldsymbol{A}\boldsymbol{I} = \boldsymbol{x}^{\mathrm{T}}\boldsymbol{A} \tag{A2.4}$$

If the vector $\boldsymbol{y} = \boldsymbol{x}$ and $\boldsymbol{A}$ is square, then U is known as the *quadratic form* given by

$$U = \boldsymbol{x}^{\mathrm{T}}\boldsymbol{A}\boldsymbol{x} \tag{A2.5}$$

Then, by the chain rule, we have the total differential, given by (A2.3) and (A2.4)

$$\frac{\partial U}{\partial \boldsymbol{x}} = \boldsymbol{x}^{\mathrm{T}}\boldsymbol{A}^{\mathrm{T}} + \boldsymbol{x}^{\mathrm{T}}\boldsymbol{A}$$

If the matrix $\boldsymbol{A}$ is symmetric, then $\boldsymbol{A} = \boldsymbol{A}^{\mathrm{T}}$ and we have

$$\frac{\partial U}{\partial \boldsymbol{x}} = \boldsymbol{x}^{\mathrm{T}}\boldsymbol{A} + \boldsymbol{x}^{\mathrm{T}}\boldsymbol{A} = 2\boldsymbol{x}^{\mathrm{T}}\boldsymbol{A} \qquad \text{(A2.6)}$$

In Least Squares we are concerned with the quadratic form

$$\Omega = \boldsymbol{v}^{\mathrm{T}}\boldsymbol{W}\boldsymbol{v}$$

Hence

$$\frac{\partial \Omega}{\partial \boldsymbol{v}} = 2\boldsymbol{v}^{\mathrm{T}}\boldsymbol{W}$$

For a minimum turning value of Ω,

$$\frac{\partial \Omega}{\partial \boldsymbol{x}} = 0$$

Therefore

$$\frac{\partial \Omega}{\partial \boldsymbol{x}} = \frac{\partial \Omega}{\partial \boldsymbol{v}}\frac{\partial \boldsymbol{v}}{\partial \boldsymbol{x}} = 2\boldsymbol{v}^{\mathrm{T}}\boldsymbol{W}\frac{\partial \boldsymbol{v}}{\partial \boldsymbol{x}} = 0$$

But

$$\boldsymbol{v} = \boldsymbol{A}\boldsymbol{x} + \boldsymbol{L}$$

so by (A2.4) $\partial \boldsymbol{v}/\partial \boldsymbol{x} = \boldsymbol{A}$, thus

$$2\boldsymbol{v}^{\mathrm{T}}\boldsymbol{W}\frac{\partial \boldsymbol{v}}{\partial \boldsymbol{x}} = 2\boldsymbol{v}^{\mathrm{T}}\boldsymbol{W}\boldsymbol{A} = \boldsymbol{0}$$

giving the two vital formulae of the whole Least Squares process of

$$\boldsymbol{v}^{\mathrm{T}}\boldsymbol{W}\boldsymbol{A} = \boldsymbol{0} \qquad \text{(A2.7)}$$

and by transposing and remembering that $\boldsymbol{W}$ is symmetric, also

$$\boldsymbol{A}^{\mathrm{T}}\boldsymbol{W}\boldsymbol{v} = \boldsymbol{0} \qquad \text{(A2.8)}$$

A2.3 Least Squares models for the general case

Ordinary algebra

To be consistent, the best estimates and the provisional values† must satisfy the mathematical models exactly, thus

$$F(\hat{x} : \hat{s}) = 0 \quad \text{and} \quad F(\overset{*}{x} : \overset{*}{s}) = 0 \qquad \text{(A2.9)}$$

†For an alternative approach to this treatment, see Section 13.14.

Expressing the model in terms of small differential changes, we have

$$F(\hat{x}:\hat{s}) = F(\overset{*}{x}:\overset{*}{s}) + \left(\frac{\partial F}{\partial \overset{*}{x}}\right)\delta x + \left(\frac{\partial F}{\partial \overset{*}{s}}\right)\delta s$$

therefore

$$\left(\frac{\partial F}{\partial \overset{*}{x}}\right)\delta x + \left(\frac{\partial F}{\partial \overset{*}{s}}\right)\delta s = 0$$

and since

$$\delta s = L + v$$

we have

$$\left(\frac{\partial F}{\partial \overset{*}{x}}\right)\delta x + \left(\frac{\partial F}{\partial \overset{*}{s}}\right)v + \left(\frac{\partial F}{\partial \overset{*}{s}}\right)L = 0 \qquad \text{(A2.10)}$$

Matrix algebra

The partial differentials are gathered together into a matrix, called the *Jacobian* matrix. We adopt the convention that there are m observations, having a weight matrix $\boldsymbol{W}$, in n parameters $(m > n)$ linked together by r equations of type (A2.10), which in matrix form, with their dimensions, are

$$\underset{(r \times n)}{\boldsymbol{A}}\ \underset{(n \times 1)}{\boldsymbol{x}} + \underset{(r \times m)}{\boldsymbol{C}}\ \underset{(m \times 1)}{\boldsymbol{v}} + \underset{(r \times m)}{\boldsymbol{C}}\ \underset{(m \times 1)}{\boldsymbol{L}} = \underset{(r \times 1)}{\boldsymbol{0}} \qquad \text{(A2.11)}$$

$\boldsymbol{L} = \overset{o}{\boldsymbol{s}} - \overset{*}{\boldsymbol{s}}$ is known, and the design matrices $\boldsymbol{A}$ and $\boldsymbol{C}$ are also known, thus the product $\boldsymbol{CL}$ is also known. Letting

$$\boldsymbol{CL} = -\boldsymbol{b}$$

equation (A2.11) becomes

$$\boldsymbol{Ax} + \boldsymbol{Cv} = \boldsymbol{b} \qquad \text{(A2.12)}$$

To obtain a unique solution for $\boldsymbol{x}$ we invoke the principle of Least Squares, making the sum of weighted squares of the residuals

$$\boldsymbol{v}^{\mathrm{T}}\boldsymbol{W}\boldsymbol{v} = \Omega$$

a minimum. Introducing a vector $\boldsymbol{\lambda}$ of Lagrangian multipliers, of dimension $(r \times 1)$, we may put

$$\Omega = \boldsymbol{v}^{\mathrm{T}}\boldsymbol{W}\boldsymbol{v} - 2\boldsymbol{\lambda}^{\mathrm{T}}(\boldsymbol{Ax} + \boldsymbol{Cv} - \boldsymbol{b}) \qquad \text{(A2.13)}$$

This vector $\boldsymbol{\lambda}$ of *correlates* or *undetermined multipliers* is an ingenious device invented by Lagrange to treat problems of conditioned minima. We can see that equation (A2.13) is valid because by equation (A2.12) the second term is zero.

Notice that the dimension of Ω is (1 × 1), i.e. a single number.

The ($m \times m$) weight matrix $\boldsymbol{W}$ is formed from the inverse of the dispersion matrix of the observations estimated in some way. We now partially differentiate the function Ω with respect to the variables $\boldsymbol{x}$ and $\boldsymbol{s}$, and equate them to zero. Before doing so, however, we will establish that differentiation with respect to the observed parameter $\boldsymbol{s}$ is the same as differentiation with respect to $\boldsymbol{v}$.

$$\frac{\partial \Omega}{\partial \boldsymbol{s}} = \frac{\partial \Omega}{\partial \boldsymbol{v}} \frac{\partial \boldsymbol{v}}{\partial \boldsymbol{s}}$$

Because $\boldsymbol{s} = \boldsymbol{L} + \boldsymbol{v}$, and $\boldsymbol{L}$ has constant elements,

$$\frac{\partial \boldsymbol{v}}{\partial \boldsymbol{s}} = \boldsymbol{I} \text{ (the unit matrix)}$$

and thus

$$\frac{\partial \Omega}{\partial \boldsymbol{s}} = \frac{\partial \Omega}{\partial \boldsymbol{v}}$$

For a minimum value of Ω we have

$$\frac{\partial \Omega}{\partial \boldsymbol{v}} = \boldsymbol{0} \quad \text{and} \quad \frac{\partial \Omega}{\partial \boldsymbol{x}} = \boldsymbol{0}$$

Applying these expressions to equation (A2.13) gives

$$\frac{\partial \Omega}{\partial \boldsymbol{v}} = \boldsymbol{0} = 2\boldsymbol{v}^{\mathrm{T}}\boldsymbol{W} - 2\boldsymbol{\lambda}^{\mathrm{T}}\boldsymbol{C}$$

Since $\boldsymbol{W}$ is symmetric

$$\boldsymbol{W}^{\mathrm{T}}\boldsymbol{v} = \boldsymbol{C}^{\mathrm{T}}\boldsymbol{\lambda} = \boldsymbol{W}\boldsymbol{v}$$

Thus we have a most important equation

$$\boldsymbol{v} = \boldsymbol{W}^{-1}\boldsymbol{C}^{\mathrm{T}}\boldsymbol{\lambda} \qquad \text{(A2.14)}$$

And again

$$\frac{\partial \Omega}{\partial \boldsymbol{x}} = \boldsymbol{0} = 2\boldsymbol{\lambda}^{\mathrm{T}}\boldsymbol{A}$$

therefore

$$\boldsymbol{A}^{\mathrm{T}}\boldsymbol{\lambda} = \boldsymbol{0} \qquad \text{(A2.15)}$$

Substituting for $\boldsymbol{v}$ from (A2.15) in equation (A2.12) gives

$$\boldsymbol{A}\boldsymbol{x} + \boldsymbol{C}\boldsymbol{W}^{-1}\boldsymbol{C}^{\mathrm{T}}\boldsymbol{\lambda} = \boldsymbol{b}$$

Putting

$$\boldsymbol{N} = \boldsymbol{C}\boldsymbol{W}^{-1}\boldsymbol{C}^{\mathrm{T}}$$

we have

$$\boldsymbol{Ax} + \boldsymbol{N\lambda} = \boldsymbol{b}$$

and

$$\boldsymbol{\lambda} = \boldsymbol{N}^{-1}(\boldsymbol{b} - \boldsymbol{Ax}) \tag{A2.16}$$

Substituting for $\boldsymbol{\lambda}$ from (A2.16) in equation (A2.15) gives

$$\boldsymbol{A}^{\mathrm{T}}\boldsymbol{N}^{-1}(\boldsymbol{b} - \boldsymbol{Ax}) = \boldsymbol{0} \tag{A2.17}$$

$$\boldsymbol{A}^{\mathrm{T}}\boldsymbol{N}^{-1}\boldsymbol{Ax} = \boldsymbol{A}^{\mathrm{T}}\boldsymbol{N}^{-1}\boldsymbol{b}$$

And putting

$$\boldsymbol{N}_1 = \boldsymbol{A}^{\mathrm{T}}\boldsymbol{N}^{-1}\boldsymbol{A} \quad \text{and} \quad \boldsymbol{b}_1 = \boldsymbol{A}^{\mathrm{T}}\boldsymbol{N}^{-1}\boldsymbol{b}$$

we have the final equations

$$\boldsymbol{N}_1\boldsymbol{x} = \boldsymbol{b}_1 \tag{A2.18}$$

The matrix $\boldsymbol{N}_1$ is square and symmetric. Traditionally the equations (A2.18) were called 'normal equations'. Their solution gives the required parameters.

From (A2.15), (A2.16) and (A2.12) we obtain another very important result

$$\boldsymbol{A}^{\mathrm{T}}\boldsymbol{N}^{-1}\boldsymbol{Cv} = \boldsymbol{0} \tag{A2.19}$$

It is so important that we remind readers that

$$\boldsymbol{N} = \boldsymbol{CW}^{-1}\boldsymbol{C}^{\mathrm{T}}$$

A2.4 Mathematical model with added constraints

It sometimes happens that there are certain constraints among the parameters which need to be allowed for. For example we may need to hold a length fixed between two points whose positions may be allowed to vary. A constraint is written as a linear equation which has to be satisfied *exactly*, i.e. as an equation, which has no residual, of the form

$$\boldsymbol{Ex} + \boldsymbol{d} = \boldsymbol{0} \tag{A2.20}$$

It is often possible, and thoroughly desirable for simplicity, to eliminate one parameter for each constraint, right at the outset. Not only does this simplify the solution but it reduces the size of the matrix to be solved. The simplest application of this procedure is to eliminate the variables for fixed points in a network.

Another practical way to solve the problem is to assign a very high weight to the constraint equations, which are then treated as observations, thus ensuring that the constraints are almost satisfied.

A theoretically correct and standard procedure is to employ further Lagrangian multipliers $\boldsymbol{k}$, one for each constraint equation, and proceed as before, and as follows.

Let $$\Omega = \boldsymbol{v}^{\mathrm{T}}\Omega\boldsymbol{v} - 2\boldsymbol{\lambda}^{\mathrm{T}}(\boldsymbol{Ax} + \boldsymbol{Cv} - \boldsymbol{b}) - 2\boldsymbol{k}^{\mathrm{T}}(\boldsymbol{Ex} + \boldsymbol{d})$$

Putting $$\left(\frac{\partial\Omega}{\partial\boldsymbol{x}}\right) = \mathbf{0}$$

gives $$\mathbf{0} = 2\boldsymbol{\lambda}^{\mathrm{T}}\boldsymbol{A} + 2\boldsymbol{k}^{\mathrm{T}}\boldsymbol{E}$$

therefore $$\boldsymbol{A}^{\mathrm{T}}\boldsymbol{\lambda} + \boldsymbol{E}^{\mathrm{T}}\boldsymbol{k} = \mathbf{0}$$

Since $\boldsymbol{\lambda}$ is the same as before (A2.13), we obtain

$$\boldsymbol{N}_1\boldsymbol{x} + \boldsymbol{E}^{\mathrm{T}}\boldsymbol{k} = \boldsymbol{b}_1 \qquad \text{(A2.21)}$$

$$\boldsymbol{Ex} + \boldsymbol{d} = \mathbf{0} \qquad \text{(A2.20)}$$

These equations may be combined into a *hypermatrix* as follows

$$\begin{bmatrix} \boldsymbol{N}_1 & \boldsymbol{E}^{\mathrm{T}} \\ \boldsymbol{E} & \mathbf{0} \end{bmatrix}\begin{bmatrix} \boldsymbol{x} \\ \boldsymbol{k} \end{bmatrix} = \begin{bmatrix} \boldsymbol{b}_1 \\ -\boldsymbol{d} \end{bmatrix}$$

(**Note:** These may be solved as they stand, although if Cholesky's method is used, the null matrix causes negative square roots which have to be flagged during the procedure to identify negative squares. When they arise, Gauss's method of solution has no problems in its general form.) This hypermatrix form is useful when making design studies to add and subtract constraint equations.

If an explicit solution for $\boldsymbol{x}$ only is desired, a smaller set of equations has to be solved. The explicit solution is obtained as follows: from (A2.21)

$$\boldsymbol{x} = \boldsymbol{N}_1^{-1}(\boldsymbol{b}_1 - \boldsymbol{E}^{\mathrm{T}}\boldsymbol{k})$$

and from (A2.20)

$$\boldsymbol{EN}_1^{-1}(\boldsymbol{b}_1 - \boldsymbol{E}^{\mathrm{T}}\boldsymbol{k}) + \boldsymbol{d} = 0$$

or

$$\boldsymbol{EN}_1^{-1}\boldsymbol{E}^{\mathrm{T}}\boldsymbol{k} = \boldsymbol{EN}_1^{-1}\boldsymbol{b}_1 + \boldsymbol{d}$$

or

$$\boldsymbol{N}_2\boldsymbol{k} = \boldsymbol{b}_2$$

and $$\boldsymbol{k} = \boldsymbol{N}_2^{-1}\boldsymbol{b}_2$$

Substituting for $\boldsymbol{k}$ in (A2.21) finally gives

$$\boldsymbol{N}_1\boldsymbol{x} = \boldsymbol{b}_1 - \boldsymbol{E}^{\mathrm{T}}\boldsymbol{N}_2^{-1}\boldsymbol{b}_2$$

or

$$\boldsymbol{N}_1\boldsymbol{x} = \boldsymbol{b}_3 \qquad \text{(A2.22)}$$

Thus the problem once again reduces to the solution of a symmetric set of equations of the form

$$\boldsymbol{Nx} = \boldsymbol{b}$$

A common simplification

Many problems are simplified at the outset so that they do not contain any constraint equations, for example by the elimination of fixed parameters when forming up the observation equations.

Also in many cases it is possible to select a mathematical model which has only one observation in each equation, giving an observation equation which contains only one residual. Thus the $\boldsymbol{C}$ matrix reduces to the unit matrix, or alternatively it has an inverse. In such cases, equations (A2.12) may be simplified by premultiplying by this inverse giving

$$\boldsymbol{C}^{-1}\boldsymbol{A}\boldsymbol{x} + \boldsymbol{C}^{-1}\boldsymbol{C}\boldsymbol{v} = \boldsymbol{C}^{-1}\boldsymbol{b}$$

which is of the form

$$\boldsymbol{A}_1\boldsymbol{x} + \boldsymbol{L}_1 + \boldsymbol{v} = \boldsymbol{0}$$

giving directly the Normal equations

$$\boldsymbol{A}_1^{\mathrm{T}}\boldsymbol{W}\boldsymbol{A}_1\boldsymbol{x} + \boldsymbol{A}_1^{\mathrm{T}}\boldsymbol{W}\boldsymbol{L}_1 = \boldsymbol{0}$$

$$\boldsymbol{N}_3\boldsymbol{x} = \boldsymbol{b}_4 \qquad \text{(A2.23)}$$

These equations are commonly found in many surveying applications. For example equations ($\boldsymbol{A}\boldsymbol{x} + \boldsymbol{C}\boldsymbol{s} = \boldsymbol{0}$) can be written

$$\boldsymbol{C}^{-1}\boldsymbol{A}\boldsymbol{x} + \boldsymbol{C}^{-1}\boldsymbol{C}\boldsymbol{s} = \boldsymbol{0}$$

$$\boldsymbol{A}_2\boldsymbol{x} + \boldsymbol{s} = \boldsymbol{0}$$

or

$$\boldsymbol{A}_2\boldsymbol{x} = \boldsymbol{L} + \boldsymbol{v}$$

A2.5 Population parameters and sample statistics

We now relate the best estimates from a sample to their theoretical population from which the sample has been drawn. If $\bar{s}$ is the population value of the observed parameter, then the population residual V is given by

$$\bar{s} = \overset{\circ}{s} + V$$

Correspondingly the vectors are written in bold notation

$$\bar{\boldsymbol{s}} = \overset{\circ}{\boldsymbol{s}} + \boldsymbol{V}$$

This compares with the vector of data from the sample

$$\hat{\boldsymbol{s}} = \overset{\circ}{\boldsymbol{s}} + \boldsymbol{v}$$

Thus

$$\mathbf{d}\boldsymbol{s} = \bar{\boldsymbol{s}} - \hat{\boldsymbol{s}} = \boldsymbol{V} - \boldsymbol{v}$$

where $\mathbf{d}\boldsymbol{s}$ is the difference between the population value and the sample estimate. Applying similar ideas to the mathematical model we have the respective sample and population models

$$\boldsymbol{A}\boldsymbol{x} + \boldsymbol{C}\boldsymbol{v} + \boldsymbol{C}\boldsymbol{L} = \boldsymbol{0}$$

and

$$\boldsymbol{A\bar{x}} + \boldsymbol{CV} + \boldsymbol{CL} = \mathbf{0}$$

Subtracting gives

$$\mathrm{A}(\bar{\boldsymbol{x}} - \boldsymbol{x}) + \boldsymbol{C}(\boldsymbol{V} - \boldsymbol{v}) = \mathbf{0}$$

$$\boldsymbol{A}\mathbf{d}\boldsymbol{x} + \boldsymbol{C}(\boldsymbol{V} - \boldsymbol{v}) = \mathbf{0} \qquad \text{(A2.24)}$$

where $\mathbf{d}\boldsymbol{x}$ is the vector of the differences between the population parameters and their corresponding values estimated from the sample by Least Squares. Now we know from equation (A2.19) that

$$\boldsymbol{A}^{\mathrm{T}}\boldsymbol{N}^{-1}\boldsymbol{C}\boldsymbol{v} = \mathbf{0}$$

thus by premultipling (A2.24) by $\boldsymbol{A}^{\mathrm{T}}\boldsymbol{N}^{-1}$ we obtain

$$\boldsymbol{A}^{\mathrm{T}}\boldsymbol{N}^{-1}\boldsymbol{A}\mathbf{d}\boldsymbol{x} + \boldsymbol{A}^{\mathrm{T}}\boldsymbol{N}^{-1}\boldsymbol{C}\boldsymbol{V} - \boldsymbol{A}^{\mathrm{T}}\boldsymbol{N}^{-1}\boldsymbol{C}\boldsymbol{v} = \mathbf{0}$$

or

$$\boldsymbol{A}^{\mathrm{T}}\boldsymbol{N}^{-1}\boldsymbol{A}\mathbf{d}\boldsymbol{x} + \boldsymbol{A}^{\mathrm{T}}\boldsymbol{N}^{-1}\boldsymbol{C}\boldsymbol{V} = \mathbf{0}$$

or say

$$\boldsymbol{N}_1\mathbf{d}\boldsymbol{x} = \boldsymbol{B}\boldsymbol{V}$$

where

$$\boldsymbol{B} = -\boldsymbol{A}^{\mathrm{T}}\boldsymbol{N}^{-1}\boldsymbol{C} \quad \text{and} \quad \boldsymbol{N}_1 = \boldsymbol{A}^{\mathrm{T}}\boldsymbol{N}^{-1}\boldsymbol{A}$$

Now

$$\boldsymbol{N}_1\mathbf{d}\boldsymbol{x}(\boldsymbol{N}_1\mathbf{d}\boldsymbol{x})^{\mathrm{T}} = \boldsymbol{B}\boldsymbol{V}(\boldsymbol{B}\boldsymbol{V})^{\mathrm{T}}$$

therefore

$$\boldsymbol{N}_1\mathbf{d}\boldsymbol{x}\mathbf{d}\boldsymbol{x}^{\mathrm{T}}\boldsymbol{N}_1 = \boldsymbol{B}\boldsymbol{V}\boldsymbol{V}^{\mathrm{T}}\boldsymbol{B}^{\mathrm{T}}$$

Taking expectations, and remembering that (See Sections 13.23 and 13.24)

$$\mathrm{E}(\mathbf{d}\boldsymbol{x}\mathbf{d}\boldsymbol{x}^{\mathrm{T}}) = \boldsymbol{D}_{\boldsymbol{x}} \quad \text{and} \quad \mathrm{E}(\boldsymbol{V}\boldsymbol{V}^{\mathrm{T}}) = \boldsymbol{D}_{\mathrm{o}}$$

we have

$$\boldsymbol{N}_1\boldsymbol{D}_{\boldsymbol{x}}\boldsymbol{N}_1 = \boldsymbol{B}\boldsymbol{D}_{\mathrm{o}}\boldsymbol{B}^{\mathrm{T}}$$

Pre and postmultiplying both sides by $\boldsymbol{N}_1^{-1}$ gives

$$\boldsymbol{N}_1^{-1}\boldsymbol{N}_1\boldsymbol{D}_{\boldsymbol{x}}\boldsymbol{N}_1\boldsymbol{N}_1^{-1} = \boldsymbol{N}_1^{-1}\boldsymbol{B}\boldsymbol{D}_{\mathrm{o}}\boldsymbol{B}^{\mathrm{T}}\boldsymbol{N}_1^{-1}$$

or

$$\boldsymbol{D}_{\boldsymbol{x}} = \boldsymbol{N}_1^{-1}\boldsymbol{B}\boldsymbol{D}_{\mathrm{o}}\boldsymbol{B}^{\mathrm{T}}\boldsymbol{N}_1^{-1}$$

but

$$\boldsymbol{B}\boldsymbol{D}_{\mathrm{o}}\boldsymbol{B}^{\mathrm{T}} = \boldsymbol{A}^{\mathrm{T}}\boldsymbol{N}^{-1}\boldsymbol{C}\boldsymbol{D}_{\mathrm{o}}(\boldsymbol{A}^{\mathrm{T}}\boldsymbol{N}^{-1}\boldsymbol{C})^{\mathrm{T}}$$

$$= \boldsymbol{A}^{\mathrm{T}}\boldsymbol{N}^{-1}\boldsymbol{C}\boldsymbol{D}_{\mathrm{o}}\boldsymbol{C}^{\mathrm{T}}\boldsymbol{N}^{-1}\boldsymbol{A}$$

and

$$\boldsymbol{C}\boldsymbol{D}_{\mathrm{o}}\boldsymbol{C}^{\mathrm{T}} = \boldsymbol{C}\sigma_{\mathrm{o}}^2\boldsymbol{W}^{-1}\boldsymbol{C}^{\mathrm{T}}$$

$$= \sigma_{\mathrm{o}}^2\boldsymbol{N}$$

giving

$$\boldsymbol{B}\boldsymbol{D}_{\mathrm{o}}\boldsymbol{B}^{\mathrm{T}} = \sigma_{\mathrm{o}}^2\boldsymbol{N}_1$$

and finally

$$\boldsymbol{D}_x = \sigma_0^2 \boldsymbol{N}_1^{-1}$$

Thus the dispersion matrix of the required parameters is the scaled inverse of the normal equations. Hence, although this inverse may not be needed to obtain the solution, it is needed for this statistical information.

Special case

When $\boldsymbol{C} = \boldsymbol{I}$ the unit matrix, the result is of the same form, namely

$$\boldsymbol{D}_x = \sigma_0^2 \boldsymbol{N}^{-1} \tag{A2.25}$$

because in full $\boldsymbol{D}_x = \sigma_0^2 (\boldsymbol{A}^{\mathrm{T}} (\boldsymbol{C}\boldsymbol{W}^{-1}\boldsymbol{C}^{\mathrm{T}})^{-1} \boldsymbol{A})^{-1}$

which reduces to equation (A2.25).

A2.6 Estimation of σ_0^2

We now show how σ_0^2 is estimated from the sample itself, as a bi-product of the Least Squares computation, using the expression:

$$\sigma_0^2 = \mathrm{E}\left(\frac{\boldsymbol{v}^{\mathrm{T}}\boldsymbol{W}\boldsymbol{v}}{m-n}\right) \tag{A2.26}$$

For economy of space, the derivations are not given in full, with some work left to the reader, who is once again reminded of the symmetric nature of many of the matrices.
From equation (A2.16) we have

$$\boldsymbol{\lambda} = \boldsymbol{N}^{-1}(\boldsymbol{b} - \boldsymbol{A}\boldsymbol{x}) \tag{A2.16}$$

and from equation (A2.14)

$$\boldsymbol{v} = \boldsymbol{W}^{-1}\boldsymbol{C}^{\mathrm{T}}\boldsymbol{\lambda} \tag{A2.14}$$

Thus we have the sample residuals given by

$$\boldsymbol{v} = \boldsymbol{W}^{-1}\boldsymbol{C}^{\mathrm{T}}\boldsymbol{N}^{-1}(\boldsymbol{b} - \boldsymbol{A}\boldsymbol{x}) \tag{A2.27}$$

The population residuals are therefore given by

$$\boldsymbol{V} = \boldsymbol{W}^{-1}\boldsymbol{C}^{\mathrm{T}}\boldsymbol{N}^{-1}(\boldsymbol{b} - \boldsymbol{A}\bar{\boldsymbol{x}}) \tag{A2.28}$$

giving by subtraction

$$\boldsymbol{V} - \boldsymbol{v} = \boldsymbol{W}^{-1}\boldsymbol{C}^{\mathrm{T}}\boldsymbol{N}^{-1}(\boldsymbol{b} - \boldsymbol{A}\bar{\boldsymbol{x}} - \boldsymbol{b} + \boldsymbol{A}\boldsymbol{x})$$

$$= -\boldsymbol{W}^{-1}\boldsymbol{C}^{\mathrm{T}}\boldsymbol{N}^{-1}\boldsymbol{A}(\bar{\boldsymbol{x}} - \boldsymbol{x})$$

$$= -\boldsymbol{W}^{-1}\boldsymbol{C}^{\mathrm{T}}\boldsymbol{N}^{-1}\boldsymbol{A}\mathbf{d}\boldsymbol{x}$$

$$\boldsymbol{V} = \boldsymbol{v} - \boldsymbol{W}^{-1}\boldsymbol{C}^{\mathrm{T}}\boldsymbol{N}^{-1}\boldsymbol{A}\mathbf{d}\boldsymbol{x}$$

or
$$\boldsymbol{V} = \boldsymbol{v} - \boldsymbol{K}\mathbf{d}\boldsymbol{x}$$

where

$$\boldsymbol{K} = \boldsymbol{W}^{-1}\boldsymbol{C}^{\mathrm{T}}\boldsymbol{N}^{-1}\boldsymbol{A} \tag{A2.29}$$

Now from equation (A2.19)

$$\boldsymbol{A}^{\mathrm{T}}\boldsymbol{N}^{-1}\boldsymbol{C}\boldsymbol{v} = \mathbf{0} \tag{A2.19}$$

Thus we have the important result

$$\boldsymbol{K}^{\mathrm{T}}\boldsymbol{W}\boldsymbol{v} = \mathbf{0} \tag{A2.30}$$

Now

$$\begin{aligned}\boldsymbol{V}^{\mathrm{T}}\boldsymbol{W}\boldsymbol{V} &= (\boldsymbol{v} - \boldsymbol{K}\mathbf{d}\boldsymbol{x})^{\mathrm{T}}\boldsymbol{W}(\boldsymbol{v} - \boldsymbol{K}\mathbf{d}\boldsymbol{x}) \\ &= \boldsymbol{v}^{\mathrm{T}}\boldsymbol{W}\boldsymbol{v} + \mathbf{d}\boldsymbol{x}^{\mathrm{T}}\boldsymbol{K}^{\mathrm{T}}\boldsymbol{W}\boldsymbol{K}\mathbf{d}\boldsymbol{x} \qquad \text{(A2.31)}\end{aligned}$$

because $\boldsymbol{K}^{\mathrm{T}}\boldsymbol{W}\boldsymbol{v} = \mathbf{0}$ and $\boldsymbol{v}^{\mathrm{T}}\boldsymbol{W}\boldsymbol{K} = \mathbf{0}$.

Now, remembering that the *trace* of a matrix is the sum of its diagonal terms, we have

$$\boldsymbol{V}^{\mathrm{T}}\boldsymbol{W}\boldsymbol{V} = \mathrm{trace}(\boldsymbol{V}\boldsymbol{V}^{\mathrm{T}}\boldsymbol{W})$$

$$\mathbf{d}\boldsymbol{x}^{\mathrm{T}}\boldsymbol{K}^{\mathrm{T}}\boldsymbol{W}\boldsymbol{K}\mathbf{d}\boldsymbol{x} = \mathrm{trace}(\mathbf{d}\boldsymbol{x}\mathbf{d}\boldsymbol{x}^{\mathrm{T}}\boldsymbol{K}^{\mathrm{T}}\boldsymbol{W}\boldsymbol{K})$$

Thus equation (A2.31) becomes

$$\mathrm{trace}(\boldsymbol{V}\boldsymbol{V}^{\mathrm{T}}\boldsymbol{W}) = \boldsymbol{v}^{\mathrm{T}}\boldsymbol{W}\boldsymbol{v} + \mathrm{trace}(\mathbf{d}\boldsymbol{x}\mathbf{d}\boldsymbol{x}^{\mathrm{T}}\boldsymbol{K}^{\mathrm{T}}\boldsymbol{W}\boldsymbol{K})$$

and taking expectations of both sides we have

$$\mathrm{E}(\mathrm{trace}(\boldsymbol{V}\boldsymbol{V}^{\mathrm{T}})\boldsymbol{W}) = \mathrm{E}(\boldsymbol{v}^{\mathrm{T}}\boldsymbol{W}\boldsymbol{v}) + \mathrm{E}(\mathrm{trace}(\mathbf{d}\boldsymbol{x}\mathbf{d}\boldsymbol{x}^{\mathrm{T}})\boldsymbol{K}^{\mathrm{T}}\boldsymbol{W}\boldsymbol{K})$$

Now

$$\mathrm{E}(\mathbf{d}\boldsymbol{x}\mathbf{d}\boldsymbol{x}^{\mathrm{T}}) = \boldsymbol{D}_{\boldsymbol{x}} = \sigma_{\mathrm{o}}^{2}\boldsymbol{N}_{1}^{-1}$$

Also

$$\mathrm{E}(\mathrm{trace}(\boldsymbol{V}\boldsymbol{V}^{\mathrm{T}})\boldsymbol{W}) = \mathrm{trace}(\boldsymbol{D}_{\mathrm{o}}\boldsymbol{W}) = \mathrm{trace}(\sigma_{\mathrm{o}}^{2}\boldsymbol{W}^{-1}\boldsymbol{W})$$

Therefore

$$\begin{aligned}\sigma_{\mathrm{o}}^{2}\ \mathrm{trace}(\boldsymbol{I}_{m}) &= \mathrm{E}(\boldsymbol{v}^{\mathrm{T}}\boldsymbol{W}\boldsymbol{v}) + \mathrm{trace}(\sigma_{\mathrm{o}}^{2}\boldsymbol{N}_{1}^{-1}\boldsymbol{K}^{\mathrm{T}}\boldsymbol{W}\boldsymbol{K}) \\ &= \mathrm{E}(\boldsymbol{v}^{\mathrm{T}}\boldsymbol{W}\boldsymbol{v}) + \sigma_{\mathrm{o}}^{2}\ \mathrm{trace}(\boldsymbol{I}_{n})\end{aligned}$$

because

$$\boldsymbol{K}^{\mathrm{T}}\boldsymbol{W}\boldsymbol{K} = \boldsymbol{N}_{1}$$

Now $\boldsymbol{I}_m$ and $\boldsymbol{I}_n$ are unit matrices of dimensions m and n, thus

$$m\sigma_{\mathrm{o}}^{2} = \mathrm{E}(\boldsymbol{v}^{\mathrm{T}}\boldsymbol{W}\boldsymbol{v}) + n\sigma_{\mathrm{o}}^{2}$$

and finally $$\sigma_o^2 = \mathrm{E}\left(\frac{\boldsymbol{v}^{\mathrm{T}}\boldsymbol{W}\boldsymbol{v}}{m-n}\right)$$

If we put

$$s_o^2 = \frac{\boldsymbol{v}^{\mathrm{T}}\boldsymbol{W}\boldsymbol{v}}{m-n}$$

then $\mathrm{E}(s_o^2) = \sigma_o^2$ and we say that s_o^2 is an unbiased estimator of σ_o^2. The denominator $m - n$ is the number of degrees of freedom in the problem, or the number of redundant observations which are not essential for a solution to be obtained.

Special case of one parameter

In the special case of one parameter estimated from m observations, the above expression for s_o^2 reduces to

$$\mathrm{E}\left(s_o^2\right) = \mathrm{E}\left(\frac{\boldsymbol{v}^{\mathrm{T}}\boldsymbol{W}\boldsymbol{v}}{m-1}\right)$$

$$= \sigma_o^2$$

and

$$s_o^2 = \frac{m}{(m \quad 1)}\frac{\boldsymbol{v}^{T}\boldsymbol{W}\boldsymbol{v}}{m}$$

Also

$$s_o^2 = \frac{m}{(m \quad 1)} \times \textit{sample variance}$$

an *unbiased estimator* of the population variance.

A2.7 Dispersion matrix of the sample residuals $\boldsymbol{D}_v$

Although the quality of the derived parameters, as expressed by their dispersion matrix $\boldsymbol{D}_x$, is generally of most interest, there are several statistical reasons for calculating the quality of the *residuals* derived in the Least Squares process. From equation (A2.27) we have

$$\boldsymbol{v} = \boldsymbol{W}^{-1}\boldsymbol{C}^{\mathrm{T}}\boldsymbol{N}^{-1}(\boldsymbol{b} - \boldsymbol{A}\boldsymbol{x}) \qquad \text{(A2.27)}$$

$$\boldsymbol{v} = -\boldsymbol{W}^{-1}\boldsymbol{C}^{\mathrm{T}}\boldsymbol{N}^{-1}\boldsymbol{C}\boldsymbol{L} - \boldsymbol{W}^{-1}\boldsymbol{C}^{\mathrm{T}}\boldsymbol{N}^{-1}\boldsymbol{A}\boldsymbol{x}$$

$$\boldsymbol{v} = -\boldsymbol{G}\boldsymbol{L} - \boldsymbol{K}\boldsymbol{x}$$

where

$$\boldsymbol{G} = \boldsymbol{W}^{-1}\boldsymbol{C}^{\mathrm{T}}\boldsymbol{N}^{-1}\boldsymbol{C} \quad \text{and} \quad \boldsymbol{K} = \boldsymbol{W}^{-1}\boldsymbol{C}^{\mathrm{T}}\boldsymbol{N}^{-1}\boldsymbol{A}$$

Differentiating gives

$$\mathbf{d}\boldsymbol{v} = -\boldsymbol{G}\mathbf{d}\boldsymbol{L} - \boldsymbol{K}\mathbf{d}\boldsymbol{x} \tag{A2.32}$$

$$\mathbf{d}\boldsymbol{v}\mathbf{d}\boldsymbol{v}^{\mathrm{T}} = (\boldsymbol{G}\mathbf{d}\boldsymbol{L} + \boldsymbol{K}\mathbf{d}\boldsymbol{x})\,(\boldsymbol{G}\mathbf{d}\boldsymbol{L} + \boldsymbol{K}\mathbf{d}\boldsymbol{x})^{\mathrm{T}}$$

Now

$$\boldsymbol{L} = \overset{\circ}{\boldsymbol{s}} - \overset{*}{\boldsymbol{s}}$$

therefore

$$\mathbf{d}\boldsymbol{L} = \mathbf{d}\overset{\circ}{\boldsymbol{s}} - \mathbf{d}\overset{*}{\boldsymbol{s}}$$

but $\mathbf{d}\overset{\circ}{\boldsymbol{s}}$ is the error vector of the observed parameters, i.e. $\mathbf{d}\overset{\circ}{\boldsymbol{s}} = \boldsymbol{V}$ the population residuals, and $\mathbf{d}\overset{*}{\boldsymbol{s}} = \mathbf{0}$, thus

$$\mathbf{d}\boldsymbol{v}\mathbf{d}\boldsymbol{v}^{\mathrm{T}} = (\boldsymbol{G}\boldsymbol{V} + \boldsymbol{K}\mathbf{d}\boldsymbol{x})\,(\boldsymbol{G}\boldsymbol{V} + \boldsymbol{K}\mathbf{d}\boldsymbol{x})^{\mathrm{T}}$$

$$\mathbf{d}\boldsymbol{v}\mathbf{d}\boldsymbol{v}^{\mathrm{T}} = \boldsymbol{G}\boldsymbol{V}\boldsymbol{V}^{\mathrm{T}}\boldsymbol{G}^{\mathrm{T}} + \boldsymbol{K}\mathbf{d}\boldsymbol{x}\mathbf{d}\boldsymbol{x}^{\mathrm{T}}\boldsymbol{K}^{\mathrm{T}} + \boldsymbol{G}\boldsymbol{V}\mathbf{d}\boldsymbol{x}^{\mathrm{T}}\boldsymbol{K}^{\mathrm{T}} + \boldsymbol{K}\mathbf{d}\boldsymbol{x}\boldsymbol{V}^{\mathrm{T}}\boldsymbol{G}^{\mathrm{T}}$$

Taking expectations we have

$$\boldsymbol{D}_v = \boldsymbol{G}\boldsymbol{D}_{\mathrm{o}}\boldsymbol{G}^{\mathrm{T}} - \boldsymbol{K}\boldsymbol{D}_x\boldsymbol{K}^{\mathrm{T}} + \boldsymbol{G}\,\mathrm{E}(\boldsymbol{V}\mathbf{d}\boldsymbol{x}^{\mathrm{T}})\boldsymbol{K}^{\mathrm{T}} + \boldsymbol{K}\,\mathrm{E}(\mathbf{d}\boldsymbol{x}\boldsymbol{V}^{\mathrm{T}})\boldsymbol{G}^{\mathrm{T}}$$

Since, as we will show below, the third and fourth terms of this expression are both equal to

$$-\boldsymbol{K}\boldsymbol{D}_x\boldsymbol{K}^{\mathrm{T}}$$

we have

$$\boldsymbol{D}_v = \boldsymbol{G}\boldsymbol{D}_{\mathrm{o}}\boldsymbol{G}^{\mathrm{T}} - \boldsymbol{K}\boldsymbol{D}_x\boldsymbol{K}^{\mathrm{T}} \tag{A2.33}$$

Note: The following is a brief outline proof that terms three and four above are both equal to

$$-\boldsymbol{K}\boldsymbol{D}_x\boldsymbol{K}^{\mathrm{T}}$$

From (A2.18)

$$\boldsymbol{x} = \boldsymbol{N}_1^{-1}\boldsymbol{b}_1$$

therefore

$$\mathbf{d}\boldsymbol{x} = \boldsymbol{N}_1^{-1}\mathbf{d}\boldsymbol{b}_1$$

$$\mathbf{d}\boldsymbol{x} = -\boldsymbol{N}_1^{-1}\,\boldsymbol{A}^{\mathrm{T}}\boldsymbol{N}^{-1}\boldsymbol{C}\boldsymbol{V}$$

The third term then is

$$\boldsymbol{G}\,\mathrm{E}(\boldsymbol{V}\mathbf{d}\boldsymbol{x}^{\mathrm{T}})\boldsymbol{K}^{\mathrm{T}} = \boldsymbol{G}\,\mathrm{E}(\boldsymbol{V}(-\boldsymbol{N}_1^{-1}\boldsymbol{A}^{\mathrm{T}}\boldsymbol{N}^{-1}\boldsymbol{C}\boldsymbol{V})^{\mathrm{T}})\boldsymbol{K}^{\mathrm{T}}$$

$$\boldsymbol{G}\,\mathrm{E}(\boldsymbol{V}\mathbf{d}\boldsymbol{x}^{\mathrm{T}})\boldsymbol{K}^{\mathrm{T}} = -\boldsymbol{G}\boldsymbol{D}_{\mathrm{o}}\boldsymbol{C}^{\mathrm{T}}\boldsymbol{N}^{-1}\boldsymbol{A}\boldsymbol{D}_x\boldsymbol{K}^{\mathrm{T}}$$

which reduces to

$$-\boldsymbol{K}\boldsymbol{D}_x\boldsymbol{K}^{\mathrm{T}}$$

This is symmetric and equals the fourth term transposed.

A common simplification

In the case of the simple model in which the $\boldsymbol{C}$ matrix is the unit matrix, it is easy to show that the general result in (A2.33) reduces to the much simpler expression

$$\boldsymbol{D}_v = \boldsymbol{D}_o - \boldsymbol{A}\boldsymbol{D}_x\boldsymbol{A}^{\mathrm{T}} \tag{A2.34}$$

A2.8 Dispersion matrix of the estimated observed parameters $\boldsymbol{D}_s$

The dispersion matrix $\boldsymbol{D}_s$ of the Least Squares estimates of the observed parameters is given by the simple expression

$$\boldsymbol{D}_s = \boldsymbol{D}_o - \boldsymbol{D}_v \tag{A2.35}$$

where $\boldsymbol{D}_v$ is given by (A2.33) for the conditioned model and by (A2.34) for the simple model. In this latter case we have the even simpler expression

$$\boldsymbol{D}_s = \boldsymbol{A}\boldsymbol{D}_x\boldsymbol{A}^{\mathrm{T}} \tag{A2.36}$$

The proof of (A2.35) follows similar lines to that for (A2.33) in which some heavy, but straightforward, matrix manipulation is involved. The basic steps are as follows.

The best estimates of the observed parameters are given by

$$\boldsymbol{s} = \boldsymbol{v} + \mathring{\boldsymbol{s}}$$

$$\mathbf{d}\boldsymbol{s} = \mathbf{d}\boldsymbol{v} + \mathbf{d}\mathring{\boldsymbol{s}}$$

but from (A2.32)

$$\boldsymbol{dv} = -\boldsymbol{GdL} - \boldsymbol{Kdx} \tag{A2.32}$$

therefore

$$\mathbf{d}\boldsymbol{s} = -\boldsymbol{G}\mathbf{d}\boldsymbol{L} - \boldsymbol{K}\mathbf{d}\boldsymbol{x} + \mathbf{d}\mathring{\boldsymbol{s}}$$

but

$$\mathbf{d}\boldsymbol{L} = \mathbf{d}\mathring{\boldsymbol{s}} = \boldsymbol{V}$$

therefore

$$\mathbf{d}\boldsymbol{s} = (\boldsymbol{I} - \boldsymbol{G})\boldsymbol{V} - \boldsymbol{K}\mathbf{d}\boldsymbol{x}$$

Substituting for

$$\mathbf{d}\boldsymbol{x} = -\boldsymbol{N}_1^{-1}\boldsymbol{A}^{\mathrm{T}}\boldsymbol{N}^{-1}\boldsymbol{C}\boldsymbol{V}$$

gives the expression

$$\mathbf{d}\boldsymbol{s} = (\boldsymbol{I} - \boldsymbol{G} + \boldsymbol{K}\boldsymbol{N}_1^{-1}\boldsymbol{A}^{\mathrm{T}}\boldsymbol{N}^{-1}\boldsymbol{C})\boldsymbol{V}$$

or say

$$\mathbf{d}\boldsymbol{s} = \boldsymbol{F}\boldsymbol{V}$$

where

$$\boldsymbol{F} = (\boldsymbol{I} - \boldsymbol{G} + \boldsymbol{K}\boldsymbol{N}_1^{-1}\boldsymbol{A}^{\mathrm{T}}\boldsymbol{N}^{-1}\boldsymbol{C})$$

Now

$$\mathbf{dsds}^{\mathrm{T}} = \boldsymbol{FVV}^{\mathrm{T}}\boldsymbol{F}^{\mathrm{T}}$$

Taking expectations

$$\boldsymbol{D}_s = \boldsymbol{FD}_0\boldsymbol{F}^{\mathrm{T}}$$

and substituting for $\boldsymbol{F}$ we obtain, after some heavy algebra,

$$\boldsymbol{D}_s = \boldsymbol{D}_0 - \boldsymbol{GD}_0\boldsymbol{G}^{\mathrm{T}} + \boldsymbol{KD}_x\boldsymbol{K}^{\mathrm{T}}$$

which from (A2.33) gives

$$\boldsymbol{D}_s = \boldsymbol{D}_0 - \boldsymbol{D}_v \qquad \text{(A2.35)}$$

SUMMARY OF FORMULAE

$$\boldsymbol{V}^{\mathrm{T}}\boldsymbol{WV} = \mathrm{trace}(\boldsymbol{VV}^{\mathrm{T}}\boldsymbol{W}) \qquad \text{(A2.1)}$$

$$\boldsymbol{v}^{\mathrm{T}}\boldsymbol{WA} = \mathbf{0} \qquad \text{(A2.7)}$$

$$\boldsymbol{A}^{\mathrm{T}}\boldsymbol{Wv} = \mathbf{0} \qquad \text{(A2.8)}$$

Observations $$\boldsymbol{Ax} + \boldsymbol{Cv} + \boldsymbol{CL} = \mathbf{0} \qquad \text{(A2.11)}$$

$$\boldsymbol{Ax} + \boldsymbol{Cv} = \boldsymbol{b} \qquad \text{(A2.12)}$$

where $\boldsymbol{b} = -\boldsymbol{CL}$

If $$\boldsymbol{N} = \boldsymbol{CW}^{-1}\boldsymbol{C}^{\mathrm{T}} \quad \text{then} \quad \boldsymbol{A}^{\mathrm{T}}\boldsymbol{N}^{-1}\boldsymbol{Cv} = \mathbf{0} \qquad \text{(A2.19)}$$

If $$\boldsymbol{N}_1 = \boldsymbol{A}^{\mathrm{T}}\boldsymbol{N}^{-1}\boldsymbol{A} = \boldsymbol{A}^{\mathrm{T}}(\boldsymbol{CW}^{-1}\boldsymbol{C}^{\mathrm{T}})^{-1}\boldsymbol{A}$$

and $$\boldsymbol{b}_1 = \boldsymbol{A}^{\mathrm{T}}\boldsymbol{N}^{-1}\boldsymbol{b} = \boldsymbol{A}^{\mathrm{T}}(\boldsymbol{CW}^{-1}\boldsymbol{C}^{\mathrm{T}})^{-1}\boldsymbol{b}$$

Normals $$\boldsymbol{N}_1\boldsymbol{x} = \boldsymbol{b}_1 \qquad \text{(A2.18)}$$

Simplification if $$\boldsymbol{C} = \boldsymbol{I} \quad \text{then} \quad \boldsymbol{A}_1\boldsymbol{x} + \boldsymbol{L}_1 = \boldsymbol{v}$$

$$\boldsymbol{A}_1^{\mathrm{T}}\boldsymbol{WA}_1\boldsymbol{x} + \boldsymbol{A}_1^{\mathrm{T}}\boldsymbol{WL}_1 = \mathbf{0}$$

$$\boldsymbol{N}_3\boldsymbol{x} = \boldsymbol{b}_4 \qquad \text{(A2.23)}$$

Constraints $$\boldsymbol{N}_1\boldsymbol{x} + \boldsymbol{E}^{\mathrm{T}}\boldsymbol{k} = \boldsymbol{b}_1 \qquad \text{(A2.21)}$$

$$\boldsymbol{Ex} + \boldsymbol{d} = \mathbf{0} \qquad \text{(A2.20)}$$

$$\boldsymbol{D}_{\boldsymbol{x}} = \sigma_o^2 \boldsymbol{N}^{-1} \tag{A2.25}$$

$$\sigma_o^2 = \mathrm{E}\left(\frac{\boldsymbol{v}^{\mathrm{T}} \boldsymbol{W} \boldsymbol{v}}{m-n}\right) \tag{A2.26}$$

$$\boldsymbol{D}_{\boldsymbol{v}} = \boldsymbol{G}\boldsymbol{D}_{\mathrm{o}}\boldsymbol{G}^{\mathrm{T}} - \boldsymbol{K}\boldsymbol{D}_{\boldsymbol{x}}\boldsymbol{K}^{\mathrm{T}} \tag{A2.33}$$

Simple case

$$\boldsymbol{D}_{\boldsymbol{v}} = \boldsymbol{D}_{\mathrm{o}} - \boldsymbol{A}\boldsymbol{D}_{\boldsymbol{x}}\boldsymbol{A}^{\mathrm{T}} \tag{A2.34}$$

$$\boldsymbol{D}_{\boldsymbol{s}} = \boldsymbol{D}_{\mathrm{o}} - \boldsymbol{D}_{\boldsymbol{v}} \tag{A2.35}$$

Appendix A3: Notation for Least Squares

Author	Observations matrix	Weight matrix	Normal equations matrix	Dispersion matrix
Allan	$\boldsymbol{Ax + L = v}$ $\boldsymbol{Ax + Cv + L = 0}$	$\boldsymbol{W}$	$\boldsymbol{Nx = b}$	$\boldsymbol{D}$
Cooper (1)	$\boldsymbol{Ax = b + v}$ $\boldsymbol{Ax + Bv = b}$	$\boldsymbol{W}$	$\boldsymbol{A^tWx = A^tb}$ Note the lower case t	$\boldsymbol{Q}$
American Manual of Photogrammetry (2)	$\boldsymbol{B\Delta - l = v}$ $\boldsymbol{B\Delta - l = Av}$	$\boldsymbol{P}$	$\boldsymbol{B^TP\Delta = B^TPl}$	$\boldsymbol{Q}$
Mikhail (3)	$\boldsymbol{A(l + v) = d}$ $\boldsymbol{Av + B\Delta = f}$ $\boldsymbol{C\Delta = g}$	$\boldsymbol{W}$	$\boldsymbol{N}$	$\boldsymbol{Q}$
Wolf (4)	$\boldsymbol{AX = L + V}$	$\boldsymbol{W}$	$\boldsymbol{Nx = A^TL}$	$\boldsymbol{Q}$
Koch (5)	$\boldsymbol{X\beta = y + e}$	$\boldsymbol{P}$	$\boldsymbol{X^TX\beta = X^Ty}$	$\boldsymbol{D}$

References

(1) M.A.R. Cooper, 1987, *Control Surveys in Civil Engineering*, Collins (ISBN 0-00-383183-3, 381 pages)

(2) American Society of Photogrammetry, 1966, *Manual of Photogrammetry* Library of Congress Catalog No 65–20813 Vol 1

(3) E.M. Mikhail, 1976, *Observations and Least Squares*, Dun-Donnelley (ISBN 0-7002-2481-5, 497 pages)

(4) P.R. Wolf and C.D. Ghilani, 1997, *Adjustment Computations*, Wiley (ISBN 0-471-16833-5, 564 pages)

(5) K.R. Koch, 1997, *Parameter Estimation and Hypothesis Testing in Linear Models*, Springer (ISBN 3–540–65257–4, 325 pages)

(6) M.A.R. Cooper and P.A. Cross, 1988, Statistical Concepts and their Application in Photogrammetry and Surveying, *Photogrammetric Record*, Vol XIII, No 73, 645–678

(7) A.L. Allan and N. Atkinson, 'Back to Basics' Series Nos 14 to 24 – Least Squares Statistics and all that, *Survey Review*, Vols 35 and 36 Nos 272 to 282

Appendix A4: The Error Ellipse and its Pedal Curve

particularly useful application of an ellipse and its pedal curve is to display me results of an error analysis in two dimensions. The pedal curve, which ows graphically the size and direction of an error function, in the vicinity a point, is useful in presenting results to clients, and in design studies, pecially of networks.

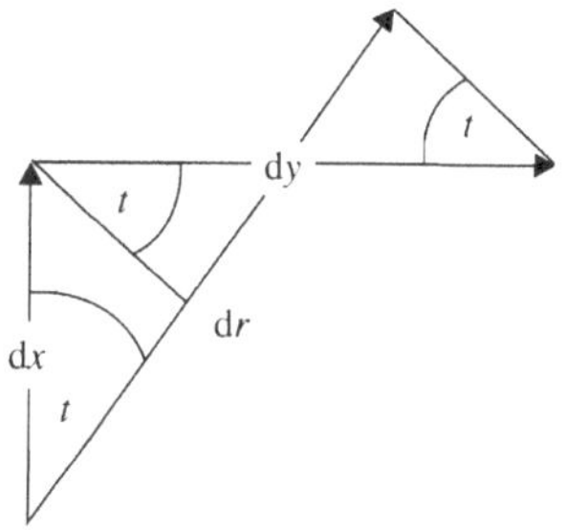

Figure A4.1

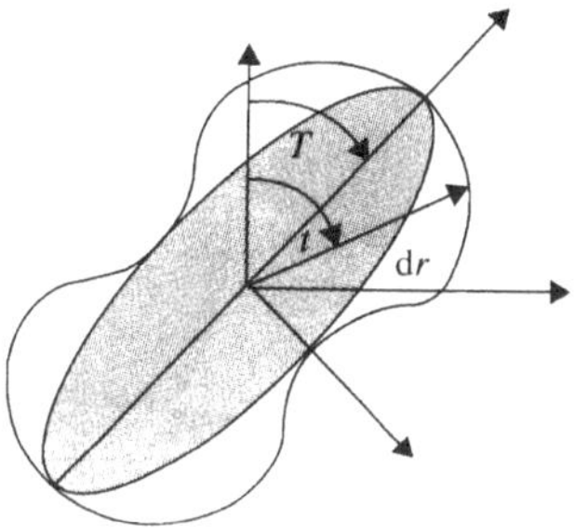

Figure A4.2

Figure A4.1 shows displacements dx and dy and their compounded effect at a bearing t. If dr rotates through a full circle about one end, the locus of e free end is a doughnut-shaped curve, the pedal curve to an ellipse. This is own in figure A4.2. As will be seen shortly, these displacements dx and dy n also be represented by standard errors derived from the dispersion matrix tained in a Least Squares analysis. By inspection, from the figure we have

$$\mathrm{d}r = \mathrm{d}x \cos t + \mathrm{d}y \sin t$$

erefore

$$\mathrm{d}r^2 = \mathrm{d}x^2 \cos^2 t + \mathrm{d}y^2 \sin^2 t + 2\mathrm{d}x\mathrm{d}y \sin t \cos t \qquad \text{(A4.1)}$$

ow dr can be taken to represent the component in direction t of the siduals dx = v_x and dy = v_y, so taking expectations we have

$$\sigma_r^2 = \sigma_x^2 \cos^2 t + \sigma_y^2 \sin^2 t + 2\sigma_{xy} \sin t \cos t \qquad \text{(A4.2)}$$

which may be recast as

$$2\sigma_r^2 = \sigma_x^2 + \sigma_y^2 + (\sigma_x^2 - \sigma_y^2)\cos 2t + 2\sigma_{xy} \sin 2t \qquad \text{(A4.3)}$$

We now wish to find the turning values (maximum and minimum) of this function. Differentiating (A4.3) with respect to t we have, putting $F = 2\sigma_r^2$,

$$\frac{dF}{dt} = -2(\sigma_x^2 - \sigma_y^2)\sin 2t + 4\sigma_{xy} \cos 2t \qquad \text{(A4.4)}$$

The turning values are obtained when $t = T$, and $dF/dt = 0$, i.e. when

$$0 = -2(\sigma_x^2 - \sigma_y^2)\sin 2T + 4\sigma_{xy} \cos 2T$$

$$\frac{\sin 2T}{\cos 2T} = \frac{2\sigma_{xy}}{\sigma_x^2 - \sigma_y^2}$$

$$\frac{\Delta \sin 2T}{\Delta \cos 2T} = \frac{2\sigma_{xy}}{\sigma_x^2 - \sigma_y^2}$$

where Δ is a positive constant. Therefore we may separate the numerator and the denominator as follows

$$\Delta \sin 2T = 2\sigma_{xy} \qquad \text{(A4.5)}$$

$$\Delta \cos 2T = \sigma_x^2 - \sigma_y^2 \qquad \text{(A4.6)}$$

Substituting from (A4.5) and (A4.6) in (A4.3) we have, after some rearranging,

$$F = 2\sigma_r^2 = \sigma_x^2 + \sigma_y^2 + \Delta \cos 2(t - T) \qquad \text{(A4.7)}$$

F is a maximum and a minimum respectively when

$$\cos 2(t - T) = +1 \text{ and } \cos 2(t - T) = -1$$

That is σ_r has maximum and minimum values given by

$$F_{max} = 2\sigma_{max}^2 = \sigma_x^2 + \sigma_y^2 + \Delta \qquad \text{(A4.8)}$$

$$F_{min} = 2\sigma_{min}^2 = \sigma_x^2 + \sigma_y^2 - \Delta \qquad \text{(A4.9)}$$

Adding and subtracting gives

$$\sigma^2_{max} + \sigma^2_{min} = \sigma^2_x + \sigma^2_y \qquad \text{(A4.10)}$$

$$\sigma^2_{max} - \sigma^2_{min} = \Delta \qquad \text{(A4.11)}$$

$$F = 2\sigma^2_r = \sigma^2_x + \sigma^2_y + \Delta \cos 2(t - T) \qquad \text{(A4.7)}$$

$$= \sigma^2_{max} + \sigma^2_{min} + (\sigma^2_{max} - \sigma^2_{min}) \cos 2(t - T) \qquad \text{(A4.12)}$$

This is the equation of the pedal curve to the ellipse whose semi-axes are respectively σ_{max} and σ_{min} with its axes oriented with respect to the original axes by the bearing T. From Chapter 10 equation (10.27) the pedal distance p to an ellipse whose semi-axes are a and b is

$$p^2 = a^2 \cos^2 U + b^2 \sin^2 U = \frac{1}{2} a^2 (1 + \cos 2U) + \frac{1}{2} b^2 (1 - \cos 2U)$$
$$= \frac{1}{2}(a^2 + b^2) + \frac{1}{2}(a^2 - b^2) \cos 2U$$

Hence we see the analogy between σ_r and p, U and $(t - T)$. From (A4.12), we have

$$U = (t - T) \quad \text{and} \quad a = \sigma_{max} \quad \text{and} \quad b = \sigma_{min}$$

The calculations are shown in Table A4.1 using the data from Chapter 13. Cells A1 to B2 are the elements of the inverse of N, its variances and covariance, which are used to calculate T from

$$\tan 2T = \frac{2\sigma_{EN}}{\sigma^2_N - \sigma^2_E} = \frac{(2x - 1.014\text{E}{-}05)}{(2.9512\text{E}{-}05 - 0.00025508)} = 0.089906$$
$$2T = 185.137^\circ, \quad T = 92.57^\circ \text{ or } -87.43^\circ$$

($2T$ is in the third quadrant because both its sine and cosine are negative.) Then we calculate Δ from either of the two equations

$$\Delta = 2\sigma_{EN} \operatorname{cosec} 2T = 0.00022648$$

or

$$\Delta = (\sigma^2_N - \sigma^2_E) \sec 2T = 0.00022648$$

and the semi-major and semi-minor axes, a and b, of the ellipse from (A4.8) and (A4.9). The calculations are shown in Table A4.1 giving a major axis a = 0.016 m oriented at a bearing of 95°, b = 0.005 m.

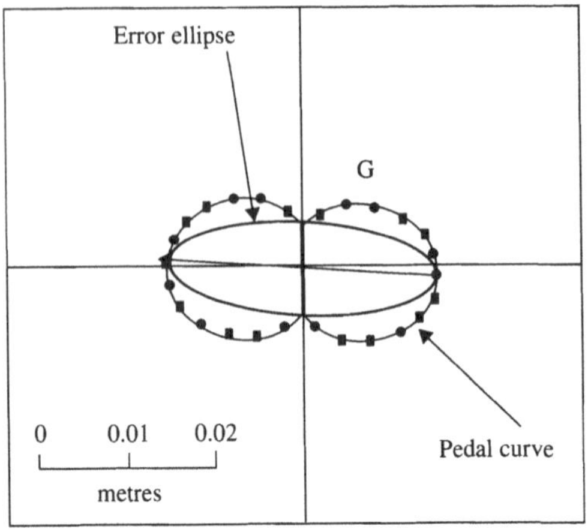

Figure A4.3

Table A4.1 Error ellipse

	A	B	C
1	0.00025508	–1.014E–05	Bearing
2	–1.014E–05	2.9512E–05	
3	Delta	0.00022648	$T°$
			–87.430461
4	Max	0.00051107	$a = 0.016$
5	Min	5.8114E–05	$b = 0.005$
6	0.01597116		
7		0.00543249	

It is quite tricky to plot the ellipse and its pedal curve by hand. It is easier to use a computer plotting routine for this. But a rough sketch is really all that is needed, as shown in Figure A4.3. This was plotted using the *MacDraw package* as if plotting by hand. To obtain the standard error at any direction all we need to do is scale it from the figure. For example, we can verify that the pedal curve cuts the axes at the correct values: in this case of

$\sigma_E = 0.016$, from cell A6 the square root of A2

and

$\sigma_N = 0.005$, from cell B7 the square root of B3.

In this case the ellipse is so aligned that this check is weak.

As with all Least Squares estimation, care is needed when using error ellipses, because they are influenced by the datum selected for the calculation of coordinates. Ellipses showing the relative errors between pairs of points often convey more useful information.

SUMMARY OF FORMULAE

$$F = 2\sigma_r^2 = \sigma_x^2 + \sigma_y^2 + \Delta \cos 2(t - T) \qquad \text{(A4.7)}$$

$$= \sigma_{\max}^2 + \sigma_{\min}^2 + (\sigma_{\max}^2 - \sigma_{\min}^2)\cos 2(t - T) \qquad \text{(A4.12)}$$

$$F_{\max} = 2\sigma_{\max}^2 = \sigma_x^2 + \sigma_y^2 + \Delta \qquad \text{(A4.8)}$$

$$F_{\min} = 2\sigma_{\min}^2 = \sigma_x^2 + \sigma_y^2 - \Delta \qquad \text{(A4.9)}$$

$$\tan 2T = \frac{2\sigma_{EN}}{\sigma_N^2 - \sigma_E^2}$$

$$\Delta = 2\sigma_{EN}\operatorname{cosec}2T$$

or

$$\Delta = (\sigma_N^2 - \sigma_E^2)\sec 2T$$

Summary of Formulae

Miscellaneous functions

Indices: $$A^n \times A^m = A^{n+m} \quad (1.2)$$

Ratios: $$\text{If } \frac{A}{B} = \frac{D}{E} \text{ then } \frac{A+B}{B} = \frac{D+E}{E} \quad (1.10)$$

Logarithms: natural base, e = 'Ln'; Briggs base 10 = 'Log'

$$\log NM = \log N + \log M \quad (1.3)$$

$$\log \frac{N}{M} = \log N - \log M \quad (1.5)$$

$$\log_B P^k = k \log_B P \quad (1.6)$$

Normal distribution function: $$f(x) = y = \frac{h}{\sqrt{\pi}} e^{-h^2 x^2} \quad (9.47)$$

Stereographic projection: $$r_F = 2R \tan \tfrac{1}{2} p_F \quad (11.16)$$

$$K_2 = \sec^2 \tfrac{1}{2} p \quad (11.25)$$

Series

Arithmetic progression: $$S = \frac{n}{2}\left[2a + (n-1)d\right] \quad (1.7)$$

Sum of first n integers: $$S = \frac{n}{2}(n+1) \quad (1.8)$$

Sum of squares of integers: $$S = \frac{n}{6}(n+1)(2n+1) \quad (1.9)$$

$$(a + b)^2 = a^2 + 2ab + b^2 \quad (1.13)$$

$$(a - b)^2 = a^2 - 2ab + b^2 \quad (1.15)$$

$$(a + b)^3 = a^3 + 3a^2b + 3ab^2 + b^3 \quad (1.17)$$

Binomial theorem

$$(a + b)^n = a^n + C_1a^{n-1}b + C_2a^{n-2}b^2 + \ldots + C_ra^{n-r}b^r + \ldots + C_{n-1}ab^{n-1} + b^n \quad (1.19)$$

$$\text{where } C_r = \frac{n(n-1)(n-2)\ldots(n-r+1)}{r!} = \frac{n!}{r!(n-r)!} \quad (1.20)$$

$$(1+x)^n = 1 + nx + \frac{n(n-1)}{2!}x^2 + \frac{n(n-1)(n-2)}{3!}x^3 + \ldots \quad (1.22)$$

$$e^x = 1 + x + \frac{x^2}{2!} + \frac{x^3}{3!} + \ldots + \frac{x^n}{n!} \quad (1.4)$$

$$\sin x = x - \frac{x^3}{3!} + \frac{x^5}{5!} - \frac{x^7}{7!} - \ldots \quad (9.12)$$

$$\cos x = 1 - \frac{x^2}{2!} + \frac{x^4}{4!} - \frac{x^6}{6!} + \ldots \quad (9.13)$$

$$\sec x = 1 + \frac{1}{2}x^2 + \frac{5}{24}x^4 + \frac{61}{720}x^6 + \ldots \quad (9.31)$$

$$\ln(1+x) = x - \frac{x^2}{2} + \frac{x^3}{3} - \frac{x^4}{4} \ldots \quad (9.18)$$

Maclaurin's theorem

$$y = F(x)$$

$$= F(0) + F'(0)x + \frac{F''(0)}{2!}x^2 + \frac{F'''(0)}{3!}x^3 + \frac{F''''(0)}{4!}x^4 + \ldots \frac{F^n(0)}{n!}x^n \quad (9.7)$$

Taylor's theorem

$$y = F(a) + \frac{F'(a)}{1!}h + \frac{F''(a)}{2!}h^2 + \frac{F'''(a)}{3!}h^3 + \frac{F''''(a)}{4!}h^4 + \ldots \frac{F^n(a)}{n!}h^n \quad (9.20)$$

Trigonometry

Pythagoras's theorem: in triangle *ABC*, right angle at *B*

$$b^2 = c^2 + a^2 \qquad (2.4)$$

Cosine formula:

$$a^2 = b^2 + c^2 - 2\,bc\cos A \qquad (3.5)$$

$$\cos^2\theta + \sin^2\theta = 1 \qquad (3.8)$$

Sine formua:

$$\frac{a}{\sin A} = \frac{b}{\sin B} = \frac{c}{\sin C} = 2R \qquad (3.9)$$

$$\sin(A + B) = \sin A\cos B + \cos A\sin B \qquad (3.12)$$

$$\cos(A + B) = \cos A\cos B - \sin A\sin B \qquad (3.13)$$

$$\sin(A - B) = \sin A\cos B - \cos A\sin B \qquad (3.14)$$

$$\cos(A - B) = \cos A\cos B + \sin A\sin B \qquad (3.15)$$

$$\tan(A+B) = \frac{\tan A + \tan B}{1 - \tan A\tan B} \qquad (3.16)$$

$$\tan(A-B) = \frac{\tan A - \tan B}{1 + \tan A\tan B} \qquad (3.17)$$

$$\sin 2A = 2\sin A\cos A \qquad (3.18)$$

$$\cos 2A = \cos^2 A - \sin^2 A \qquad (3.19)$$

$$\cos 2A = 2\cos^2 A - 1 \qquad (3.20)$$

$$\cos 2A = 1 - 2\sin^2 A \qquad (3.21)$$

$$\cos 3A = 4\cos^3 A - 3\cos A \qquad (3.22)$$

$$\sin 3A = 3\sin A - 4\sin^3 A \qquad (3.23)$$

$$\sin A + \sin B = 2\sin\tfrac{1}{2}(A+B)\cos\tfrac{1}{2}(A-B) \qquad (3.24)$$

$$\sin A - \sin B = 2\cos\tfrac{1}{2}(A+B)\sin\tfrac{1}{2}(A-B) \qquad (3.25)$$

$$\cos A + \cos B = 2\cos\tfrac{1}{2}(A+B)\cos\tfrac{1}{2}(A-B) \qquad (3.26)$$

$$\cos A - \cos B = -2\sin\tfrac{1}{2}(A+B)\sin\tfrac{1}{2}(A-B) \qquad (3.27)$$

In triangle ABC where $2s = a + b + c$

$$\sin^2\tfrac{1}{2}A = \frac{(s-b)(s-c)}{bc} \qquad (3.28)$$

$$\cos^2\tfrac{1}{2}A = \frac{s(s-a)}{bc} \qquad (3.29)$$

$$\tan^2\tfrac{1}{2}A = \frac{(s-b)(s-c)}{s(s-a)} \qquad (3.30)$$

$$\text{Area}\ \Delta ABC = \sqrt{[s(s-a)(s-b)(s-c)]} \qquad (3.31)$$

Cartesian and polar coordinates in two dimensions

Given (r, U), find x and y from

$$x = r \cos U \quad y = r \sin U \tag{4.1}$$

Given (x, y), find U and r from

$$\tan U = \frac{y}{x} \tag{4.2}$$

and

$$r = \sqrt{(x^2 + y^2)} \tag{4.3}$$

A useful check is

$$r = x \cos U + y \sin U \tag{4.4}$$

Coordinate differences in two dimensions

$$\Delta x = r \cos U \quad \Delta y = r \sin U \tag{4.6}$$

$$\tan U = \frac{\Delta y}{\Delta x} \tag{4.7}$$

$$r = \sqrt{(\Delta x^2 + \Delta y^2)} \tag{4.8}$$

$$r = \Delta x \cos U + \Delta y \sin U \tag{4.9}$$

Equations of a straight line in two dimensions

Gradient form: $$y = mx + c \tag{4.10}$$

Polar form: $$L'x + M'y = p \tag{4.13}$$

Parametric forms: $$x = a + t \cos U \quad y = b + t \sin U \tag{4.14}$$

$$x = a + tL \quad y = b + tM \tag{4.15}$$

Direction cosines in two dimensions

$$L = \cos U \qquad M = \sin U$$

$$L^2 + M^2 = 1 \tag{3.7}$$

Angle between lines: $$\cos W = L_1L_2 + M_1M_2 \tag{4.22}$$

Orthogonal lines: $$L_1L_2 + M_1M_2 = 0 \tag{4.23}$$

Parallel lines: $$L_1L_2 + M_1M_2 = 1 \tag{4.24}$$

The Circle

radius of circle $r = \frac{1}{2}d$ diameter (2.5)

circumference / diameter $\frac{c}{d} = \pi$ constant pi (2.6)

$$\alpha \text{ rad} = \alpha^\circ \times \pi/180^\circ \qquad (2.7)$$

$$\alpha^\circ = \alpha \text{ rad} \times 180^\circ/\pi \qquad (2.8)$$

Equation of a circle: $x^2 + y^2 + 2gx + 2fy + c = 0$ (4.25)

Length of tangent: $x_E{}^2 + y_E{}^2 + 2gx_E + 2fy_E + c = t^2$ (4.28)

Eqn of tangent at F: $(x_F + g)\,x + (y_F + f)\,y + (gx_F + fy_F + c) = 0$ (4.30)

Coordinate transformations in three dimensions

$$x = r \cos V \cos U \quad y = r \cos V \sin U \quad z = r \sin V \qquad (5.2)$$

$$r = \sqrt{(x^2 + y^2 + z^2)} \qquad (5.3)$$

$$\tan U = \frac{y}{x} \qquad (5.4)$$

$$\tan V = \frac{z}{\sqrt{\left(x^2 + y^2\right)}} \qquad (5.5)$$

A useful check is $r = (x \cos U + y \sin U) \cos V + z \sin V$ (5.6)

Coordinate differences in three dimensions

$$\Delta x = x_B - x_A \text{ for } x \quad \Delta y = y_B - y_A \text{ for } y \quad \Delta z = z_B - z_A \text{ for } z \qquad (5.7)$$

$$\Delta x = r \cos V \cos U \quad \Delta y = r \cos V \sin U \quad \Delta z = \sin V \qquad (5.8)$$

$$r = \sqrt{(\Delta x^2 + \Delta y^2 + \Delta z^2)} \qquad (5.9)$$

$$\tan U = \frac{\Delta y}{\Delta x} \qquad (5.10)$$

$$\tan V = \frac{\Delta z}{\sqrt{\left(\Delta x^2 + \Delta y^2\right)}} \qquad (5.11)$$

A useful check is

$$r = (\Delta x \cos U + \Delta y \sin U) \cos V + \Delta z \sin V \qquad (5.12)$$

$$AB^2 = (X_B - X_A)^2 + (Y_B - Y_A)^2 + (Z_B - Z_A)^2 \qquad (5.1)$$

Direction cosines and lines in three dimensions

$$L^2 + M^2 + N^2 = 1 \quad (5.14)$$

$$X_P = X_A + t.L$$
$$Y_P = Y_A + t.M$$
$$Z_P = Z_A + t.N \quad (5.18)$$

$$\cos\theta = LL' + MM' + NN' \quad (5.19)$$

$$AD = (X_B - X_A)L' + (Y_B - Y_A)M' + (Z_B - Z_A)N' \quad (5.21)$$

$$BD = AB\sqrt{((LM' - ML')^2 + (MN' - NM')^2 + (NL' - LN')^2)} \quad (5.22)$$

Equation of a plane in three dimensions

$$LX + MY + NZ = P \quad (5.28)$$

Equation of a sphere

$$x^2 + y^2 + z^2 + 2ex + 2fy + 2gz + h = 0 \quad (5.30)$$

Areas and volumes

$$\text{Area}\,\Delta\,ABC = \sqrt{[s(s-a)(s-b)(s-c)]} \quad (3.31)$$
$$\Delta = \tfrac{1}{2}bc\sin A \quad (6.1)$$

Simpson's rule: $A = \frac{1}{6}(y_1 + 4y_m + y_2)(x_2 - x_1)$ (9.42)

Area of sector of circle: $A_S = \frac{1}{2}R^2\theta$ (9. 43)

Surface area of sphere $A = 4\pi R^2$ (9.44)

Volume of sphere: $V = \frac{4}{3}\pi R^3$ (9.45)

Area by determinants

$$D = x_A y_B - x_B y_A \quad (6.4)$$

$$D = \begin{vmatrix} x_A & y_A & 1 \\ x_B & y_B & 1 \\ x_C & y_C & 1 \end{vmatrix}$$

$$D = y_A x_B + y_B x_C + y_C x_A - x_C y_A - x_A y_B - x_B y_C \qquad (6.5)$$

$$D = x_A D_1 - y_A D_2 + 1.D_3$$

where

$$D_1 = \begin{vmatrix} y_B & 1 \\ y_C & 1 \end{vmatrix} \qquad D_2 = \begin{vmatrix} x_B & 1 \\ x_C & 1 \end{vmatrix} \qquad D_3 = \begin{vmatrix} x_B & y_B \\ x_C & y_C \end{vmatrix}$$

Volume: $$V = \tfrac{1}{3}\Delta H \qquad (6.7)$$

$$6V = -\begin{vmatrix} x_A & y_A & z_A & 1 \\ x_B & y_B & z_B & 1 \\ x_C & y_C & z_C & 1 \\ x_D & y_D & z_D & 1 \end{vmatrix} \qquad (6.15)$$

Matrices

Addition: $$\boldsymbol{C} = \boldsymbol{A} + \boldsymbol{B} \qquad (7.11)$$

Subtraction: $$\boldsymbol{D} = \boldsymbol{A} - \boldsymbol{B} \qquad (7.12)$$

Associative law: $$\boldsymbol{A}(\boldsymbol{B} + \boldsymbol{C}) = \boldsymbol{AB} + \boldsymbol{AC} \qquad (7.13)$$

Multiplication: $$\boldsymbol{Ax} = \boldsymbol{b} \qquad (7.3)$$

Division: $$\boldsymbol{x} = \boldsymbol{A}^{-1}.\boldsymbol{b} \qquad (7.4)$$

Inversion: $$\boldsymbol{A}.\boldsymbol{A}^{-1} = \boldsymbol{A}^{-1}.\boldsymbol{A} = \boldsymbol{I} \qquad (7.5)$$

$$\boldsymbol{C} = \boldsymbol{AB} \text{ then } \boldsymbol{C}^{-1} = \boldsymbol{B}^{-1}\boldsymbol{A}^{-1} \qquad (7.15)$$

$$\boldsymbol{D} = \boldsymbol{ABC} \text{ then } \boldsymbol{D}^{-1} = \boldsymbol{C}^{-1}\boldsymbol{B}^{-1}\boldsymbol{A}^{-1} \qquad (7.17)$$

Transpose: $$\boldsymbol{C} = \boldsymbol{AB} \text{ then } \boldsymbol{C}^{\mathrm{T}} = \boldsymbol{B}^{\mathrm{T}}\boldsymbol{A}^{\mathrm{T}} \qquad (7.14)$$

$$\boldsymbol{D} = \boldsymbol{ABC} \text{ then } \boldsymbol{D}^{\mathrm{T}} = \boldsymbol{C}^{\mathrm{T}}\boldsymbol{B}^{\mathrm{T}}\boldsymbol{A}^{\mathrm{T}} \qquad (7.16)$$

Length of vector: $$S^2 = \boldsymbol{x}^{\mathrm{T}}\boldsymbol{x} \qquad (7.18)$$

Orthogonal matrix $$\boldsymbol{R}^{\mathrm{T}} = \boldsymbol{R}^{-1} \qquad (7.26)$$

Vectors

$$\boldsymbol{a} = |\boldsymbol{a}|\bar{\boldsymbol{a}} \qquad (8.1)$$

Components: $$\boldsymbol{a} = \boldsymbol{a}_x + \boldsymbol{a}_y + \boldsymbol{a}_z \qquad (8.2)$$

$$= |\boldsymbol{a}_x|\bar{\boldsymbol{a}}_x + |\boldsymbol{a}_y|\bar{\boldsymbol{a}}_y + |\boldsymbol{a}_z|\bar{\boldsymbol{a}}_z$$

Unit vectors: $$\bar{\boldsymbol{a}}_x = \mathbf{i} \quad \bar{\boldsymbol{a}}_y = \mathbf{j} \quad \bar{\boldsymbol{a}}_z = \mathbf{k} \qquad (8.3)$$

$$\boldsymbol{a}_x = |\boldsymbol{a}|\cos\alpha \quad \boldsymbol{a}_y = |\boldsymbol{a}|\cos\beta \quad \boldsymbol{a}_z = |\boldsymbol{a}|\cos\gamma \qquad (8.4)$$

$$\boldsymbol{a}_x = |\boldsymbol{a}|L\mathbf{i} \quad \boldsymbol{a}_y = |\boldsymbol{a}|M\mathbf{j} \quad \boldsymbol{a}_z = |\boldsymbol{a}|N\mathbf{k} \qquad (8.5)$$

Length of vector: $|\boldsymbol{a}| = \sqrt{(a_x a_x + a_y a_y + a_z a_z)}$ (8.6)

Dot or scalar product: $\boldsymbol{a} \cdot \boldsymbol{b} = |\boldsymbol{a}|\ |\boldsymbol{b}| \cos U$ (8.7)

Orthogonal unit vectors:

$$\mathbf{i} \cdot \mathbf{i} = 1 \quad \mathbf{j} \cdot \mathbf{j} = 1 \quad \mathbf{k} \cdot \mathbf{k} = 1 \qquad (8.8)$$

$$\mathbf{i} \cdot \mathbf{j} = 0 \quad \mathbf{i} \cdot \mathbf{k} = 0 \quad \mathbf{j} \cdot \mathbf{k} = 0 \qquad (8.9)$$

$$\mathbf{j} \cdot \mathbf{i} = 0 \quad \mathbf{k} \cdot \mathbf{i} = 0 \quad \mathbf{k} \cdot \mathbf{j} = 0 \qquad (8.10)$$

$$\mathbf{a} \cdot \mathbf{b} = (a_x \mathbf{i} + a_y \mathbf{j} + a_z \mathbf{k}) \cdot (b_x \mathbf{i} + b_y \mathbf{j} + b_z \mathbf{k}) \qquad (8.11)$$

$$\mathbf{a} \cdot \mathbf{b} = a_x b_x + a_y b_y + a_z b_z \qquad (8.12)$$

Angle between vectors: $\cos A = \dfrac{\mathbf{p} \cdot \mathbf{q}}{|\boldsymbol{p}||\boldsymbol{q}|}$ (8.14)

Cross or vector product: $\boldsymbol{a} \times \boldsymbol{b} = |\boldsymbol{a}|\ |\boldsymbol{b}| \sin U\ \bar{\boldsymbol{c}}$ (8.15)

$$\boldsymbol{b} \times \boldsymbol{a} = -\boldsymbol{a} \times \boldsymbol{b} \qquad (8.16)$$

Orthogonal unit vectors:

$$\mathbf{i} \times \mathbf{i} = 0 \quad \mathbf{j} \times \mathbf{j} = 0 \quad \mathbf{k} \times \mathbf{k} = 0 \qquad (8.17)$$

$$\mathbf{i} \times \mathbf{j} = \mathbf{k} \quad \mathbf{j} \times \mathbf{k} = \mathbf{i} \quad \mathbf{k} \times \mathbf{i} = \mathbf{j} \qquad (8.18)$$

$$\mathbf{j} \times \mathbf{i} = -\mathbf{k} \quad \mathbf{k} \times \mathbf{j} = -\mathbf{i} \quad \mathbf{i} \times \mathbf{k} = -\mathbf{j} \qquad (8.19)$$

Area of parallelogram: $|\boldsymbol{a} \times \boldsymbol{b}| = |\boldsymbol{a}|\ |\boldsymbol{b}| \sin U$ (8.21)

Scalar triple product vol. of parallelepiped

$$V = \mathbf{a} \cdot \mathbf{d} = \mathbf{a} \cdot (\mathbf{b} \times \mathbf{c}) \qquad (8.22)$$

Calculus:Differentiation

If $y = x^n$ $\quad \dfrac{dy}{dx} = nx^{n-1}$ (9.5)

If $y = F(x) = Ax^n$ $\quad \dfrac{dy}{dx} = F'(x) = nAx^{n-1}$ (9.6)

Product rule: If $y = uv$ $\quad \dfrac{dy}{dx} = u\dfrac{dv}{dx} + v\dfrac{du}{dx}$ (9.22)

Quotient rule: If $y = \frac{u}{v}$ $$\frac{dy}{dx} = \frac{v\frac{du}{dx} - u\frac{dv}{dx}}{v^2} \qquad (9.24)$$

$$\frac{d(\sin x)}{dx} = \cos x \qquad (9.9)$$

$$\frac{d(\cos x)}{dx} = -\sin x \qquad (9.10)$$

$$\frac{d(e^x)}{dx} = e^x \qquad (9.16)$$

$$\frac{d(\ln x)}{dx} = \frac{1}{x} \qquad (9.17)$$

$$\frac{d(\tan x)}{dx} = \sec^2 x \qquad (9.23)$$

Partial differentiation

If $F(x, y, R) = 0$ $$\frac{\partial F}{\partial x}\delta x + \frac{\partial F}{\partial y}\delta y + \frac{\partial F}{\partial R}\delta R = 0 \qquad (9.33)$$

If $a = \frac{b \sin A}{\sin B}$ $$\frac{\delta a}{a} = \frac{\delta b}{b} + \cot A\delta A - \cot B\delta B \qquad (9.35)$$

Curvature of plane curve

$$R = \frac{\left[1 + \left(\frac{dy}{dx}\right)^2\right]^{3/2}}{\frac{d^2y}{dx^2}} \qquad (9.37)$$

Calculus: Integration

$$x = \int dx \qquad (9.41)$$

$$\int x^n dx = \frac{1}{n+1}x^{n+1} + k \qquad (9.39)$$

$$\int \sec x \, dx = \ln \tan\left(\frac{\pi}{4} + \frac{x}{2}\right) + k \qquad (9.40)$$

Conic sections

Circle: $X^2 + Y^2 + 2fX + 2gY + c = 0$ (10.8)

Parabola: $y^2 = 4ax$ (10.14)

Hyperbola: $$\frac{x^2}{a^2} - \frac{y^2}{b^2} = 1 \quad (10.38)$$

Ellipse: $$\frac{x^2}{a^2} + \frac{y^2}{b^2} = 1 \quad (10.19)$$

where $b^2 = a^2(1 - e^2)$

Area of the ellipse: $A = \pi ab$ (10.21)

Pedal distance to ellipse: $p^2 = a^2 (1 - e^2 \sin^2 U)$ (10.27)

Radius of curvature in plane of ellipse:

$$R = \frac{a(1 - e^2)}{(1 - e^2 \sin^2 U)^{3/2}} \quad (10.32)$$

Length of normal to ellipse: $$PC = \frac{a^2}{P} \quad (10.33)$$

Spherical trigonometry

$$\cos a = \cos b \cos c + \sin b \sin c \cos A \quad (11.1)$$

$$\frac{\sin a}{\sin A} = \frac{\sin b}{\sin B} = \frac{\sin c}{\sin C} \quad (11.2)$$

$$\cos A \cos c = \cot b \sin c - \sin A \cot B \quad (11.3)$$

Spherical excess $A + B + C - 180° = e = 11.19°$ (11.11)

$$e'' = \frac{\Delta}{R^2} \times 206\,265 \quad (11.12)$$

Lune $$e = \frac{\Delta}{R^2} \quad (11.13)$$

Solution of equations

Linear equations: $\boldsymbol{N x} = \boldsymbol{b}$ (12.2)

$\boldsymbol{x} = \boldsymbol{N}^{-1}\boldsymbol{b}$ (12.3)

$\boldsymbol{L\,U}$ decomposition: $\boldsymbol{N} = \boldsymbol{L\,U}$ (12.5)

$\boldsymbol{L\,U\,x} = \boldsymbol{b}$ (12.6)

Intermediate vector: $\boldsymbol{U x} = \boldsymbol{f}$ (12.7)

$\boldsymbol{L f} = \boldsymbol{b}$ (12.8)

Quadratic equation: $ax^2 + bx + c = 0$ (12.11)

$$x = \frac{-b \pm \sqrt{\left(b^2 - 4ac\right)}}{2a} \qquad (12.12)$$

Cubic equation: $ax^3 + bx^2 + cx + d = 0$ (12.19)

Newton's method: $F(x, y) \approx F(x', y') + \dfrac{\partial F}{\partial x}\delta x + \dfrac{\partial F}{\partial y}\delta y \approx 0$ (12.29)

Least Squares estimation

Functional models: $F(\boldsymbol{x}: \boldsymbol{s}) = 0$ (13.5)

Normal equations: $\boldsymbol{Nx} = \boldsymbol{b}$ (13.13)

$$\Omega = \boldsymbol{v}^{\mathrm{T}}\boldsymbol{v} \quad (13.21)$$

$$\boldsymbol{A}^{\mathrm{T}}\boldsymbol{v} = \boldsymbol{0}$$

Observation equations: $\boldsymbol{Ax} + \boldsymbol{L} = \boldsymbol{v}$ (13.26)

Normal equations: $\boldsymbol{A}^{\mathrm{T}}\boldsymbol{Ax} + \boldsymbol{A}^{\mathrm{T}}\boldsymbol{L} = \boldsymbol{0}$ (13.27)

Weighted normal equations: $\boldsymbol{A}^{\mathrm{T}}\boldsymbol{WAx} + \boldsymbol{A}^{\mathrm{T}}\boldsymbol{WL} = \boldsymbol{0}$ (13.41)

Sample variance: $s^2 = \dfrac{\Sigma v^2}{m}$ (13.42)

Expectation: $\mathrm{E}(\boldsymbol{V}^{\mathrm{T}}\boldsymbol{V}) = \sigma^2$

Covariance: $\mathrm{covar} = \mathrm{E}(V_A V_B) = \sigma_{AB}$

$$s_{\bar{A}}^2 = \mathrm{E}\left(\frac{V_S^2}{m^2}\right) = \frac{m s_A^2}{m^2} = \frac{s_A^2}{m} \quad (13.44)$$

Weights: $w_A = \dfrac{1}{s_A^2}$ (13.45)

Dispersion matrix: $\boldsymbol{D}_{\mathrm{o}}^{-1} = \boldsymbol{W}$ and $\boldsymbol{D}_{\mathrm{o}} = \boldsymbol{W}^{-1}$ (13.49)

$$\boldsymbol{Adx} = \boldsymbol{V} - \boldsymbol{v} \quad (13.51)$$

$$D_x = s_{\mathrm{o}}^2 N^{-1} \quad (13.53)$$

$$\boldsymbol{VV}^{\mathrm{T}} = \boldsymbol{Av}(\boldsymbol{Av})^{\mathrm{T}} = \boldsymbol{Avv}^{\mathrm{T}}\boldsymbol{A}^{\mathrm{T}} \quad (13.54)$$

Variance: $\sigma_{\mathrm{o}}^2 = \mathrm{E}\left(\dfrac{\boldsymbol{v}^{\mathrm{T}}\boldsymbol{Wv}}{m-n}\right)$ (A2.26)

Special case: $s_{\mathrm{o}}^2 = \dfrac{\boldsymbol{v}^{\mathrm{T}}\boldsymbol{Wv}}{m-1}$ (13.55)

Dispersion matrix of residuals: $\boldsymbol{D}_v = \boldsymbol{D}_{\mathrm{o}} - \boldsymbol{D}_s$ (13.56)

where $\boldsymbol{D}_s = \boldsymbol{AD}_x\boldsymbol{A}^{\mathrm{T}}$ (13.57)

$$\boldsymbol{D}_s = \boldsymbol{A}\boldsymbol{D}_x\boldsymbol{A}^{\mathrm{T}} \tag{13.58}$$

Test ratio:

$$t = \frac{v}{\sigma_v}$$

$$\delta_i^{\mathrm{u}} = \frac{\Delta_i^{\mathrm{u}}}{\sigma_i} \tag{13.59}$$

$$\mathrm{MDE} = \Delta_i^{\mathrm{u}} = \delta_i^{\mathrm{u}}\sigma_i = (a+b)\sigma_i \tag{13.60}$$

$$\mathbf{d}\boldsymbol{x} = \boldsymbol{N}^{-1}\mathbf{A}^{\mathrm{T}}\mathbf{W}\mathbf{d}\boldsymbol{L} \tag{13.61}$$

F test :

$$t = \frac{s_1^2}{s_2^2}$$

General Least Squares:

$$\boldsymbol{Bx} + \boldsymbol{C}\delta\boldsymbol{S} = 0 \tag{13.65}$$

$$\boldsymbol{C}^{-1}\boldsymbol{Bx} + \delta\boldsymbol{S} = 0 \tag{13.66}$$

$$\boldsymbol{Ax} + \boldsymbol{Cv} + \boldsymbol{CL} = 0 \tag{A2.11}$$

$$\boldsymbol{N}_1\boldsymbol{x} = \boldsymbol{b}_1 \tag{A2.18}$$

where $\boldsymbol{N}_1 = \boldsymbol{A}^{\mathrm{T}}\boldsymbol{N}^{-1}\boldsymbol{A}$

$$\boldsymbol{b}_1 = \boldsymbol{A}^{\mathrm{T}}\boldsymbol{N}^{-1}\boldsymbol{b}$$

$$\boldsymbol{N} = \boldsymbol{C}\boldsymbol{W}^{-1}\boldsymbol{C}^{\mathrm{T}}$$

$$\boldsymbol{b} = -\boldsymbol{CL}$$

Line fitting:

$$y = mx + c \tag{13.66}$$

$$(x)\delta m + (1)\delta c = L + v \tag{13.68}$$

$$(x)\delta m + (1)\delta c + (m)\delta x - \delta y = 0 \tag{13.68}$$

$$\boldsymbol{N}_1\boldsymbol{x} + \boldsymbol{E}^{\mathrm{T}}\boldsymbol{k} = \boldsymbol{b}_1 \tag{13.69}$$

$$\boldsymbol{Ex} + \boldsymbol{d} = \mathbf{0} \tag{13.70}$$

Index